Physical, Chemical and Biological Markers in Argentine Archaeology: Theory, Methods and Applications

Edited by

Débora M. Kligmann
Marcelo R. Morales

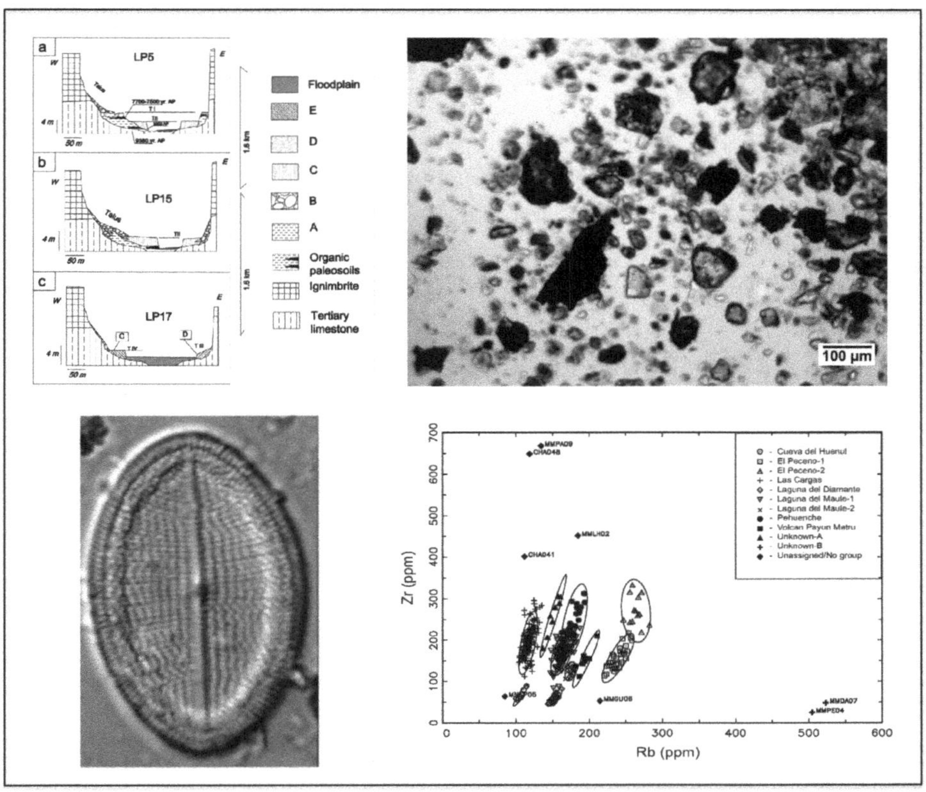

BAR International Series 2678
2014

Published in 2016 by
BAR Publishing, Oxford

BAR International Series 2678

Physical, Chemical and Biological Markers in Argentine Archaeology: Theory, Methods and Applications

ISBN 978 1 4073 1322 1

BAR Publishing is the trading name of British Archaeological Reports (Oxford) Ltd. British Archaeological Reports was first incorporated in 1974 to publish the BAR Series, International and British. In 1992 Hadrian Books Ltd became part of the BAR group. This volume was originally published by Archaeopress in conjunction with British Archaeological Reports (Oxford) Ltd / Hadrian Books Ltd, the Series principal publisher, in 2014. This present volume is published by BAR Publishing, 2016.

Printed in England

BAR
PUBLISHING

BAR titles are available from:

BAR Publishing
122 Banbury Rd, Oxford, OX2 7BP, UK
EMAIL info@barpublishing.com
PHONE +44 (0)1865 310431
FAX +44 (0)1865 316916
www.barpublishing.com

TABLE OF CONTENTS

INDICADORES FÍSICOS, QUÍMICOS Y BIOLÓGICOS EN LA ARQUEOLOGÍA ARGENTINA: TEORÍA, MÉTODOS Y APLICACIONES

RESUMEN

En las últimas dos décadas se han incrementado notablemente en la Arqueología Argentina los trabajos que recurren a métodos y técnicas de las llamadas "ciencias duras" para resolver problemas arqueológicos de distinta índole. Estos estudios involucran la participación de profesionales de disciplinas tales como la física, la química y la biología en investigaciones arqueológicas y/o la especialización por parte de los propios arqueólogos en métodos y técnicas comúnmente utilizados en dichas ciencias.

Los trabajos que aplican los métodos y técnicas recién mencionados suelen quedar desperdigados en distintas publicaciones periódicas de arqueología en nuestro país, ya que no existen revistas especializadas y son escasas en el exterior (ej. Journal of Archaeological Science, Geoarchaeology, Archaeometry, Revue d'Archaeometrie, etc.). Este tomo compila una selección de trabajos originalmente presentados en el XVII Congreso Nacional de Arqueología Argentina en el marco de un simposio que reunía especialistas en estas temáticas, intentando generar un espacio para la discusión y el intercambio de información obtenida a partir de la utilización de indicadores físicos, químicos y biológicos tales como los análisis de isótopos estables, los análisis físico-químicos de artefactos y sedimentos y los análisis de microfósiles, entre otros.

Este libro incluye once capítulos que tratan sobre temas que van desde los estudios paleoambientales multi-proxy en las provincia de Jujuy en el límite norte de nuestro país (Oxman *et al.* y Tchilinguirian *et al.*), hasta los estudios de ADN de Parmigiani en la isla de Tierra del Fuego el punto más austral de la Argentina y del continente Americano, cubriendo vastas extensiones territoriales y temáticas. Así, el trabajo de Cortegoso y colaboradores aborda el estudio de fuentes de obsidiana en torno a la región de Cuyo en el sector central de nuestro país y el de Mercuri y Restifo se aboca a una temática similar pero en la provincia de Salta en el Noroeste Argentino. Por su parte, Killian y colaboradores presentan un estudio de isótopos estables sobre maíces de los sectores Norte y Sur de la Puna Argentina, en las provincias de Jujuy y Catamarca respectivamente, orientadas a la calibración de un set de valores de referencia para estudios de dieta. En la provincia de Catamarca se desarrollan otras tres investigaciones, una en la Puna y dos en la zona de valles. En el primer caso contamos con el trabajo de Grana *et al.*, que aborda el estudio de canales de riego prehispánicos a partir del análisis de diatomeas en sedimentos. En la zona de valles, por su parte, Kligmann y Lantos presentan los resultados de estudios experimentales de campo y de laboratorio para identificar actividades vinculadas a la preparación y consumo de alimentos en los sedimentos mientras que Lindskoug aborda el análisis de micro-espículas de carbón para la determinación de eventos de incendios naturales durante el período Aguada de la región. Finalmente, Dejean y colaboradores no sólo presentan nuevos resultados de estudios de ADN sobre restos humanos procedentes de la provincia de Santa Cruz, en la Patagonia Argentina, sino que discuten una serie de temáticas que van desde lo metodológico hasta aspectos éticos en torno al manejo de este tipo de muestras en nuestro país.

Este tomo no intenta agotar la diversidad de estudios que aplican este tipo de herramientas a problemas arqueológicos concretos, ni cubrir la diversidad regional y cronológica de estos casos en la Argentina, sino más bien presentar una serie de trabajos originales que permiten ejemplificar el grado de desarrollo y la diversidad de temáticas en las que actualmente se aplican indicadores de otras disciplinas en la arqueología de nuestro país.

INDICATEURS PHYSIQUES, CHIMIQUES ET BIOLOGIQUES EN ARCHÉOLOGIE: THÉORIE, MÉTHODES ET APPLICATIONS

RÉSUMÉ

Durant les dernières deux décennies, les travaux d'Archéologie Argentine qui ont recours aux méthodes et techniques des dites « sciences dures » pour résoudre des problématiques archéologiques de différente nature, ont connu une croissance remarquable. Ces études impliquent la participation de professionnels de disciplines telles que la physique, la chimie et la biologie appliquées à la recherche archéologique et/ou la spécialisation des archéologues en méthodes et techniques fréquemment utilisées dans de telles sciences.

Les travaux qui appliquent les méthodes et techniques récemment mentionnées ont tendance à se disséminer dans différentes publications périodiques d'archéologie dans notre pays, étant donné qu'il n'existe pas de revues spécialisées, et que celles-ci sont rares à l'étranger (ex. Journal of Archaeological Science, Geoarchaeology, Archaeometry, Revue d'Archéométrie, etc.). Ce tome compile une sélection de travaux présentés à l'origine dans le XVII Congreso Nacional de Arqueología Argentina dans le cadre d'un symposium qui réunissait des spécialistes de ces thématiques, tentant de créer un espace pour la discussion et l'échange d'information obtenue à partir d'indicateurs physiques, chimiques et biologiques, telles que les analyses d'isotopes stables, les analyses physico-chimiques d'artéfacts et de sédiments, et les analyses de microfossiles entre autres.

Ce livre inclut onze chapitres qui traitent ces thèmes qui vont depuis les études paléoenvironnementales multi-proxy dans la province de Jujuy à l'extrême nord du pays (Oxman *et al.* et Tchilinguirian *et al.*), jusqu'aux études d'ADN de Parmigiani sur l'île de Terre de Feu, le point le plus austral de l'Argentine et du continent Américain, couvrant de vastes territoires et thématiques. De cette façon, le travail de Cortegoso et collaborateurs aborde l'étude de gisements d'obsidienne dans la région de Cuyo dans le secteur central de notre pays et celui de Mercuri et Restifo est consacré à une thématique similaire mais dans la province de Salta au nord de l'Argentine. De leur côté, Killian et collaborateurs présentent une étude d'isotopes stables sur des maïs des secteurs Nord et Sud de la Puna Argentine, dans les provinces de Jujuy et Catamarca respectivement, orientées à la calibration d'un ensemble de valeurs de références pour les études d'alimentation. Dans la province de Catamarca, trois autres recherches se sont développées, une dans la Puna et deux dans la zone des vallées. Dans le premier cas, le travail de Grana *et al.* aborde l'étude des canaux d'irrigation préhispaniques à partir de l'analyse des diatomées dans les sédiments. D'autre part, dans la zone des vallées, Kligmann et Lantos présentent les résultats d'études expérimentales de terrain et de laboratoire pour identifier les activités liées à la préparation et la consommation d'aliments dans les sédiments alors que Lindskoug aborde l'analyse des micro-spicules de carbone pour la détermination d'évènements incendiaires naturels durant la période Aguada dans la région. Enfin, Dejean et collaborateurs présentent non seulement les résultats d'études ADN sur restes humains provenant de la province de Santa Cruz en Patagonie Argentine, mais aussi discutent une série de problèmes qui vont depuis la méthodologie jusqu'aux aspects éthiques tournant autour de la gestion de ce type d'échantillon dans notre pays.

Ce tome n'est pas une tentative d'épuiser la diversité des études qui appliquent ce type d'outils à des problèmes archéologiques concrets, ni de couvrir la diversité régionale et chronologique de ces cas en Argentine, sinon plus une présentation de travaux originaux qui permettent d'illustrer le degré de développement et la diversité de thématiques auxquelles s'appliquent actuellement des indicateurs d'autres disciplines dans l'archéologie de notre pays.

LIST OF CONTRIBUTORS

Barberena, Ramiro
CONICET - Laboratorio de Geoarqueología, Facultad de Filosofía y Letras, Universidad Nacional de Cuyo, Mendoza, Argentina.

Bocci, Marco
Università della Calabria, Italy.

Carnese, Francisco R.
Sección Antropología Biológica, Instituto de Ciencias Antropológicas, Facultad de Filosofía y Letras, Universidad de Buenos Aires - Centro de Estudios Biomédicos, Biotecnológicos, Ambientales y Diagnóstico, Universidad Maimónides, Buenos Aires, Argentina.

Cohen, Ma. Lorena
CONICET - Instituto Superior de Estudios Sociales, Instituto de Arqueología y Museo de la Universidad Nacional de Tucumán, Tucumán, Argentina.

Cornejo, Luis
Universidad de Chile, Chile.

Cortegoso, Valeria
CONICET - Universidad Nacional de Cuyo, Mendoza, Argentina.

Crespo, Cristian M.
CONICET - Centro de Estudios Biomédicos, Biotecnológicos, Ambientales y Diagnóstico y Fundación de Historia Natural Félix Azara. Departamento de Ciencias Naturales y Antropológicas, Universidad Maimónides, Argentina.

De Francesco, Anna María
Università della Calabria, Italy.

Dejean, Cristina B.
Sección Antropología Biológica, Instituto de Ciencias Antropológicas, Facultad de Filosofía y Letras, Universidad de Buenos Aires - Centro de Estudios Biomédicos, Biotecnológicos, Ambientales y Diagnóstico y Fundación de Historia Natural Félix Azara, Departamento de Ciencias Naturales y Antropológicas, Universidad Maimónides, Buenos Aires, Argentina.

Durán, Victor
CONICET - Universidad Nacional de Cuyo, Mendoza, Argentina.

Giesso, Martín
Northeastern Illinois University, Illinois, USA.

Gil, Adolfo
CONICET - Museo de Historia Natural de San Rafael, Mendoza, Argentina.

Glascock, Michael D.
Research Reactor Center, University of Missouri, Columbia, USA.

Grana, Lorena
CONICET - Departamento de Biodiversidad y Biología Experimental, Facultad de Ciencias Exactas y Naturales, Universidad de Buenos Aires, Buenos Aires, Argentina.

Killian Galván, Violeta A.
CONICET - Instituto de Geocronología y Geología Isotópica - Universidad de Buenos Aires, Buenos Aires, Argentina.

Kligmann, Débora M.
CONICET - Instituto de Arqueología, Facultad de Filosofía y Letras, Universidad de Buenos Aires, Buenos Aires, Argentina.

Lanata, José L.
CONICET - Instituto de Investigaciones en Diversidad Cultural y Procesos de Cambio, Universidad Nacional de Río Negro, Argentina.

Lantos, Irene J.
CONICET - Museo Etnográfico, Facultad de Filosofía y Letras, Universidad de Buenos Aires, Buenos Aires, Argentina.

Lindskoug, Henrik B.
CONICET - Museo de Antropología, Facultad de Filosofía y Humanidades, Universidad Nacional de Córdoba, Argentina.

Maidana, Nora I.
CONICET - Departamento de Biodiversidad y Biología Experimental, Facultad de Ciencias Exactas y Naturales, Universidad de Buenos Aires, Buenos Aires, Argentina.

Mercuri, Cecilia
CONICET - Instituto de Arqueología, Facultad de Filosofía y Letras, Universidad de Buenos Aires, Buenos Aires, Argentina.

Morales, Marcelo R.
CONICET - Instituto de Arqueología, Facultad de Filosofía y Letras, Universidad de Buenos Aires, Buenos Aires, Argentina.

Neme, Gustavo
CONICET - Museo de Historia Natural de San Rafael, Mendoza, Argentina.

Oliszewski, Nurit
CONICET - Instituto Superior de Estudios Sociales - Universidad Nacional de Tucumán, Tucumán, Argentina.

Olivera, Daniel E.
CONICET - Instituto Nacional de Antropología y Pensamiento Latinoamericano - Universidad de Buenos Aires, Buenos Aires, Argentina.

Oxman, Brenda I.
CONICET - Instituto de Arqueología, Facultad de Filosofía y Letras, Universidad de Buenos Aires, Buenos Aires, Argentina.

Panarello, Héctor O.
CONICET - Instituto de Geocronología y Geología Isotópica - Universidad de Buenos Aires, Buenos Aires, Argentina.

Parmigiani, Vanesa
CONICET - Centro Austral de Investigaciones Científicas, Tierra del Fuego - Facultad de Ciencias Naturales y Museo, Universidad Nacional de La Plata, Buenos Aires, Argentina.

Pirola, Malena
Instituto de Arqueología, Facultad de Filosofía y Letras, Universidad de Buenos Aires, Buenos Aires, Argentina.

Restifo, Federico
CONICET - Instituto de Arqueología, Facultad de Filosofía y Letras, Universidad de Buenos Aires, Buenos Aires, Argentina.

Sanhueza, Lorena
Universidad de Chile, Chile.

Tchilinguirian, Pablo
CONICET - Instituto Nacional de Antropología y Pensamiento Latinoamericano - Universidad de Buenos Aires, Buenos Aires, Argentina.

Yacobaccio, Hugo D.
CONICET - Instituto de Arqueología, Facultad de Filosofía y Letras, Universidad de Buenos Aires, Buenos Aires, Argentina.

LIST OF REVIEWERS

Barba Pingarrón, Luis A.
Instituto de Investigaciones Antropológicas, Universidad Nacional Autónoma de México, México.

Barberena, Ramiro
CONICET - Laboratorio de Geoarqueología, Facultad de Filosofía y Letras, Universidad Nacional de Cuyo, Mendoza, Argentina.

Borrero, Luis A.
CONICET - Instituto Multidisciplinario de Historia y Ciencias Humanas, Departamento de Investigaciones Prehistóricas y Arqueológicas, Buenos Aires, Argentina.

De Nigris, Mariana
CONICET - Instituto Nacional de Antropología y Pensamiento Latinoamericano, Buenos Aires, Argentina.

Díaz Pardo, Carolina A.
Instituto de Ecología y Biodiversidad, Laboratorio de Paleoecología, Departamento de Ciencias Ecológicas, Facultad de Ciencias, Universidad de Chile, Santiago, Chile.

Escola, Patricia S.
CONICET - Escuela de Arqueología, Universidad Nacional de Catamarca, Catamarca, Argentina.

Franco, Nora V.
CONICET - Instituto Multidisciplinario de Historia y Ciencias Humanas, Departamento de Investigaciones Prehistóricas y Arqueológicas, Buenos Aires, Argentina.

Huckleberry, Gary A.
Department of Geosciences, University of Arizona, Tucson, USA.

Joly, Delphine
UMR 6566, Bât 24-25, Campus de Beaulieu, Université de Rennes 1, Rennes, France.

Kornstanje, Alejandra
CONICET - Instituto de Arqueología y Museo, Tucumán, Argentina.

Lazzari, Marisa
Archaeology Department, Exeter University, UK.

Lema, Verónica S.
CONICET - Departamento Científico de Arqueología, Laboratorio 1, Museo de Ciencias Naturales de La Plata, Buenos Aires, Argentina.

López, Gabriel
CONICET - Instituto de Arqueología, Facultad de Filosofía y Letras, Universidad de Buenos Aires, Buenos Aires, Argentina.

Luna, Leandro
CONICET - Museo Etnográfico J. B. Ambrosetti, Facultad de Filosofía y Letras, Universidad de Buenos Aires, Buenos Aires, Argentina.

Mondini, Mariana
CONICET - Laboratorio de Zooarqueología y Tafonomía de Zonas Áridas, Instituto de Antropología de Córdoba, Museo de Antropología, Universidad Nacional de Córdoba, Córdoba, Argentina.

Ratto, Norma
Museo Etnográfico J. B. Ambrosetti, Facultad de Filosofía y Letras, Universidad de Buenos Aires, Buenos Aires, Argentina.

Schittek, Karsten
Seminar für Geographie und ihre Didaktik, Universität zu Köln, Germany.

Seldes, Verónica
CONICET - Instituto Nacional de Antropología y Pensamiento Latinoamericano, Buenos Aires, Argentina.

Tripaldi, Alfonsina
CONICET - Instituto de Geociencias Básicas, Aplicadas y Ambientales, Departamento de Ciencias Geológicas, Facultad de Ciencias Exactas y Naturales, Universidad de Buenos Aires, Buenos Aires, Argentina.

Zangrando, Atilio F.J.
CONICET - Centro Austral de Investigaciones Científicas - Laboratorio de Antropología, Tierra del Fuego, Argentina.

APPLICATIONS OF PHYSICAL, CHEMICAL AND BIOLOGICAL MARKERS IN ARGENTINE ARCHAEOLOGY: A BRIEF STATE OF THE ART

Marcelo R. Morales[1] and Débora M. Kligmann[1]

[1] CONICET - Instituto de Arqueología, Facultad de Filosofía y Letras, Universidad de Buenos Aires

INTRODUCTION

Even though the beginning of archaeological studies dates back to very old times, the use and applications of disciplines such as physics, chemistry, geology, and biology in archaeology have rapidly increased during the last sixty years worldwide (e.g. Butzer 1971; Cook and Heizer 1965; Cornwall 1958; Courty *et al.* 1989; Davidson and Shackley 1976; Dincauze 2000; Drucker 1972; Evans and O'Connor 1999; Garrison 2003; Goffer 2007; Goldberg and Mcphail 2006; Goldberg *et al.* 2001; Harris 1989; Herz and Garrison 1998; Holliday 1992, 2004; Lasca and Donahue 1990; Limbrey 1975; Luff and Rowley-Conwy 1994; Pollard and Heron 1996; Rapp 2009; Rapp and Hill 1998; Sánchez Vizcaíno and Cañabate Guerrero 1998; Shackley 1975; Stein and Farrand 1985, 2001; Waters 1992; Weiner 2010, among many others).

Papers that apply methods and technics of the so-called "hard sciences" to solve diverse archaeological problems in argentine archaeology have become more popular during the last two decades. These studies involve the participation of professionals coming from several fields such as physics, chemistry, geology and biology, as well as of archaeologists technically trained in those disciplines. Papers that apply this kind of approaches can only be found as isolated contributions in argentine archaeological meetings, symposia, and in non-specific publications, because there are no local technical journals such as those internationally available (e.g. *Journal of Archaeological Method and Theory*, *Journal of Archaeological Research*, *Journal of Archaeological Science*, *Geoarchaeology*, *Archaeometry*, *Revue d'Archaeometrie*, etc.).

For this reason, we organized a Symposium at the "*XVII Congreso Nacional de Arqueología Argentina*" ("XVII National Meeting of Argentine Archaeology", October 2010, Mendoza City, Argentina) seeking to offer a specialist-oriented arena to share new information and discuss methodological and technical issues regarding the application of physical, chemical and biological tools in archaeology. This book includes some of the papers presented at that symposium, and partially illustrates the state of the art regarding the utilization of these analytical markers in Argentina. It is worth noting that about 80% of the participants were archaeologists specialized in chemical, earth and natural sciences, and also that a high number of them were trained in national and foreign institutions related to those fields of science.

We say that this book "partially" illustrates the state of the art because it neither fully represents the diversity of studies that use this kind of analytical markers, nor the regional and chronological variability of archaeological problems where they were employed. Moreover, topics such as analytical chemistry applied to ceramics, rock art or metallurgy, probably the most developed scientific lines of research in our country, are not included in this volume because they have been the subject matter of particular symposia in the same national meeting just mentioned. In the same way, other less frequently used tools such as remote sensing and GIS, artifact functional analysis, geophysics, radiocarbon dating, and soil micromorphology, among others, are also absent.

After succinctly mentioning some general theoretical and methodological issues regarding the title of this book, in the following sections of this introductory chapter we briefly summarize and review the analytical markers mentioned in this volume and the case studies in which they were applied.

A FEW THEORETICAL AND METHODOLOGICAL CONCEPTS REGARDING THE TITLE OF THIS BOOK

Why do we speak of physical, chemical and biological markers instead of using the terms geoarchaeology or archaeometry?

Geoarchaeology applies methods and technics of the earth sciences or geosciences, in a broad sense (e.g. stratigraphy, sedimentology, pedology, geomorphology, geophysics, geochemistry, geochronology, mineralogy, petrography-petrology, paleontology, climatology and geography), to solve archaeological problems such as: 1) identification of site formation processes, 2) paleoenvironmental reconstructions, 3) raw material provenance, 4) relative and absolute chronology, and 5) location of archaeological sites, among many others (Gladfelter 1981; Hassan 1979; Herz and Garrison 1998; Holliday 1991; Pollard 1999; Rapp 1987; Rapp and Gifford 1982; Rapp and Hill 1998; Thorson and Holliday 1990; Waters 1992; Zárate 1993). Geoarchaeology is archaeology: research problems are archaeological even though the methodologies to solve them can be borrowed from some other discipline. From this perspective, methods and technics coming from the earth sciences can be conceived as a means rather than as an end. For others authors, however, geoarchaeology represents a narrower field of action, applying the methods and technics of just a few geosciences (e.g. sedimentology, pedology and geomorfology) (e.g. Gladfelter 1977; Renfrew 1976; Waters 1992). Taking this limited definition into account, some of the archaeological goals (e.g. geophysical and geochemical prospections, raw material sourcing and absolute dating) fall into the scope of archaeometry (Butzer 1982; De Atley and Bishop 1991; Ehrenreich 1995; Herz and Garrison 1998; Renfrew 1976; Waters 1992).

Archaeometry, also known as archaeological science or scientific archaeology, can be defined as the relationship between archaeology and the physical, chemical and material sciences. This term was coined from archaeology and metric, meaning measurement of archaeological findings. It applies scientific methods and techniques to the analysis of archaeological materials to study the past. Research topics comprise disciplines and techniques such as: physical and chemical dating methods, which provide archaeologists with absolute and relative chronologies; analyses of artifacts, concerning provenance, technology and types of use; environmental approaches, which provide information on landscape, climate, flora, and fauna past changes; anthropological studies dealing with diet, nutrition, health, and pathologies; remote-sensing and geophysical-survey applications; mathematical methods for data treatment including the role of computers in handling, analyzing, and modeling and conservation sciences, involving the study of decay processes and the development of new methods of conservation (Butzer 1982; De Atley and Bishop 1991; Ehrenreich 1995; Herz and Garrison 1998; Olin 1982; Waters 1992).

Terms as geoarchaeology and archaeometry are too narrow or too wide, respectively, to accurately include all the analytical markers considered in this book. The term geoarchaeology is restrictive because several biological lines of evidence, for instance DNA analysis, fall outside the scope of this sub-discipline of archaeology. On the other hand, archaeometry, as defined above, consists of a myriad of particular analyses, tools and research focuses with nothing in common beyond their archaeological interest. Moreover, being archaeology a discipline dealing with material remains, almost everything in this field of research is "measurable". Another reason to exclude both of these terms to describe the kinds of approaches contained in this volume is that these concepts entail the idea of an archaeological "naturalist approach" or a "post-positivist approach" that narrow down the range of theoretical perspectives where they can be applied. Regarding the latter, we consider that the same markers or analytical tools can be applied from different theoretical and methodological perspectives and are not "theory-dependent" or limited to only one specific sub-discipline or archaeological field. Thus, we here prefer to mention the individual markers as related to their primary science or field of origin, focusing on their different scopes as well as on the potential topics for their application. We avoid referring to them as "proxies" because the term "proxy" (*sensu* Dincauze 2000) exclusively refers to those kinds of data that allow us to acquire "indirect" information about the objects being studied. Several cases involving this kind of analytical tools -some of them included in this volume, for instance DNA or soil chemistry data- usually generate "direct" information, falling outside the "proxy" definition.

Finally, we believe that some of the contributions of this book can be considered as interdisciplinary, involving the collaboration between archaeologists and chemists, geologists and biologists. However, most of the papers presented in this volume can be considered as transdisciplinary meaning those approaches that include cases studied by researchers originally formed in archaeology that have been trained in some aspects -or even have their post-graduate specialization- of a second formal discipline. Thus, they can use the methods and technics coming from this second discipline in an accurate manner to solve archaeological problems. We consider that both kinds of interactions between disciplines are essential to solve specific archaeological problems in a scientific and efficient way.

PHYSICAL, CHEMICAL AND BIOLOGICAL MARKERS USED IN THIS BOOK: A BRIEF OVERVIEW OF PAPERS

The papers in this special volume were divided into six broad categories to cover the range of theoretical and technical advances as well as their applications.

XRF and NAA in Obsidian Sourcing

As evidenced by publication dates, only a decade has made a great difference in obsidian sourcing studies due to the application of physico-chemical markers. The two main techniques involved in obsidian sourcing are: XRF (X-Ray Fluorescence) and NAA (Neutron Activation Analysis). Both of them have been extremely useful in establishing the origin of the majority of the chemical varieties of obsidian found in the argentine territory. Some broad researches such as those published by Barberena *et al.* 2010; Bellelli and Pereyra 2002; Bellelli *et al.* 2006; Durán *et al.* 2004; Escola 2004; Giesso *et al.* 2011; Gómez Otero and Stern 2005; Stern 2000; Stern *et al.* 2012; Yacobaccio *et al.* 2002, 2004, among others, must be mentioned due to their importance for the archaeology of Northwest Argentina, Cuyo and Patagonia. These papers are key for determining the origin of most of the known chemical varieties of obsidian in the areas just mentioned. However, several sources still remain unknown, particularly in Northwest Argentina, due to the diversity and quantity of volcanic events in the area. Thus, a systematic ongoing work is required.

The most evident fact of the first contribution of this book is the exceptional number of chemical determinations presented. The paper by **Cortegoso and co-authors** called "Chemical characterization of obsidian in west-central Argentina and central Chile: archaeological problems and perspectives" has involved more than a thousand determinations, including 100 from six different known sources and 911 from 106 archaeological sites located on both sides of the Andean range. This work allowed the authors to discuss the distributional patterns of obsidian in the region as well as to discuss some technical issues regarding the use of different XRF spectrometers and the NAA method in the data obtained.

The second contribution by **Mercuri and Restifo** called "Obsidian provenance analysis from Alero Cuevas site, Salta province, Argentina: application and complementarity of physico-chemical and macroscopic methods" deals with a key issue when dealing with obsidian sourcing and artifact provenance during the analysis of a lithic assemblage. As it is well known, it is virtually impossible to chemically determine the full set of artifacts and debris coming from an archaeological site. Thus, we need original approaches to build bridges between chemical obsidian sourcing and everyday lab macroscopic obsidian "type" determinations. This paper exemplifies one of these attempts, exploring the viability and efficiency of a joint analysis of physico-chemical methods such as XRF with some particular macroscopic attributes to be observed and described during lithic analysis. The authors also evaluate its utility, advantages and limitations during the study of the assemblages of different layers of the Alero Cuevas site.

Stable Isotopes

Stable isotopes (e.g. $\delta^{18}O$, $\delta^{15}N$, $\delta^{13}C$ and $\delta^{86}Sr$) offer different kinds of information depending on the sampled substrate -mineral or biological- (human remains, animals, vegetation, sediments, water, etc.) and the selected fraction of the sample (e.g. apatite or collagen in bones and bulk organic matter or cellulose macro remains in soils) of archaeological evidences or sedimentary sequences. The resulting values of isotope analysis allow us to obtain information about a huge variety of topics such as diet composition changes, photosynthetic regime of vegetation communities, hydrologic imbalance of water systems, population movements, drought periods, nutritional stress, etc. (Ambrose 1993; Codron *et al.* 2005; Schoeninger 1995, among others).

It is probable that the earliest and most frequent applications of isotope studies in argentine archaeology were related to values obtained from human tissues oriented to disentangle several diet issues in different regions of the country (Barberena 2002; Barrientos 1999, 2001; Borrero *et al.* 2001; Fernández and Panarello 1991, 1996; Guichón *et al.* 2001; Olivera and Yacobaccio 1999; Panarello *et al.* 2006; Tessone *et al.* 2005, 2009; Yesner *et al.* 1991, 2003; Zangrando *et al.* 2004, among many others). The majority of the obtained values offered limited information due to the lack of broad isotopic ecology surveys focused in obtaining reference values of different elements, of the trophic and metabolic fractionation, of the environmental effects over values, etc. In recent years, some colleagues working in several regions started to make efforts in that direction (Barberena *et al.* 2009, 2011; Fernández and Panarello 1999-2001; Tessone 2010; Tessone and Belardi 2010; Tessone *et al.* 2009; Ugan *et al.* 2012; Yacobaccio *et al.* 2009, 2010). For a complete review about the state of the art in isotopic applications in argentine archaeology, a detailed summary is available in Samec (2011).

In this volume, an important contribution has been made in the arena of agriculture related problems under the title "Intraspecific variability in $\delta^{15}N$ and $\delta^{13}C$ values of archaeological *Zea mays* samples (northwest argentine Puna)", by **Killian Galván, Oliszewski, Olivera and Panarello**. The chapter efficiently deals with the complexity related to the variability in the values detected in one of the most conspicuous cultivars in the Andes, the corn. In their paper, the authors analyze several maize races associated with different uses and discuss the implications of the utility of direct archaeological specimen values in paleodietary reconstructions.

DNA

More than two decades after the bloom of ancient DNA studies in diverse scientific fields in the 1980's, this biochemical line of inquiry, capable of dealing with problems such as population dynamics, local extinctions, peopling routes, population replacements, etc., has recently become an area of growing interest and development in argentine archaeology.

A few seminal papers involving argentine samples were published by Demarchi *et al.* 2001; Lalueza *et al.* 1993-1994 and Lalueza-Fox *et al.* 1995. Some of the few new studies in this field include the research carried out by Moraga and co-authors (2009) that involves mitochondrial DNA of more than 40 individuals from both Argentinean and Chilean Patagonia. The data obtained in this study evidenced the possibility of a founder linage (haplogroup B) that could correspond to the first peopling wave into the southern part of the continent. Other studies considering mitochondrial DNA were carried out in the central area of the country by García and Demarchi (2009) and Nores and Demarchi (2011). The first one discusses the distribution of mtDNA haplogroups while the second one evidences not only the reduction of A and B haplotypes and the increase of C and D through time but also some spatial differences in their distribution in the mentioned area. A recent contribution by some of these authors (García *et al.* 2012) deals with the Phylogeography of mitochondrial haplogroup D1 in the same area of Argentina. Other few particular issues regarding pre Columbian DNA are available in Carnese *et al.* 2009; Crespo *et al.* 2009, 2010; Demarchi 2009; and Mendisco *et al.* 2011.

The two contributions presented in this book related to DNA analysis are extremely valuable, not only because of the already mentioned scarcity of papers on this topic, but also for the potential approaches exemplified by them.

The first one, by **Dejean and co-authors**, called "History of a dialogue. Scopes and limitations of ancient

DNA analysis in archaeological research", solidly discusses different aspects of ancient DNA analysis and its potential applications in archaeology, regarding several methodological, technical and even legal aspects of this kind of studies. To complete this excellent work the authors also summarize some results recently obtained and published dealing with different individuals housed in two Museum's collections and from one archaeological site: Pampa Grande (Salta province, NW Argentina). The results from Pampa Grande not only established mitochondrial DNA lineages of the sampled individuals, but also determined nuclear DNA and chromosome Y microsatellites.

The second one, by **Parmigiani**, called "Determination of age and sex on dental pieces of *Lama guanicoe*: a methodological approach", uses DNA in a less usual context, in zooarchaeology. As explained in this paper, the extremely poor preservation conditions constitute one of the major problems in the zooarchaeology of Tierra del Fuego, being tooth almost the only recoverable remains. Thus, the only way to deal with resource utilization, selection and management by past populations is trying to deal with this taphonomic limitation. This paper makes an effort in this direction, using DNA analysis to determine sex in samples, and tooth enamel incremental lines to estimate the age of death of the camelids analyzed, trying to calibrate both techniques for future applications in archaeological contexts.

Diatoms

It is broadly known that assemblages of organisms living in aquatic environments markedly modify their characteristics due to environmental changes. For this reason, sediments of water bodies are an invaluable source of paleoenvironmental data that allow us to understand past landscapes and their modifications through time. Diatoms, in particular, are one of the most commonly used proxies and are known as excellent paleolimnological indicators due to several factors: the abundance of their valves, their usual good preservation in lake sediments, the broad variety of adaptative strategies, and their short life cycle that allows them to respond almost immediately to changes in their environment (Kumke *et al.* 2004; Smol and Cumming 2000).

These organisms were first studied in Argentina towards the 1930's in the Pampean region (Frenguelli 1938) and the early 40's in Patagonia (Frenguelli 1942; Krasske 1949). However, their use as a biological proxies (e.g. Stoermer and Smol 1999; Straub 1990, among others) in different archaeological and environmental researches only became frequent in Argentina in the last twenty years (Grana 2007; Grana and Morales 2005; Haberzettl *et al.* 2005; Hassan *et al.* 2004, 2012; Kligmann 2003, 2009; Kligmann and Maidana 1996; Lupo *et al.* 2006b, 2007; Maidana 1996; Martínez Machiavello and Díaz 1997; Mayr *et al.* 2005; Morales 2004, 2011; Yacobaccio and Morales 2005; Zolitschka *et al.* 2006). Some of these researches were particularly concerned in establishing the environmental scenarios and resource availability for human occupation in different regions (Fernández *et al.* 2012; Grana 2007; Grana and Morales 2005; Morales 2004, 2011; Yacobaccio and Morales 2005). Some other efforts have been made to discuss raw material sourcing in ceramics (De la Fuente 1997; Kligmann and Calderari 2012; Solá and Morales 2007), site formation processes (Kligmann 2003, 2009; Kligmann *et al.* 2013) or to communicate its potential applications in the archaeological field (e.g. Martínez Macchiavello *et al.* 1999).

In this volume, two chapters used diatoms as a key line of inquiry evidencing the potential diversity in terms of the problems that can be faced with this biological indicator. The first one called "Methodological proposal to recognize irrigation channels using diatom analysis as a biomarker: Peñas Coloradas (Antofagasta de la Sierra, southern Puna of Argentina)" (**Grana, Cohen and Maidana**) offers an innovative approach by applying diatom studies in the Salty Puna of Catamarca province, to test if some suspicious landforms in this extremely dry area correspond to irrigation channels. The solid methodological design and the coherent results show the utility of diatoms as indicators of palaeochannels, setting a good starting point for future investigations dealing with similar problems in the area.

The second paper, called "Paleoenvironmental data from Quebrada de Lapao, Jujuy province, Argentina (23.36° S, 66.36° W, 3650 m a.s.l.) for the 9400-7300 BP span" (**Tchilinguirian, Morales, Oxman and Pirola**) used diatoms in a more traditional way, as one of the main archives in a multiproxy paleoenvironmental study carried out in Lapao ravine in the Dry Puna of Jujuy province. This work illustrates, not only the utility of diatoms in this kind of research, but it also shows its power in detecting and magnifying weak environmental modification signals in particular localities.

Pollen

Pollen is another biomarker frequently applied in paleoenvironmental studies in Argentina. In 1918, Lennart von Post conducted the earliest quantitative analysis of pollen. This research, carried out in Swedish bogs, can be considered the seminal work of modern pollen analysis. In 1929, the same author carried out the first known pollen analysis of South America in the argentine territory. This analysis included the study of sediment samples recovered from lakes Fagnano, Baño Nuevo and Cabo Domingo in Tierra del Fuego province.

Despite the fact that the first pollen studies in Argentina took place in the late 1920's, the papers that followed were written almost four decades later in Cuyo (Vogel and Lerman 1969) and northwestern Argentina (Markgraf 1985; Pöthe de Baldis 1971, 1974; Pöthe de Baldis and Salas 1977) and more than five decades later in Patagonia (Heusser *et al.* 1988; Mancini and Trivi 1992; Markgraf 1987, 1989; Schäbitz 1989, 1991, 1994) and Pampa (Nieto and D´Antoni 1985; Romero and Fernández 1981). In the past twenty years, pollen analyses have increased noticeably their frequency and spatial coverage turning into one of the most frequent proxy in paleoenvironmental studies regularly used by archaeologists in Argentina (e.g. Bamonte and Mancini 2011; Lupo 1992, 1998; Lupo *et al.* 2006a, 2006b; Mancini 1993, 2002, 2009; Mancini *et al.* 2005; Paez and Prieto 1993; Paez *et al.* 1999, 2002; Prieto 1996; Stutz and Prieto 2003; Tchilinguirian *et al.* 2012, among many others).

Two of the contributions of this volume have applied pollen analysis in reconstructing past environments. The first one has already been mentioned in the diatom section (**Tchilinguirian *et al.***) and uses pollen as one of the proxies applied in a multi-proxy analysis of Lapao ravine's deposits. The work of **Oxman and Yacobaccio** is the other contribution involving pollen studies. In this case, the paper entitled "Pollen analysis of Pastos Chicos: paleoenvironmental and archaeological implications during the Holocene in the dry Puna of Argentina" exemplifies how paleoenvironmental data coming from pollen analysis is a useful tool to discuss archaeological processes such as the human peopling of the highlands of south-central Andes. The authors present and compare the results of the pollen analysis of Pastos Chicos sequence with those obtained in Lapao (the small "*quebrada*" of the same drainage system studied in the work of Tchilinguirian *et al.* previously mentioned). The comparison has allowed the authors to obtain paleoenvironmental signals with different spatial scope. Thus, they could achieve a better understanding of the environmental scenario of the initial peopling of the region, allowing the discussion of different hypotheses related to this topic.

Sediments

The first paper that presents the results of chemical analysis of argentine archaeological sediments was written by Ducloux, and included as a report in a general paper published by Greslebin in 1928. Between the 1950's and the 1990's, some more isolated reports of this type were published (Bayarsky *et al.* 1999; Cappannini 1950; Etchichury 1975, 1976, 1978; Etchichury and Tófalo 1980; Forzinetti 1980; Forzinetti *et al.* 1988-90; Mazzoni and Spalletti 1974; Spalletti and Salazar 1988; Teruggi 1968). It is interesting to mention that all of these authors were earth science scientists who analyzed sediment samples collected by archaeologists. Thus, they did not take part during fieldwork. While almost all of the papers cited took some general attributes into account (color, organic matter, carbonates, grain-size and mineralogy), none of them analyzed two very important attributes in current geoarchaeological studies: pH and phosphorous. Also, it is important to mention that only some of these geologists included control samples. In sum, it can be said that these reports were mere descriptions and that in most of the cases the results were not incorporated into the archaeological interpretation of the sites. Marcelo Zárate, a geologist specialized in Quaternary studies, constitutes the only exception to this trend. He became interested in fully collaborating with archaeologists in the late 1980's (Zárate 1986-87, 1993, 1994, 1997; Zárate and Flegenheimer 1991; Zárate and Prieto 1994, among others), establishing the foundations for similar collaborations that took place only several years later.

It was not until the present century that the first analyses of archaeological sediments were carried out in Argentina by archaeologists trained in the earth sciences. Also, some other geologists started to work intensively in archaeological research teams, taking part during fieldwork as well as labwork (Díaz País and Kligmann 2009; Eugenio *et al.* 2010; Favier Dubois 2000, 2001, 2003; Kligmann 1998, 2003, 2009; Kligmann and Díaz País 2010, 2012; Kligmann and Ratto 2009; Kligmann *et al.* 2012, 2013; Ortiz 2003; Ozán 2012; Ratto *et al.* 2012, 2013; Roldán *et al.* 2005, 2008; Sampietro and Vattuone 2005; Tchilinguirian and Olivera 2000a, 2000b; Villa and Kligmann 2013, among others).

Unfortunately, analysis of argentine archaeological sediments are still the exception rather that the rule. While many local archaeologists do not collect sediment samples, others collect them but do not analyze them, probably because archaeologists trained in the earth sciences are scarce and it is not easy to find geologists interested in the Quaternary who are willing to collaborate. We do hope that in the near future this type of analyses will be part of routine archaeological labwork in our country.

As far as archaeological charcoals and ashes recovered in sediment samples are concerned, very few people in Argentina have analyzed the use of vegetable resources as fuel (Jofré 2005, 2007; Lindskoug and Mors 2010; Marconetto 1993, 1996, 1999, 2002a, 2002b, 2003, 2005, 2006, 2008).

Three of the papers included in this book deal with sediment analyses but in very different ways. The first two papers use sediment data to discuss paleoenvironmental issues while the third one applies sedimentological analysis to discuss human behavior.

The first contribution, entitled "In pursuit of the fire. Contributions of microcharcoal analysis to the archaeology of the Ambato valley (Catamarca)", by **Henrik Lindskoug**, presents the preliminary results of studies carried out in order to generate information about past wildfires and construct paleoenvironmental data for the Ambato Valley, Catamarca province. In this paper, microcharcoal is used as a paleoenvironmental indicator of the presence of fires in the past in order to evaluate the last phase of the Aguada occupation of the area. The tools and methods applied in this study were developed in pedoanthracology and sedimentology (e.g. point counting technique).

The second contribution, by **Tchilinguirian, Morales, Oxman and Pirola**, has already been mentioned in both the diatom and the pollen sections of this chapter. Paleoenvironmental research carried out in the Andes highlands during the past decades has shown great variability in terms of the impact of broad scale climate changes in different environmental settings. In order to address this issue, the authors undertook systematic studies of several localities in the Puna of Argentina oriented to modeling past environmental scenarios for human adaptation in the area. In this paper they summarize the main results obtained by multi-proxy analysis carried out in the Quebrada de Lapao, located at 3650 m a.s.l. in the Dry Puna of Jujuy province.

The last contribution, entitled "On stews and sediments: contributions of experimental field and lab archaeology to the study of sedimentological modifications", by **Kligmann and Lantos**, presents the results of a geoarchaeological experiment carried out to examine human modifications of sediments resulting from food preparation and consumption activities. The experiment involved the deposition of organic matter coming from food remains in sediments of different texture, both in the field and in the lab, as well as the measurement of its postdepositional decomposition. The main goals were to reproduce organic matter decay in sedimentary contexts, and to create reference data that will enable archaeologists to reconstruct past human activities through sedimentological analyses.

SOME CONCLUIDING REMARKS

The papers in this special volume highlight a variety of aspects and approaches related to the study of archaeological materials in Argentina, applying methods and technics coming from the chemical, earth and natural sciences. As we have already mentioned, the contributions presented here do not intent to summarize the current situation in our country, although the wide range of case studies exemplifies the archaeological advantages of using analytical markers borrowed from those fields of science. We believe that this volume illustrates well the great development that some kinds of studies have had in our country, such as obsidian sourcing or paleoenvironmental research. At the same time, it shows that other particular tools -such as DNA, microcharcoal analysis or soil chemistry- are less frequent, and given that they have proved useful in many other parts of the world, they disserve more attention in the future research agenda of argentine teams. This volume also illustrates some innovative approaches to particular archaeological problems, such as the use of diatoms in the detection of archaeological irrigation channels or the development of calibration datasets or reference value databases for archaeological interpretations, such as those related to isotopic studies. In summary, we think that this book presents a set of good examples to showcase the relevance of this kind of studies in argentine archaeology, the current research trends and the future research needs. We hope that this book will encourage many other colleagues to join this tendency, cooperating with other disciplines that have proved to be largely useful in solving archaeological problems.

ACKNOWLEDGEMENTS

We would like to thank the authors who participated in this volume as well as the peer reviewers for their generous collaboration. We would also like to thank Guti Tessone for reading a preliminary version of this chapter and David Davison (BAR) for publishing this volume as part of the British Archaeological Reports, International Series.

REFERENCES

AMBROSE, S.H.
1993. Isotopic analysis of paleodiets: Methodological and interpretive considerations. In *Investigations of Ancient Human Tissue. Chemical Analysis in Anthropology*, edited by M.K. Sandford, pp. 59-130. Pensylvania, Gordon and Breach Science Publishers.

BAMONTE, F.P. AND M.V. MANCINI
2011. Palaeoenvironmental changes since Pleistocene-Holocene transition: Pollen analysis from a wetland in southwestern Patagonia (Argentina). *Review of Palaeobotany and Palynology* 165(1): 103-110.

BARBERENA, R.
2002. *Los Límites del Mar: Isótopos Estables en Patagonia Meridional*. Sociedad Argentina de Antropología, Buenos Aires.

BARBERENA, R., A. HAJDUK, A.F. GIL, G.A. NEME, V. DURÁN, M.D. GLASCOCK AND A.A. RUGHINI
2010. Obsidian in the south-central Andes: Geological, geochemical, and archaeological assessment of north Patagonian sources (Argentina). *Quaternary International* 245(1): 25-36.

BARBERENA, R., C. MÉNDEZ, F. MENA AND O. REYES
2011. Endangered species, archaeology, and stable isotopes: huemul (*Hippocamelus bisulcus*) isotopic ecology in central-western Patagonia (South America). *Journal of Archaeological Science* 38(9): 2313-2323.

BARBERENA, R., A.F. ZANGRANDO, A.F. GIL, G.A. MARTÍNEZ, G.G. POLITIS, L.A. BORRERO AND G.A. NEME
2009. Guanaco *Lama guanicoe* isotopic ecology in southern South America: spatial and temporal tendencies, and archaeological implications. *Journal of Archaeological Science* 36(12): 2666-2675.

BARRIENTOS, G.
1999. Composición isotópica ($\delta^{13}C$) de muestras de restos óseos humanos del sitio Arroyo Seco 2 (provincia de Buenos Aires): inferencias paleodietarias. *Relaciones de la Sociedad Argentina de Antropología* 24: 81-94.

2001. Estudios de isótopos estables del carbono y nitrógeno en Pampa, Patagonia continental y Tierra del Fuego: una evaluación. *Revista Argentina de Antropología Biológica* 3(2): 35.

BAYARSKY, A., A. MONTI AND M.A. GONZÁLEZ
1999. Sedimentología del Alero Pintoscayoc 1, departamento Humahuaca, Jujuy. In *Pintoscayoc, arqueología de quebradas altas en Humahuaca*, by M.I. Hernández Llosas. Unpublished PhD Dissertation, Facultad de Filosofía y Letras, Universidad de Buenos Aires.

BELLELLI, C. AND F.X. PEREYRA
2002 Análisis geoquímicos de obsidiana. Distribuciones de fuentes y artefactos arqueológicos en el Noroeste del Chubut (Patagonia argentina). *Werken* 3: 99-118.

BELLELLI, C., F.X. PEREYRA AND M. CARBALLIDO
2006. Obsidian localization and circulation in northwestern Patagonia (Argentina): sources and archaeological record. *Special Publications of the Geological Society of London* 257(1): 241-255.

BORRERO, L.A., R. GUICHÓN, R. TYKOT, J. KELLY, A. PRIETO AND P. CÁRDENAS
2001. Dieta a partir de isótopos estables en restos óseos humanos de Patagonia Austral. Estado actual y perspectivas. *Anales del Instituto de la Patagonia, Serie Ciencias Humanas* 29: 119-127.

BUTZER, K.W.
1971. *Environment and Archaeology. An Ecological Approach to Prehistory*. Second edition. Aldine, Chicago.

1982. *Archaeology as Human Ecology. Method and Theory for a Contextual Approach*. Cambridge University Press, New York.

CAPPANNINI, D.A.
1950. Informe sobre la investigación de las tierras de la Gruta del Oro. *Runa* III: 35-36.

CARNESE, F.R., F. MENDISCO, C. KEYSER, C.B. DEJEAN, J.M. DUGOUJON, C.M. BRAVI, B. LUDES AND E. CRUBÉZY
2009. Paleogenetical study of pre Columbian samples from Pampa Grande (Salta, Argentina). *American Journal of Physical Anthropology* 141(3): 452-462.

CODRON, J., D. CODRON, J.A. LEE-THORP, M. SPONHEIMER, W.J. BOND, D. DE RUITER AND R. GRANT
2005. Taxonomic, anatomical, and spatio-temporal variations in the stable carbon and nitrogen isotopic compositions of plants from an African savanna. *Journal of Archaeological Science* 32(12): 1757-1772.

COOK, S.F. AND R.F. HEIZER
1965. *Studies on the Chemical Analysis of Archaeological Sites*. University of California Press, Berkeley.

CORNWALL, J.W.
1958. *Soils for the Archaeologist*. Phoenix House, London.

COURTY, M.A., P. GOLDBERG AND R. MACPHAIL
1989. *Soils and Micromorphology in Archaeology*. Cambridge University Press, Cambridge.

CRESPO, C.M., C.B. DEJEAN, F.R. CARNESE AND J.L. LANATA
2009. Extracción y tipificación de ADN a partir de piezas dentales de muestras arqueológicas de la prov. de Santa Cruz. In *Actas de las IX Jornadas Nacionales de Antropología Biológica*, pp. 97.

CRESPO, C.M., C.B. DEJEAN, M.B. POSTILLONE, J.L. LANATA AND F.R. CARNESE
2010. Historias en código genético: Los aportes de los estudios de ADN antiguo en antropología y sus implicancias éticas. *Runa 31*(2): 153-174.

DAVIDSON, D.A. AND M.L. SHACKLEY (EDITORS)
1976. *Geoarchaeology. Earth Science and the Past*. Duckworth, London.

DE ATLEY, S.P. AND R.L. BISHOP
1991. Toward an integrated interface for archaeology and archaeometry. In *The Ceramic Legacy of Anna O. Shepard*, edited by R.L. Bishop and F.W. Lange, pp. 358-380. University Press of Colorado, Niwot.

DE LA FUENTE, G.
1997. Aplicación de un bioindicador arqueológico (diatomeas) para el estudio de fuentes de aprovisionamiento de arcillas. Unpublished Graduate Thesis, Escuela de Arqueología, Universidad Nacional de Catamarca.

DEMARCHI, D.
2009. Microsatélites, distancias genéticas y estructura de poblaciones nativas sudamericanas. *Revista Argentina de Antropología Biológica* 1(1): 73-88.

DEMARCHI, D.A., G.M. PANZETTA-DURTARI, S. COLANTONIO AND A.J. MARCELLINO
2001. Absence of the 9-bp deletion of mitochondrial DNA in pre-Hispanic inhabitants of Argentina. *Human Biology* 73(4): 575-582.

DÍAZ PAÍS, E. AND D.M. KLIGMANN
2009. Comparación de cuatro métodos para medir fósforo en sedimentos arqueológicos. In *Arqueometría Latinoamericana: 2do. Congreso Argentino y 1ro. Latinoamericano*, Volumen 1, edited by O.M. Palacios, C. Vázquez, T. Palacios and E. Cabanillas, pp. 190-198. Comisión Nacional de Energía Atómica (CNEA), Buenos Aires.

DINCAUZE, D.F.
2000. *Environmental Archaeology. Principles and Practice*. Cambridge University Press, Cambridge.

DRUCKER, P.
1972. *Stratigraphy in Archaeology: An Introduction*. Modules in Anthropology 30, Addison-Wesley, Massachusetts.

Durán, V., M. Giesso, M. Glascock, G. Neme, A. Gil and R.L. Sanhueza
2004. Estudios de redes de aprovisionamiento y redes de distribución de obsidiana durante el Holoceno Tardío en el sur de Mendoza (Argentina). *Estudios Atacameños* 28: 25-43.

Ehrenreich, R.M.
1995. Archaeometry into Archaeology. *Journal of Archaeological Method and Theory* 2(1): 1-6.

Escola, P.S.
2004. Variabilidad en la explotación y distribución de obsidianas en la Puna Meridional argentina. *Estudios Atacameños* 28: 9-24.

Etchichury, M.C.
1975. Sedimentología del perfil de el alero de las manos pintadas. *Relaciones* N.S. IX: 177-185.

1976. Sedimentología de la cueva de "las manos", Estancia Alto Río Pinturas (provincia de Santa Cruz). *Relaciones* N.S. X: 251-260.

1978. Análisis de sedimentos del abrigo de los Chelcos (departamento San Alberto, Córdoba). *Relaciones* N.S. XII: 207-208.

Etchichury, M.C. and O.R. Tófalo
1980. Sedimentología de los depósitos de la Cueva Grande del Cañadón Quesada Arroyo Feo, provincia de Santa Cruz. *Revista de la Asociación Argentina de Mineralogía, Petrología y Sedimentología* 11(1-2): 42-67.

Eugenio, E., A. Murgo and A. García
2010. Análisis físico-químicos de sedimentos para evaluar el impacto de la acción antrópica en sitios arqueológicos de cazadores recolectores pescadores. El caso de El Divisadero Monte 6 (General Lavalle, Buenos Aires). In *Mamül Mapu: Pasado y Presente desde la Arqueología Pampeana*, edited by M. Berón, L. Luna, M. Bonomo, C. Montalvo, C. Aranda and M. Carrera Aizpitarte, Tomo I, pp. 53-64. Libros del Espinillo, Ayacucho.

Evans, J. and T. O'Connor
1999. *Environmental Archaeology: Principles and Methods*. Sutton Publishing, Gloucestershire.

Favier Dubois, C.M.
2000. Pedogénesis y formación de registros en Bahía San Sebastián (Tierra del Fuego) y Lago Roca (Santa Cruz). In *Soplando en el Viento... Actas de las II Jornadas de Arqueología de la Patagonia*, pp. 319-332. San Carlos de Bariloche.

2001. Análisis geoarqueológico de los procesos de formación del registro, cronología y paleoambientes en sitios arqueológicos de Fuego-Patagonia. Unpublished PhD Dissertation, Facultad de Ciencias Exactas y Naturales, Universidad de Buenos Aires.

2003. Late Holocene climatic fluctuations and soil genesis in southern Patagonia: effects on the archaeological record. *Journal of Archaeological Science* 30(12): 1657-1664.

Fernández J. and H. Panarello
1991. Paleodietas y patrones de movilidad de cazadores recolectores: su estimación en base a los isótopos estables del carbono. *Anales de Arqueología y Etnología* 43-45: 599-611.

1996. El Rodeo. Análisis de las relaciones isotópicas 13C/12C en el colágeno. In *Contribución a la Arqueología del Río Pinturas, Provincia de Santa Cruz*, edited by C. Gradin and A.M. Aguerre, pp. 300-310. Ediciones Búsqueda del Ayllu, Concepción del Uruguay.

1999-2001. Isótopos del carbono en la dieta de herbívoros y carnívoros de los Andes Jujeños. Xama 12-14: 71-85.

Fernández, M., N. Maidana and J. Rabassa
2012. Palaeoenvironmental conditions during the Middle Holocene at Isla de los Estados (Staaten Island, Tierra del Fuego, 54° S, Argentina) and their influence on the possibilities for human exploration. *Quaternary International* 256: 78-87.

Forzinetti, M.E.
1980. Análisis sedimentológico de muestras de la Cueva del Oro. *Relaciones* N.S. XIV(1): 98-101.

Forzinetti, M.E., M. Assaro and E. Turco Greco
1988-90. Estudio sedimentológico del depósito. In *La Cueva de Haichol. Arqueología de los Pinares Cordilleranos del Neuquén*, edited by J. Fernández, pp. 45-73. Anales de Arqueología y Etnología 43/45 (I-III), Facultad de Filosofía y Letras, Universidad Nacional de Cuyo, Mendoza.

Frenguelli, J.
1938. Diatomeas de la Bahia de San Blas. *Revista del Museo de La Plata, NS, 1*.

1942. *Diatomeas del Neuquén (Patagonia)*. Instituto del Museo de la Universidad de La Plata, La Plata.

García, A. and D.A. Demarchi
2009. Incidence and distribution of Native American mtDNA haplogroups in central Argentina. *Human Biology* 81(1): 59-69.

García, A., M. Pauro, R. Nores, C.M. Bravi and D.A. Demarchi
2012. Phylogeography of mitochondrial haplogroup D1: An early spread of subhaplogroup D1j from Central Argentina. *American Journal of Physical Anthropology* 149(4): 583-590.

Garrison, E.G.
2003. *Techniques in Archaeological Geology*. Springer, Berlin.

GIESSO, M., V. DURÁN, G. NEME, M.D. GLASCOCK, V. CORTEGOSO, A. GIL AND L. SANHUEZA
2011. A study of obsidian source usage in the Central Andes of Argentina and Chile. *Archaeometry 53*(1): 1-21.

GLADFELTER, B.G.
1977. Geoarchaeology: the geomorphologist and archaeology. *American Antiquity* 42(4): 519-538.

1981. Developments and directions in geoarchaeology. In *Advances in Archaeological Method and Theory* 4, edited by M.B. Schiffer, pp. 343-364. Academic Press, New York.

GOFFER, Z.
2007. *Archaeological Chemistry*. Second Edition. John Wiley & Sons, Inc., New Jersey.

GOLDBERG, P., V.T. HOLLIDAY AND C.R. FERRING
2001. *Earth Sciences and Archaeology*. Klumer Academic, New York.

GOLDBERG, P. AND R.I. MCPHAIL
2006. *Practical and Theoretical Geoarchaeology*. Blackwell Publishing, Oxford.

GÓMEZ OTERO, J. AND C.R. STERN
2005. Circulación, intercambio y uso de obsidianas en la costa de la provincia del Chubut (Patagonia argentina), durante el Holoceno tardío. *Intersecciones en Antropología* 6: 93-108.

GRANA, L.
2007. Diatomeas fósiles: un acercamiento a la relación entre Sociedad y el Ambiente durante el Holoceno en Antofagasta de la Sierra, Puna Meridional Argentina. Unpublished Graduate Thesis. Facultad de Filosofía y Letras, Universidad de Buenos Aires.

GRANA, L. AND M.R. MORALES
2005. Primeros resultados paleoambientales del análisis de diatomeas fósiles del Holoceno Medio y Tardío de la cuenca del Rio Miriguaca, Antofagasta de la Sierra, Puna Catamarqueña. In *Entre Pasados y Presentes II*, edited by T. Bourlot, D. Bozzuto, C. Crespo, A.C. Hecht and N. Kuperszmit pp. 392-409. Fundación de Historia Natural Felix de Azara. Buenos Aires.

GRESLEBIN, H.
1928. Las llamadas "botijas" o "tinajas" de la provincia de San Luis (República Argentina). *Physis* IX(32): 46-71.

GUICHÓN, R.A., L.A. BORRERO, A.I. PRIETO, P. CÁRDENAS AND R. TYKOT
2001. Nuevas determinaciones de isótopos estables para Tierra del Fuego. *Revista Argentina de Antropología Biológica* 3(1): 113-126.

HABERZETTL, T., M. FEY, A. LÜCKE, N. MAIDANA, C. MAYR, C. OHLENDORF AND B. ZOLITSCHKA
2005. Climatically induced lake level changes during the last two millennia as reflected in sediments of Laguna Potrok Aike, southern Patagonia (Santa Cruz, Argentina). *Journal of Paleolimnology* 33(3): 283-302.

HARRIS, E.C.
1989. *Principles of Archaeological Stratigraphy*, Second Edition. Academic Press, San Diego.

HASSAN, F.A.
1979. Geoarchaeology: the geologist and archaeology. *American Antiquity* 44(2): 267-270.

HASSAN, G., M.A. ESPINOSA AND F.I. ISLA
2004. Análisis paleoambiental de una secuencia del Holoceno tardío del arroyo Claromecó (Provincia de Buenos Aires, Argentina) basado en diatomeas. *Ameghiniana* 41(2): 161-170.

2012. Fluctuaciones de salinidad durante el Holoceno en la laguna costera de Mar Chiquita (provincia de Buenos Aires): una aproximación cuantitativa basada en diatomeas. *Ameghiniana* 48(4): 496-507.

HERZ, N. AND E.G. GARRISON
1998. *Geological Methods for Archaeology*. Oxford University Press, New York.

HEUSSER, C.J., J. RABASSA, A. BRANDANI AND R. STUCKENRATH
1988. Late-Holocene vegetation of the Andean Araucaria region, province of Neuquén, Argentina. *Mountain Research and Development* 8(1): 53-63.

HOLLIDAY, V.T.
1991. Earth science and archaeology. *Reviews in Anthropology* 16: 35-45.

2004. *Soils in Archaeological Research*. Oxford University Press, New York.

HOLLIDAY, V.T. (EDITOR)
1992. *Soils in Archaeology. Landscape Evolution and Human Occupation*. Smithsonian Institution Press, Washington D.C.

JOFRÉ, I.C.
2005. La formación del registro antracológico: Estudio estadístico de los efectos de las técnicas arqueológicas de recuperación sobre carbón vegetal. *La Zaranda de Ideas* 1: 21-41.

2007. Estudio antracológico en Tebenquiche Chico (Dpto. Antofagasta de la Sierra, Provincia de Catamarca). In *Paleoetnobotánica del Cono Sur: Estudios de Casos y Propuestas Metodológicas*, compiled by B. Marconetto, M.P. Babot y N. Oliszewski, pp. 153-178. Ferreyra Editor para el Museo de Antropología, Facultad de Filosofía y Humanidades, Universidad Nacional de Córdoba, Córdoba.

KLIGMANN, D.M.
1998. Formation processes of the archaeological record: a case study from the southern Puna, Catamarca, Argentina. In *Proceedings of the XIII Congress of the International Union of Prehistoric and Protohistoric Sciences 1996* (Sections 1-3, Volume 1), edited by C. Arias, A. Bietti, L. Castelletti and C. Peretto, pgs. 377-382, A.B.A.C.O. Srl, Forlì.

2003. Procesos de Formación de Sitios Arqueológicos: Tres Casos de Estudio en la Puna Meridional Catamarqueña Argentina. Unpublished PhD Dissertation, Facultad de Filosofía y Letras, Universidad de Buenos Aires.

2009. Procesos de Formación de Sitios Arqueológicos: Tres Casos de Estudio en la Puna Meridional Catamarqueña Argentina. BAR Publishing - British Archaeological Reports (BAR) International Series 1949, Oxford.

KLIGMANN, D.M., A.M. ALBINO AND E. DÍAZ PAÍS
2012. Anfisbenas para los dioses: Zooarqueología de una ofrenda animal. In *De la Puna a las Sierras: Avances y Perspectivas en Zooarqueología Andina*, edited by A.D. Izeta and G.L. Mengoni Goñalons. South American Archaeology Series, British Archaeological Reports (BAR), International Series, BAR Publishing, Oxford. In press.

KLIGMANN, D.M. AND M. CALDERARI
2012. Diatoms and ceramic provenance: A cautionary tale. *Archaeometry* 54(1): 129-143.

KLIGMANN, D.M. AND E. DÍAZ PAÍS
2010. Caracterización físico-química de sedimentos provenientes de fogones experimentales. In *La Arqueometría en Argentina y Latinoamérica*, edited by S. Bertolino, R. Cattáneo and A. D. Izeta, pp. 305-310. Facultad de Filosofía y Humanidades, Universidad Nacional de Córdoba, Córdoba.

2012. Mirando lo invisible. Primeros análisis físico-químicos de los sedimentos de Tolombón y del Pukara de Angastaco (provincia de Salta). In *Al Borde del Imperio. Paisajes Sociales en Áreas Periféricas del Qollasuyu*, edited by V. Williams and B. Cremonte. In press.

KLIGMANN, D.M. AND N.I. MAIDANA
1996. Informe del sitio arqueológico "Alero de las Circunferencias" (Humahuaca, Jujuy): análisis de asociaciones de diatomeas. In Pintoscayoc, Arqueología de Quebradas Altas en Humahuaca, by M.I. Hernández Llosas. Unpublished PhD Dissertation, Facultad de Filosofía y Letras, Universidad de Buenos Aires.

KLIGMANN, D.M. AND N. RATTO
2009. Aportes de la sedimentología a la arqueología: El caso del sitio Fiambalá 1 (Puna meridional catamarqueña). In *Arqueometría Latinoamericana: 2do. Congreso Argentino y 1ro. Latinoamericano*, Volumen 1, edited by O.M. Palacios, C. Vázquez, T. Palacios and E. Cabanillas, pp. 199-208. Comisión Nacional de Energía Atómica (CNEA), Buenos Aires.

KLIGMANN, D.M., N. RATTO AND N.I. MAIDANA
2013. Microarqueología y procesos de formación del registro: análisis mineralógicos y diatomológicos del sitio El Zorro (Puna transicional, Departamento de Tinogasta, Catamarca). In *Arqueometría Argentina: Estudios Pluridisciplinarios*, edited by M. Ramos, M. Lanza, V. Helfer, V. Pernicone, F. Bognanni, C. Landa, V. Aldazabal and M. Fernández, pp. 113-136. Programa de Arqueología Histórica y Estudios Pluridisciplinarios, Departamento de Ciencias Sociales, Universidad Nacional de Luján. Aspha Ediciones, Buenos Aires.

KRASSKE, G.
1949. Subfossile diatomeen aus den Mooren Patagoniens und Feuerlands. *Ann. Aead. Sci. Fenn.*, A, 14: 1-94.

KUMKE, T., U. KIENEL, J. WECKSTRÖM, A. KORHOLA AND K.-H. HUBBERTEN
2004. Inferred Holocene paleotemperatures from diatoms at Lake Lama, Central Siberia. *Arctic, Antartic, and Alpine Research* 35(4): 624-634.

LALUEZA, C., A. PÉREZ-PÉREZ, E. PRATS, P. MORENO, J. PONS AND D. TURBÓN
1993-1994. Ausencia de la delección de 9 bp COII/tRNA Lys en aborígenes de Fuego-Patagonia mediante el análisis de DNA antiguo. *Anales del Instituto de la Patagonia, Serie Ciencias Humanas* 22: 181-191.

LALUEZA-FOX, C., A. PÉREZ-PÉREZ, E. PRATS and D. TURBÓN
1995. Linajes mitocondriales de los aborígenes de Tierra del Fuego y Patagonia. *Anales del Instituto de la Patagonia, Serie Ciencias Humanas* 23: 75-86.

LASCA, N.P. AND J. DONAHUE (EDITORS)
1990. *Archaeological Geology of North America.* Geological Society of America, Centennial Special Volume 4, Boulder.

LIMBREY, S.
1975. *Soil Science and Archaeology.* Academic Press, London.

LINDSKOUG, H.B. AND V.A. MORS
2010. Donde hubo fuego, ¿cenizas quedan? Residuos de combustión en el sitio Piedras Blancas, Dpto. Ambato, Catamarca. In *La Arqueometría en Argentina y Latinoamérica*, edited by S. Bertolino, R. Cattáneo and A. D. Izeta, pp. 243-250. Facultad de Filosofía y Humanidades, Universidad Nacional de Córdoba, Córdoba.

LUFF, R. AND P. ROWLEY-CONWY
1994. *Whither Environmental Archaeology?* Oxbow Monographs 38, Oxbow Books, Oxford.

LUPO, L.C.

LUPO, L.C.
1992. El Sitio ICc-4. Reconstrucción paleoambiental para los últimos 10 000 años. Noroeste argentino. Unpublished report to Consejo Nacional de Investigaciones Científicas y Técnicas.

1998. Estudio sobre la lluvia polínica actual y la evolución del paisaje a través de la vegetación durante el Holoceno en la cuenca del Río Yavi. Borde Oriental de la Puna, Noroeste argentino. Unpublished PhD Dissertation, Bamberg University, Germany.

LUPO, L.C., M.M. BIANCHI, E. ARAOZ, R. GRAU, C. LUCAS, R. KERN AND M. GROSJEAN
2006a. Climate and human impact during the past 2000 years as recorded in the Lagunas de Yala, Jujuy, northwestern Argentina. *Quaternary international* 158(1): 30-43.

LUPO, L.C., M. MORALES, A. MALDONADO AND M. GROSJEAN
2006b. A high-resolution pollen and diatom record from Laguna Los Pululos (22° 36' S / 66° 44' W / 4500 m asl), NW Argentinean Puna, since *ca.* 800 AD. *Abstracts' Book of the International Symposium: Reconstructing Past Regional Climate Variations in South America over the Late Holocene: A New PAGES Initiative*, pp. 24.

LUPO, L.C., M.R. MORALES, H.D. YACOBACCIO, A. MALDONADO AND M. GROSJEAN
2007. Cambios Ambientales en la Puna Jujeña durante los últimos 1200 años: explorando su impacto en la economía pastoril. Número Especial de la *Revista Pacarina*, Tomo III, Resúmenes Ampliados del XVI CNAA, pp. 151-156.

MAIDANA, N.I.
1996. Asociación de diatomeas fósiles del sitio arqueológico Potrero del Caballo Muerto (Puna de Jujuy, Argentina). In *Actas y Memorias del XI Congreso Nacional de Arqueología Argentina* (13ª Parte), Tomo XXV (1/4): 53-60. San Rafael, Mendoza.

MANCINI, M.V.
1993. Recent pollen spectra from forest and steppe of South Argentina: a comparison with vegetation and climate data. *Review of Palaeobotany and Palynology* 77(1): 129-142.

2002. Vegetation and climate during the Holocene in Southwest Patagonia, Argentina. *Review of Palaeobotany and Palynology* 122(3): 101-115.

2009. Holocene vegetation and climate changes from a peat pollen record of the forest-steppe ecotone, Southwest of Patagonia (Argentina). *Quaternary Science Reviews* 28(15): 1490-1497.

MANCINI, M.V., M.M. PAEZ, A.R. PRIETO, S. STUTZ, M. TONELLO AND I. VILANOVA
2005. Mid-Holocene climatic variability reconstruction from pollen records (32–52 S, Argentina). *Quaternary International* 132(1): 47-59.

MANCINI, M.V. AND M. TRIVI
1992. Búsqueda de análogos modernos en el sistema polen del Alero Cárdenas (Provincia de Santa Cruz). *Publicaciones Especiales de la Asociación Paleontológica Argentina* 2: 81-84.

MARCONETTO, M.B.
1993. *Aprovisionamiento y consumo de elementos combustibles en el sector medio de la Quebrada de Humahuaca (Pcia. de Jujuy)*. Unpublished Graduate Thesis, Facultad de Filosofía y Letras, Universidad de Buenos Aires.

1996. Qué quemaba Don Santiago. Un ensayo de identificación de carbón vegetal arqueológico del sitio Alero Don Santiago (Pcia. de Chubut). Paper presented at the *III Jornadas de Arqueología de la Patagonia*, San Carlos de Bariloche, Argentina.

1999. Las leñas del Jaguar. In *En Los Tres Reinos: Prácticas de Recolección en el Cono Sur de América*, edited by C.A. Aschero, M.A. Korstanje and P.M. Vuoto, pp. 179-186. Ediciones Magna Publicaciones, Facultad de Ciencias Naturales e Instituto Miguel Lillo, Universidad Nacional de Tucumán.

2002a. Análisis de los vestigios de combustión de los sitios Alero Don Santiago y Campo Moncada (Pcia del Chubut). In *Plantas y Cazadores de la Patagonia*, compiled by C. Pérez de Micou, pp. 33-54. Facultad de Filosofía y Letras, Universidad de Buenos Aires, Buenos Aires.

2002b. Analysis of burnt building structures of the Ambato valley (Catamarca, Argentina). In *Charcoal Analysis. Methodological Approaches, Palaeo-ecological Results and Wood Uses*, edited by S. Thiebault, pp. 267-271. BAR International Series 1063, Oxford.

2003. Recursos forestales: Oferta y disponibilidad en contextos Aguada. Paper presented at the *V Jornadas La Cultura de la Aguada y su Dispersión*, La Rioja, Argentina.

2005. *Recursos forestales y el proceso de diferenciación social en tiempos prehispánicos en el valle de Ambato, Catamarca*. Unpublished PhD Dissertation, Facultad de Ciencias Naturales y Museo, Universidad Nacional de La Plata.

2006. La gente, la leña, el monte. In *El Modo de Hacer las Cosas*, edited by C. Pérez de Micou, pp. 101-128. Facultad de Filosofía y Letras, Universidad de Buenos Aires, Buenos Aires.

2008. *Recursos Forestales y el Proceso de Diferenciación Social en Tiempos Prehispánicos en el Valle de Ambato, Catamarca, Argentina*. South American Archaeology Series 1785, No. 3, edited by A.D. Izeta. British Archaeological Reports (BAR) - Oxford.

MARKGRAF, V.
1985. Paleoenvironmental history of the last 10,000 years in northwestern Argentina. *Zentralblatt für Geologie und Palaontologie* 11: 1739-1748.

1987. Paleoenvironmental changes at the northern limit of the subantarctic *Nothofagus* forest, lat 37° S, Argentina. *Quaternary Research* 28(1): 119-129.

1989. Palaeoclimates in Central and South America since 18,000 BP based on pollen and lake-level records. *Quaternary Science Reviews* 8(1): 1-24.

MARTÍNEZ MACCHIAVELLO, J.C., G. DE LA FUENTE AND N. RATTO
1999. La utilización de diatomeas (Bacillariophyta) en la investigación arqueológica: una perspectiva interdisciplinaria. *Arqueología* 9: 49-69.

MARTÍNEZ MACCHIAVELLO, J.C. AND L. DÍAZ
1997. Diatomeas (chrysophita-Bacillariophyceae) de un perfil de una turbera de Milluyoc, Provincia de Jujuy, Argentina. *Insula* 26: 29-44.

MAYR, C., M. FEY, T. HABERZETTL, S. JANSSEN, A. LÜCKE, N.I. MAIDANA AND B. ZOLITSCHKA
2005. Palaeoenvironmental changes in southern Patagonia during the last millennium recorded in lake sediments from Laguna Azul (Argentina). *Palaeogeography, Palaeoclimatology, Palaeo-ecology* 228(3): 203-227.

MAZZONI, M.M. AND L.A. SPALLETTI
1974. Sedimentología de la Cueva de Los Toldos, provincia de Santa Cruz. *Revista de la Asociación Argentina de Mineralogía, Petrología y Sedimentología* 5(1-2): 15-34.

MENDISCO, F., C. KEYSER, C. HOLLARD, V. SELDES, A.E. NIELSEN, E. CRUBÉZY AND B. LUDES
2011. Application of the iPLEXTM Gold SNP genotyping method for the analysis of Amerindian ancient DNA samples: Benefits for ancient population studies. *Electrophoresis* 32(3-4): 386-393.

MORAGA, M., F. MENA, O. REYES, G. BARRIENTOS, R. GOÑI, N. FRANCO AND L.A. BORRERO
2009. Identificación de linajes mitocondriales fundadores en restos humanos prehistóricos de Patagonia y Tierra del Fuego. Paper presented at the Novenas Jornadas Nacionales de Antropología Biológica, Puerto Madryn, Chubut, Argentina.

MORALES, M.R.
2004. Casi Invisibles. Diatomeas, ambientes locales y estrategias cazadoras-recolectoras durante la primera mitad del Holoceno en la Puna desértica. Unpublished Graduate Thesis, Facultad de Filosofía y Letras. Universidad de Buenos Aires.

2011. *Arqueología Ambiental del Holoceno Temprano y Medio en la Puna Seca Argentina. Modelos Paleoambientales Multi-escalas y sus Implicancias para la Arqueología de Cazadores-Recolectores*. British Archaeological Research 2295, South American Archaeology Series, vol. 15. BAR Publishing, Oxford, UK.

NIETO, M.A. AND H.L. D'ANTONI
1985. Pollen analysis of sediments of the Atlantic shore at Mar Chiquita (Buenos Aires province, Argentina). *Zentralbl. Geol. Palaontol* 1: 1731-1738.

NORES, R. AND D.A. DEMARCHI
2011. Análisis de haplogrupos mitocondriales en restos humanos de sitios arqueológicos de la provincia de Córdoba. *Revista Argentina de Antropología Biológica* 13(1): 43-54.

OLIN, J.
1982. *Future Directions in Archaeometry: A Roundtable*. Smithsonian Institution Press, Washington, DC.

OLIVERA, D.E. AND H.D. YACOBACCIO
1999. Estudios de paleodieta en poblaciones humanas de los Andes del Sur a través de isótopos estables. In *Actas del V Congreso Nacional de Paleopatología*, pp. 190-211. Asociación Española de Paleopatología, Alcalá la Real.

ORTIZ, G.
2003. La aplicación de análisis químicos para la interpretación de áreas de actividad en la región subandina de Jujuy. *Cuadernos de la Facultad de Humanidades y Ciencias Sociales* 20: 291-321. Universidad Nacional de Jujuy, San Salvador de Jujuy.

OZÁN, I.L.
2012. Intensidad de ocupación humana en cazadores-recolectores: una mirada microestrati-gráfica desde las ciencias del suelo. *La Zaranda de Ideas* 8(1): 65-81.

PAEZ, M.M. AND A.R. PRIETO
1993. Paleoenvironmental reconstruction by pollen analysis from loess sequences of the southeast of Buenos Aires (Argentina). *Quaternary International* 17: 21-26.

PAEZ, M.M., A.R. PRIETO AND M.V. MANCINI
1999. Fossil pollen from Los Toldos locality: A record of the Late-glacial transition in the Extra-Andean Patagonia. *Quaternary International* 53: 69-75.

PAEZ, M.M., F. SCHÄBITZ AND S. STUTZ
2002. Modern pollen-vegetation and isopoll maps in southern Argentina. *Journal of Biogeography* 28(8): 997-1021.

PANARELLO, H., F. ZANGRANDO, A. TESSONE, L. KOZAMEH AND N. TESTA
2006. Análisis comparativo de paleodietas humanas entre la región del Canal Beagle y Península Mitre: perspectivas desde los isótopos estables. *Magallania* 34(2): 37-46.

POLLARD, A.M.
1999. Geoarchaeology: an introduction. In Geoarchaeology: *Exploration, Environments, Resources*, edited by A.M. Pollard, pp. 7-14. Geological Society of America Special Publication 165, London.

POLLARD, A.M. AND C. HERON
1996. *Archaeological Chemistry*. The Royal Society of Chemistry, Cambridge.

PÖTHE DE BALDIS, E.D.
1971. Microplancton del Silúrico superior de la provincia de Santiago del Estero, República Argentina. *Ameghiniana*: 8(3-4): 282- 290.

1974. Microplancton adicional del Silúrico superior de Santiago del Estero, República Argentina. *Ameghiniana* 11(4): 313-327.

PÖTHE DE BALDIS, E.D. AND E. SALAS
1977. Análisis palinológico de la Formación Tarija. *IGCP, Proy. Pal. Sup. América del Sur, Boletín* 2(7).

PRIETO, A.R.
1996. Late Quaternary vegetational and climatic changes in the Pampa grassland of Argentina. *Quaternary Research* 45(1): 73-88.

RAPP, G., JR.
1987. Geoarchaeology. *Annual Review of Earth and Planetary Sciences* 15: 97-113.

2009. *Archaeomineralogy*. Segunda edición. Springer.

RAPP, G., JR. AND J.A. GIFFORD
1982. Archaeological geology. *American Scientist* 70(1): 45-53.

RAPP, G., JR. AND C.L. HILL
1998. *Geoarchaeology. The Earth-Science Approach to Archaeological Interpretation*. Yale University Press, New Haven.

RATTO, N., D.M. KLIGMANN, C. MONTERO, N. BONOMO, F. HONGN, B. VALERO GARCÉS, A. OSELLA AND N. RUSSO
2012. Historia ambiental, geo-arqueología y arqueo-geofísica del oeste tinogasteño (Catamarca): su impacto y consecuencias en las sociedades del primer milenio. Paper presented at the Workshop: Arqueología del Período Formativo en Argentina, Tafí del Valle, Tucumán, Argentina.

RATTO, N., D.M. KLIGMANN AND N.G. RUSSO
2013. Geoarqueología de sitios de altura (4000-5000 msnm) del oeste tinogasteño (Catamarca). In *Delineando Prácticas de la Gente del Pasado: Los Procesos Socio-Históricos del Oeste Catamarqueño*, compiled by N. Ratto, pp. 357-377. Sociedad Argentina de Antropología, Altuna Impresores, Buenos Aires.

RENFREW, C.
1976. Archaeology and the earth sciences. In *Geoarchaeology. Earth Science and the Past*, edited by Davidson, D.A. and M.L. Shackley, pp. 1-5. Duckworth, London.

ROLDÁN, J., M.M. SAMPIETRO VATTUONE, L. DEL VALLE NEDER AND M.A. VATTUONE
2008. Efectos antrópicos de uso de suelos durante el Formativo en el Valle de Tafí (Tucumán-Argentina). *Chungara* 40(2): 161-172.

ROLDÁN, J., M.M. SAMPIETRO VATTUONE AND M.A. VATTUONE
2005. Técnicas analíticas para el estudio de los suelos de sitios agrícolas prehispánicos. In *Actas del Primer Congreso Argentino de Arqueometría*, pp. 60-69. Universidad Nacional de Rosario, Santa Fe.

ROMERO, E.J. AND C.A. FERNÁNDEZ
1981. Palinología de paleosuelos del Cuaternario de los alrededores de Lobería, Provincia de Buenos Aires, Argentina. *Ameghiniana* 18(3-4): 273-285.

SAMEC, C.T.
2011. Perspectiva isotópica sobre la alimentación de camélidos domésticos y silvestres de la Puna Jujeña: construyendo un marco de referencia para estudios arqueológicos. Unpublished Graduate Thesis, Facultad de Filosofía y Letras, Universidad de Buenos Aires.

SAMPIETRO, M.M. AND M.A. VATTUONE
2005. Reconstruction of activity areas at a Formative household in Northwest Argentina. *Geoarchaeology* 20(4): 337-354.

SÁNCHEZ VIZCAÍNO, A. AND M.L. CAÑABATE GUERRERO
1998. *Indicadores Químicos para la Arqueología*. Centro Andaluz de Arqueología Ibérica, Universidad de Jaén, Baeza.

SCHÄBITZ, F.
1989. Untersuchungen zum aktuellen Pollenniederschlag und zur holozänen Klima- und Vegetationsentwicklung in den Anden Nord-Neuquéns, Argentinien. *Bamberger Geographische Schriften*, [Bd.] 8.

1991. Holocene vegetation and climate in southern Santa Cruz, Argentina. *Bamberger Geographische Schriften* 11: 235-244.

1994. Holocene climatic variations in northern Patagonia, Argentina. *Palaeogeography, Palaeoclimatology, Palaeoecology 109*(2): 287-294.

SCHOENINGER, M.J.
1995. Stable isotopes studies in human evolution. *Evolutionary Anthropology* 4(3): 83-98.

SHACKLEY, M.L.
1975. *Archaeological Sediments. A Survey of Analytical Methods*. John Wiley and Sons, New York.

SMOL, J.P. AND B.F. CUMMING
2000. Tracking long-term changes in climate using algal indicators in lake sediments. *Journal of Phycology* 36(6): 986-1011.

SOLÁ, P. AND M.R. MORALES
2007. Caracterización petrográfica y biológica de un tiesto cerámico estilo "incaico" hallado en Susques (Jujuy). *Intersecciones en Antropología* 8: 361-364.

SPALLETTI, L.A. AND J.A. SALAZAR
1988. Sedimentología de una unidad arqueológica de Tafí del Valle (Tucumán, R. Argentina) y análisis composicional de las cerámicas asociadas. In *Sistemas de Asentamiento Prehispánicos en el Valle de Tafí*, edited by E.E. Berberián, pp. 145-167. Editorial Comechingonia, Córdoba.

STEIN, J.K. AND W.R. FARRAND (EDITORS)
1985. *Archaeological Sediments in Context*. University of Maine, Orono.

STEIN, J.K. AND W.R. FARRAND (EDITORS)
2001. *Sediments in Archaeological Context*. The University of Utah Press, Salt Lake City.

STERN, C.
2000. Sources of obsidian artefacts from the Pali Aike, Fells Cave and Cañadón La Leona archaeological sites in southernmost Patagonia. In *Desde el País de los Gigantes. Perspectivas Arqueológicas en Patagonia* 2: 43-55. Universidad Nacional de la Patagonia Austral, Río Gallegos.

STERN, C., I. PEREDA AND A. AGUERRE
2012. Multiple primary and secondary sources for chemically similar obsidians from the area of Portada Covunco, West-Central Neuquén, Argentina. *Archaeometry* 54(3): 442-453.

STOERMER, E. AND J. SMOL
1999. *The Diatoms: Applications for the Environmental and Earth Science*. Cambridge University Press, London.

STRAUB, F.
1990. Diatomeés et reconstitution des environnements préhistoriques. *Archéologie Neuchathloise* 10. Hauterive-Chapréveyers 4.

STUTZ, S. AND A.R. PRIETO
2003. Modern pollen and vegetation relationships in Mar Chiquita coastal lagoon area, southeastern Pampa grasslands, Argentina. *Review of Palaeobotany and Palynology* 126(3): 183-195.

TCHILINGUIRIAN, P., M.R. MORALES, B. OXMAN, L.C. LUPO, D.E. OLIVERA AND H.D. YACOBACCIO
2012. Early to Middle Holocene transition in the Pastos Chicos record, dry Puna of Argentina. *Quaternary International*. http://dx.doi.org/10.1016/j.quaint.2012.03.006.

TCHILINGUIRIAN, P. AND D. OLIVERA
2000a. De aguas y tierras: aportes para la reactivación de campos agrícolas arqueológicos en la Puna argentina. *Relaciones de la Sociedad Argentina de Antropología* XXV: 99-118.

2000b. Aportes edáficos para el estudio de campos agrícolas arqueológicos Bajo del Coypar, provincia de Catamarca. *Cuaternario y Ciencias Ambientales* 4: 51-55.

TERUGGI, M.E.
1968. Geología y sedimentología de las cuevas de la Cuchilla de las Aguilas (Sierras de Tandil, Prov. de Buenos Aires). *Etnía* 7 (Enero-Julio): 13-21.

TESSONE, A.
2010. Arqueología y ecología isotópica. Estudio de isótopos estables de restos humanos del Holoceno tardío en Patagonia meridional. Unpublished PhD Dissertation, Facultad de Filosofía y Letras, Universidad de Buenos Aires.

TESSONE, A. AND J. BELARDI
2010. Evaluación de variaciones temporales del $\delta^{13}C$ y $\delta^{15}N$ en el colágeno de herbívoros de los lagos Tar y San Martín (provincia de Santa Cruz, Patagonia). In *Zooarqueología a Principios del Siglo XXI, Aspectos Teóricos, Metodológicos y Casos de Estudio*, edited by M. Gutiérrez, M. De Nigris, P. Fernández, M. Giardina, A. Gil, A. Izeta, G. Neme and H. Yacobaccio, pp. 345-357. Ediciones del Espinillo, Buenos Aires.

TESSONE, A., A.F. ZANGRANDO, G. BARRIENTOS, R. GOÑI, H. PANARELLO AND M. CAGNONI
2009. Stable isotope studies in the Salitroso Lake Basin (southern Patagonia, Argentina): assessing diet of Late Holocene hunter-gatherers. *International Journal of Osteoarchaeology* 19(2): 297-308.

TESSONE, A., A. ZANGRANDO, G. BARRIENTOS, S. VALENCIO, H. PANARELLO AND R. GOÑI
2005. Isótopos estables del carbono en Patagonia meridional: datos de la cuenca del lago Salitroso (Provincia de Santa Cruz, República Argentina). *Magallania* 33(2): 21-28.

THORSON, R.M. AND V.T. HOLLIDAY
1990. Just what is geoarchaeology. *Geotimes* 35(7): 19-20.

UGAN, A., G. NEME, A. GIL, J. COLTRAIN, R. TYKOT AND P. NOVELLINO
2012. Geographic variation in bone carbonate and water $\delta^{18}O$ values in Mendoza, Argentina and their relationship to prehistoric economy and settlement. *Journal of Archaeological Science* 39(8): 2752-2763.

VILLA, M.V. AND D.M. KLIGMANN
2013. Primeros resultados de los análisis geoarqueológicos del sitio El Churcal (Quebrada de Humahuaca, Jujuy). In *Arqueometría Argentina: Estudios Pluridisciplinarios*, edited by M. Ramos, M.

Lanza, V. Helfer, V. Pernicone, F. Bognanni, C. Landa, V. Aldazabal and M. Fernández, pp. 137-158. Programa de Arqueología Histórica y Estudios Pluridisciplinarios, Departamento de Ciencias Sociales, Universidad Nacional de Luján. Aspha Ediciones, Buenos Aires.

VOGEL, J.C. AND J.C. LERMAN
1969. Groningen radiocarbon dates VIII. *Radiocarbon* 11: 351-387.

VON POST, L.
1929. Die zeichenschrift der pollenstatistik. *Geol. Fören. Förh.* 51: 543-565.

WATERS, M.R.
1992. *Principles of Geoarchaeology. A North American Perspective*. The University of Arizona Press, Tucson.

WEINER, S.
2010. *Microarchaeology. Beyond the Visible Archaeological Record*. Cambridge University Press, New York.

YACOBACCIO, H., P. ESCOLA, M. LAZZARI AND F. PEREYRA
2002. Long distance obsidian traffic in Northwestern Argentina. In *Geochemical Evidence for Long Distance Exchange*, edited by M.D. Glascock, pp. 167-203. Bergin and Garvey, Westport.

YACOBACCIO, H. AND M. MORALES
2005. Mid-Holocene environment and human occupation of the Puna (Susques, Argentina). *Quaternary International* 132(1): 5-14.

YACOBACCIO, H.D., M.R. MORALES AND C.T. SAMEC
2009. Towards an isotopic ecology of herbivory in the Puna ecosystem: New results and patterns on *Lama glama*. *International Journal of Osteoarchaeology* 19(2): 144-155.

YACOBACCIO, H.D., F.X. PEREYRA, M. LAZZARI AND M.D. GLASCOCK
2004. Quest for ancient routes: obsidian sourcing research in Northwestern Argentina. *Journal of Archaeological Science* 31(2): 193-204.

YACOBACCIO, H.D., C.T. SAMEC AND M.P. CATÁ
2010. Isótopos estables y zooarqueología de camélidos en contextos pastoriles de la puna (Jujuy, Argentina). In *Zooarqueología a Principios del Siglo XXI. Aportes Teóricos, Metodológicos y Casos de Estudio*, edited by M. Gutiérrez, M. De Nigris, P. Fernández, M. Giardina, A. Gil, A. Izeta, G. Neme and H. Yacobaccio, pp. 77-86. Editorial del Espinillo, Buenos Aires.

YESNER, D.R., M.J. FIGUERERO TORRES, R.A. GUICHÓN AND L.A. BORRERO
1991. Análisis de isótopos estables en esqueletos humanos: confirmación de patrones de subsistencia etnográficos para Tierra del Fuego. *Shincal* 3(2): 182-191.

2003. Stable isotope analysis of human bone and ethnohistoric subsistence patterns in Tierra del Fuego. *Journal of Anthropological Archaeology* 22(3): 279-291.

ZANGRANDO A.F., A. TESSONE, S. VALENCIO, H. PANARELLO, M.E. MANSUR AND M. SALEMME
2004. Isótopos estables y dietas humanas en ambientes costeros. In *Avances en Arqueometría 2003*, pp. 91-97. Universidad de Cádiz, España.

ZÁRATE, M.A.
1986-87. Marco estratigráfico y geológico del sitio 3 de la localidad arqueológica Cerro La China (provincia de Buenos Aires). *Relaciones de la Sociedad Argentina de Antropología* N.S. XVII(1): 29-35.

1993. La geología del cuaternario y la arqueología. In *Actas del Curso de Postgrado "Introducción al Estudio del Cuaternario"*, edited by A.P. Calmels, J.O. Tullio and O.C. Carballo, pp. 313-321. Santa Rosa.

1994. Geoarqueología. En *Jornadas de Arqueología e Interdisciplinas*, pp. 21-33. PREP - CONICET, Buenos Aires.

1997. Late Pleistocene geoarchaeology of the southern Pampas, Buenos Aires province, Argentina. *Anthropologie* XXXV(2): 197-205.

ZÁRATE, M.A. AND N. FLEGENHEIMER
1991. Geoarchaeology of the Cerro La China locality (Buenos Aires, Argentina): site 2 and site 3. *Geoarchaeology* 6(3): 273-294.

ZÁRATE, M.A. AND A.P. PRIETO
1994. Role of the geological sciences in the archaeological research of Buenos Aires province, Argentina. *Quaternary of South America and Antarctic Peninsula* 10: 311-320.

ZOLITSCHKA, B., F. SCHÄBITZ, A. LÜCKE, H. CORBELLA, B. ERCOLANO, M. FEY, T. HABERZETTL, S. JANSSEN, N. MAIDANA, C. MAYR, C. OHLENDORF, G. OLIVA, M.M. PAEZ, G.H. SCHLESER, J. SOTO, P. TIBER AND M. WILLE
2006. Crater lakes of the Pali Aike Volcanic Field as key sites for paleoclimatic and paleoecological reconstructions in southern Patagonia, Argentina. *Journal of South American Earth Sciences* 21(3): 294-309.

CHEMICAL CHARACTERIZATION OF OBSIDIAN IN CENTRAL WESTERN ARGENTINA AND CENTRAL CHILE: ARCHAEOLOGICAL PROBLEMS AND PERSPECTIVES

Valeria Cortegoso[1], Michael D. Glascock[5], Anna María De Francesco[4], Víctor Durán[1], Gustavo Neme[2], Adolfo Gil[2], Martín Giesso[3], Lorena Sanhueza[6], Luis Cornejo[6], Ramiro Barberena[1] and Marco Bocci[4]

[1] Universidad Nacional de Cuyo-CONICET, [2] Museo de Historia Natural de San Rafael-CONICET, [3] Northeastern Illinois University, [4] Università della Calabria, [5] Research Reactor Center-University of Missouri, Columbia, [6] Universidad de Chile

Abstract

Trace element analysis has been performed on 911 samples from 106 archaeological sites and more than 100 samples from six sources. The samples come from archaeological sites located in all environments within a large area located between 32° and 37° south and 67° and 72° west, and in a chronological range of 9000 to 300 years BP. Analyses were performed to determine the spatial distribution of sources and to investigate the potential of these data to test models of mobility and exchange proposed for different regions in that area. Three different X-ray fluorescence (XRF) spectrometers were used: two energy dispersive and one wavelength dispersive. This paper discusses some problems of interpretation of archaeological and geological records related to the application of different methods and equipment. This experience shows the need to improve the geochemical sampling of sources and to confront different methods of analysis and calibration. The results provide an adequate resolution for discriminating the use of different sources throughout the Holocene, indicating variable spatial and temporal distributions. The general trend shows a preferential use of the sources located in the mountainous regions. Las Cargas source, used since the early Holocene (*ca.* 9000-8000 BP), together with Laguna del Maule source, were the most important sources in the entire area (including eastern and western slopes of the Andes).

Key words: *Chemical characterization - Obsidian - Archaeological problems*

INTRODUCTION

Using different geochemical methods, several characterization studies of obsidian sources in Chile and Argentina have taken place in recent decades. The results show a large chronological range of exploitation and an increasingly precise knowledge of the spatial distribution of obsidian sources throughout different areas in both countries (Durán *et al.* 2004; Seelenfreund *et al.* 1996; Stern *et al.* 2000; Yacobaccio *et al.* 2002).

During the past twenty years, a number of research projects have generated questions related to the exploitation, circulation and use of raw materials in the regions we are studying (Durán 2000; Durán *et al.* 2004, 2006; Gil 2006; Lagiglia 1997; Neme 2002). Until the mid-1990s, little was known about the procurement and distribution of obsidian in the Cuyo region (i.e., Central western Argentina) and central Chile. In 2002, a systematic survey of obsidian sources in the region located between 32 and 37 degrees latitude south was initiated. Several obsidian sources have been located and chemically characterized by means of neutron activation analysis (Durán *et al.* 2004) and X-ray fluorescence (De Francesco *et al.* 2006a; Giesso *et al.* 2011).

This paper discusses some methodological problems regarding interpretation of the archaeological and geological records related to the application of different methods and equipment. Trace element analysis has been performed on 911 samples from 106 archaeological sites and more than 100 samples from six sources throughout the South Central Andes (Durán *et al.* 2012; Giesso *et al.* 2011). The samples come from archaeological sites located in all environments within the area and in a chronological range extending from 9000 to 300 years BP. Three different XRF spectrometers were used to perform the measurements: two energy-dispersive XRF (i.e., an Elva-X table top and a Bruker III-V portable from the University of Missouri Research Reactor - MURR), and a wavelength-dispersive XRF (i.e., Philips PW 1480 from the University of Calabria). The data for the archaeological sites was compared with the results of source analysis performed with XRF and with previous results obtained with Neutron Activation Analysis (NAA). The analysis was performed to investigate the spatial distribution of obsidian from six sources. Three of them are high-altitude cordilleran sources: Laguna del Diamante, Las Cargas, and Laguna del Maule. The remaining three are low-altitude extra cordilleran sources: El Peceño, Cerro Huenul and Payún Matrú. The potential of these data to test models of mobility and exchange proposed for different regions in the area are evaluated.

The purpose of this paper is to present the lessons learned during the years of studying the sources of archaeological obsidian in the area in order to assess methodological insights. The large number of samples that have been analyzed using different methods, as well as consultation with different laboratories, enabled the detection and correction of inaccuracies during the process. The primary problem identified in recent years is related to the characterization and assignment of one of the sub-sources in the Laguna del Diamante source, designed as Arroyo Paramillos, which is geochemically similar to Las Cargas source, separated by more than 100 km. Among the previously published results two unexpected patterns appeared: at first, it was noted that there were unexpected distribution patterns of Laguna del Diamante source, particularly conspicuous by its absence in the archaeological sites in the vicinity of this source in both slopes of the Andes (Durán *et al.* 2004; Giesso *et al.* 2011). The second unexpected pattern was the broad distribution of Las Cargas obsidian source that had a dominant presence even in the Maipo valley, located in western slope of the Andes, much closer to Laguna del Diamante source. The following discussion will rectify these results recently presented in the light of a reallocation of the sub-source Arroyo Paramillos of Laguna del Diamante. New data is also presented for the dispersion of obsidian in the archaeological sites located nearby this source in the region of Central western Argentina, and in the upper Maipo, Chile (Durán *et al.* 2012). By providing a review and critique of the results, some of the possible difficulties in these kind of research will be identified, providing information that can be useful for those beginning to study the distribution of obsidian in other regions using comparable methods.

ENVIRONMENT AND OBSIDIAN SOURCES IN SOUTHWESTERN SOUTH AMERICA

The study area covers a large part of central Chile and Argentina (32° to 37° south latitude and 67 ° to 72 ° west longitude). This large territory extends from the coast of the Pacific Ocean on the west to the Mendoza plains on the east. The region is divided by the Andean mountains, which have an average width of 150 km, and heights reaching up to 6900 m a.s.l. The eastern plains include some prominent geological features, such as the southern volcanic fields reaching heights up to 3500 m a.s.l. The region as a whole is a mosaic of heterogeneous environments influenced by the Pacific Ocean, the Cordillera de la Costa, the Andes and the eastern plains. From a west-to-east perspective, the environments located to the west of the Andes are more humid than those in the east and the vegetation is more abundant. The eastern slopes of the Andes are arid to semiarid. The rivers that drain to the Pacific are shorter but carry a greater volume of water than those that drain into the Atlantic. On both slopes of the Andes, the altitude causes significant differences in the distribution of precipitation in the local environments.

In order to organize the information, seven areas have been distinguished (Figure 1): 1) The northern Cordillera which includes the upper basin of the Tunuyán and Mendoza rivers in Argentina; 2) The western Andean slopes which includes the basins of the Maipo and Cachapoal rivers in their lower valleys, and the Pacific coast; these areas do not have obsidian sources; 3) The central Chilean Cordillera which includes the upper Cachapoal and Maipo river basins; this area has not been systematically surveyed yet, but unpublished information suggests the existence of obsidian in the region; 4) The central Argentinean Cordillera which includes the upper Atuel and Diamante river basins; the sources located here are Las Cargas and Laguna del Diamante; 5) The central eastern plains, between the Atuel and Diamante rivers, and the Llancanelo basin; the El Peceño source is found here; 6) The southern Cordillera which includes the Río Grande basin; the Maule sources are located in this area; 7) The southern eastern plains which includes the Barrancas-Colorado river and the Payunia volcanic region; Payún Matrú and Cerro Huenul sources are found here.

Numerous volcanoes, some still active, are located in the central and southern portions of the region. Volcanic activity decreases to the north of 34 degrees, but it continues to the south in Patagonia. Extra-Andean volcanism (*volcanismo de retroarco*) is very abundant in southern Mendoza, an extension of several thousand square km of volcanoes that were active during the Holocene in the Payunia region. Some of these volcanoes rise to heights of more than 3500 m a.s.l. The distribution of volcanoes and silicic lavas conditions the presence and absence of obsidian sources throughout the region (Durán and Mikkan 2009).

The Sources

As shown in Figure 1, a total of six sources were located. The first three are in the Cordillera region: (a) Laguna del Diamante with two sub-sources: Arroyo Las Numeradas and Arroyo Paramillos; (b) Las Cargas, and (c) Laguna del Maule with three sub-sources: Laguna del Maule, Arroyo El Pehuenche and Laguna Negra. The remaining three sources are located in the oriental plains: (d) Cerro Huenul; (e) Cerro Peceño, and (f) Payún Matrú. The large amount of obsidian knapping debris recorded in Laguna del Maule - Laguna Negra, El Peceño, Cerro Huenul and Las Cargas, suggests that they were major obsidian quarries.

Laguna del Diamante: is a lagoon in an old volcanic caldera of 300 km^2 located at 3200 m a.s.l. in the upper valley of Diamante river near the border between Argentina and Chile. The access to this source is seasonally restricted from both sides. The area includes ignimbrites and pyroclastic deposits. The size of nodules is more than 30 cm in diameter on the edge of the caldera, high ground above 3800 m a.s.l., progressively decreasing along the streams that drain into the lagoon. Nodules found on the beaches are 2 to 3 cm. Two sub-sources can be geochemically distinguished in the same volcanic complex: Arroyo Las Numeradas and Arroyo Paramillos.

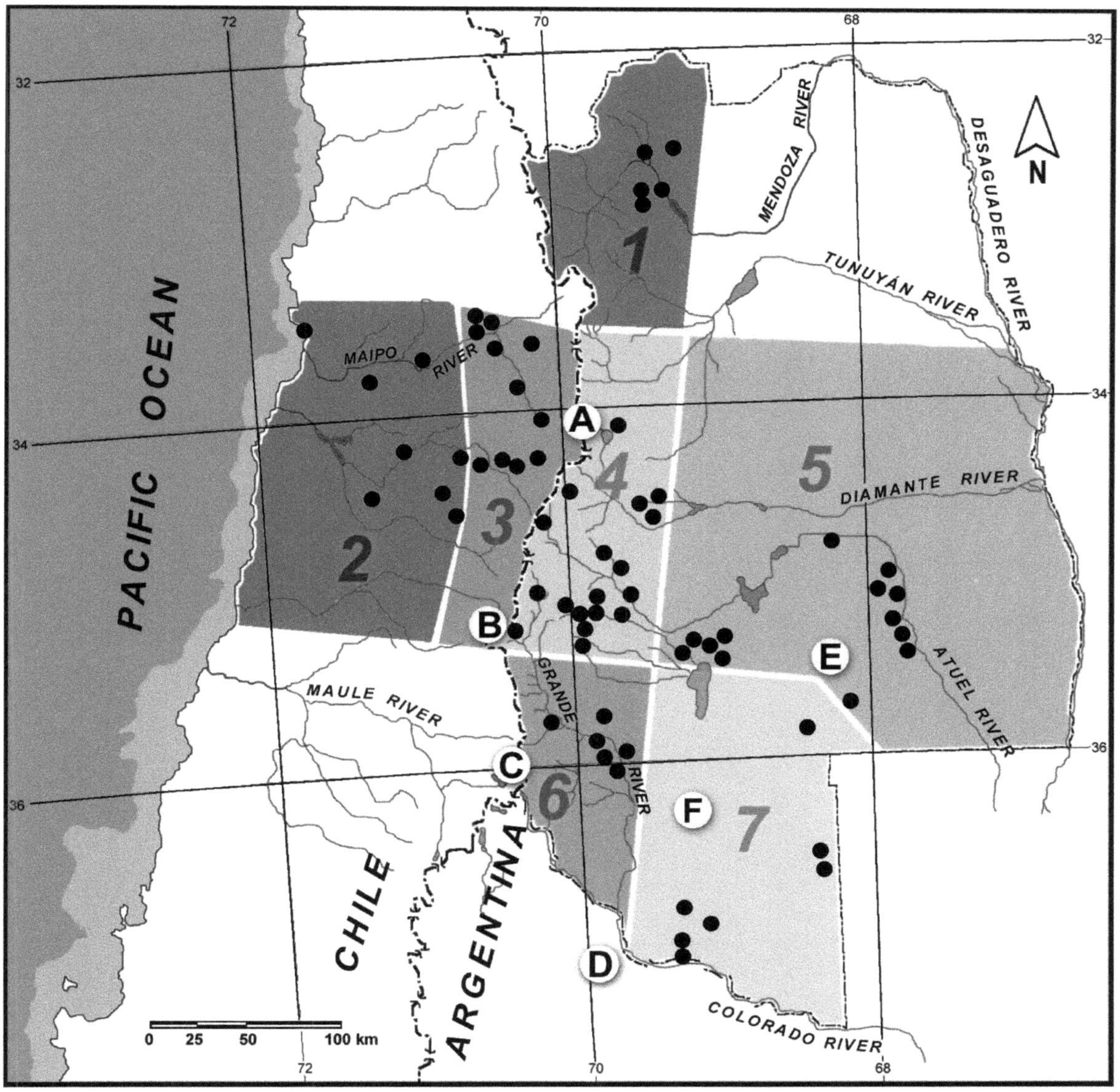

Figure 1: The seven areas in which the region was divided and the archaeological sites and sources analyzed by this paper. Sources: A) Laguna del Diamante; B) Las Cargas; C) Laguna del Maule); D) Cerro Huenul; E) El Peceño; F) Payún Matrú.

Las Cargas: the source is located on the border between Argentina and Chile, at 2350 m a.s.l in the volcanic complex Planchón-Peteroa. Access to the source is seasonally restricted from both sides. The total surveyed area of the primary source is around 8 km^2, but according to information from local inhabitants, the distribution of the obsidian in the area should be greater. The obsidian appears to be associated with volcanic tuff with glass inclusions and blocks that can reach 0.5 m^3. It is a source of good-quality obsidian, and includes a great amount of knapping debris over large surfaces.

Laguna del Maule - Laguna Negra - Arroyo El Pehuenche: the Laguna del Maule area is a vast volcanic complex located in the high Cordillera, between Argentina and Chile, at altitudes around 2400 m a.s.l. It is the largest obsidian source in the study region, covering an area of approximately 900 km^2. The source has several outcrops in Laguna Negra (Argentina) including blocks that can reach 1 m^3, dispersed in a 20-km^2 area. In other sectors, like Arroyo Pehuenche, the nodules are smaller, between 2 and 5 cm in diameter. It is very variable, but has high quality obsidian, and the access is seasonally restricted. The three sectors have a relatively homogeneous chemical signature.

El Peceño: is located at around 1450 m a.s.l., on the northwestern flank of the El Nevado volcano, in eastern Mendoza (Durán *et al.* 2004). Raw materials are dispersed over a radius of ~1000 m around the cone. The nodules vary from large (30 cm in diameter) to very small (2 cm). The raw material availability is good, their quality is variable, and is not a massive outcrop like the other sources. The source is available year round.

Payún Matrú: is a large-scale volcanic complex, located on the eastern plains (1800 m a.s.l), being available year round. There are two main volcanoes (Payún Matrú and Payún Liso) with heights over 3600 m a.s.l. Surveys were conducted on the majority of the crater and slopes of Payún Matrú. Lava flows that contain blocks of obsidian were located in these sectors, but they contain a low quality raw material for knapping. Therefore, it is still necessary to locate the precise provisioning area. It is possible that the source has been hidden by recent volcanic events (Durán and Mikkan 2009).

Cerro Huenul: this source is located at 900 m a.s.l., on a plateau on the right margin of the Colorado River (Barberena *et al.* 2011; Durán *et al.* 2004; Seelenfreund *et al.* 1996). Access is easy and year round, and the obsidian is of high quality. The source includes a lot of ravines that have scattered obsidian over several square kilometers. The nodules are of medium to small size, usually not larger than 10 cm in diameter. Raw material suitable for knapping is quite abundant (Durán *et al.* 2004).

MATERIALS AND METHODS

ED-XRFs from the University of Missouri Research Reactor

Source characterizations are based on neutron activation analysis; the first study included 37 pieces from four sources (Durán *et al.* 2004). Currently there are 154 samples analyzed by NAA from all the sources (Giesso *et al.* 2011). Nondestructive analysis of 428 obsidian artifacts from 68 archaeological sites was performed in 2007 with an *ElvaX* desktop energy-dispersive x-ray fluorescence (ED-XRF) spectrometer (Giesso *et al.* 2011) transported from the University of Missouri to the Universidad Nacional de Cuyo - CONICET (Figure 2). The instrument consists of an x-ray generator, an x-ray detector, and a multi-channel analyzer (MCA). The detector is a solid-state Si-pin-diode with an area of 30 mm^2 and a resolution of 180 eV at 6.4 keV (at a count-rate of 6000 counts per second). The *ElvaX* uses thermo-electric cooling instead of liquid nitrogen to cool the solid-state detector. The output signal of the detector is formed by a time-variant time-processor with pile-up rejector, base line restorer, and automatic adaptation of shaping time to the input count rate. The MCA consists of a fast shaping amplifier (FSA) and a 4096-channel spectrometric analog-to-digital converter (ADC), built as a successive approximation ADC with channels, a 32-bit per channel buffer RAM, "sliding scale" linearization of differential non-linearity, and dead time correction circuit. The x-ray tube has a tungsten anode with a 140 μm Be window.

Analysis using the *Elva-X* was conducted at 35 kV using a tube current of 45 μA and operating time of 400 seconds. Concentrations were calculated in parts per million using the *ElvaX Regression* program based on the quadratic regression model from a series obsidian reference samples previously characterized by XRF and neutron activation analysis (NAA). The analysis permits quantification of the following eleven elements: potassium (K), titanium (Ti), manganese (Mn), iron (Fe), zinc (Zn), gallium (Ga), rubidium (Rb), strontium (Sr), yttrium (Y), zirconium (Zr), and niobium (Nb). The equipment projects individual spectra on the computer screen, allowing visual comparisons with previous samples from sources and artifacts to perform preliminary source identification.

The semi-portable XRF equipment provides a precise, fast and inexpensive approach to this vast obsidian collection, including the analysis of complete tools without having to transport them outside of Argentina. Very few samples could not be assigned with certainty to any of the sources, usually due to their very small size, and these were subsequently taken to the University of Missouri Research Reactor (MURR) for analysis by NAA. To identify the obsidian sources, the MURR NAA database was used and 20 additional samples were added to the analysis.

In 2009, a Bruker Tracer III-V portable ED-XRF was taken to the Universidad Nacional de Cuyo-CONICET where analysis of 41 samples of sources and 509 artifacts from 46 sites was completed. The Bruker instrument that has a rhodium anode was operated at 40kV and 17 uA. Count times were 180 seconds. The analysis allowed the quantification of 13 elements: potassium (K), titanium (Ti), manganese (Mn), iron (Fe), zinc (Zn), gallium (Ga), rubidium (Rb), strontium (Sr), yttrium (Y), zirconium (Zr), niobium (Nb), lead (Pb) and thorium (Th). The equipment also displayed a spectrum on the computer screen that allowed comparisons between artifacts and samples from sources and enabled a very quick visual identification. These analyses, which are still being processed, are based on reallocation of the sub-sources and will improve the results previously obtained.

WD-XRF in the Laboratory of Earth Sciences, University of Calabria

Based on the proposals of Crisci *et al.* (1994) and De Francesco *et al.* (2008), 101 obsidian artifacts from 12 archaeological sites in Argentina and three in Chile were tested using the non-destructive method of X-ray fluorescence, but using a different equipment and statistical analysis (Durán *et al.* 2012). The equipment was a Philips PW 1480 wavelength dispersive XRF located in the Laboratory of Earth Sciences at the University of Calabria. Non-destructive analytical methodology was chosen to define the origin of obsidian used in the manufacture of artifacts. While this method may be considered less accurate than others (for example: NAA), it has the advantage of preserving the integrity of the samples. In routine analysis by XRF and NAA, the samples are reduced to powder, preventing the study artifacts with heritage value. Moreover, this methodology has been successfully used in numerous archaeological sites in Italian and European Neolithic (Biagi *et al.* 2007; De Francesco *et al.* 2005, 2006b, 2008, 2011; Marini *et al.* 2007) and in the study region (De Francesco *et al.* 2006a).

Fragments with shapes and volumes similar to those of the archaeological samples were extracted from the nodules of obsidian sources (27 samples of 7 sources and sub-sources were analyzed). These fragments were then analyzed by WD-XRF. Intensity data obtained was compared with those from powders using the standard method, to verify the sensitivity for discriminating between the different geochemical types. Using the standard method, parts of the source samples were reduced to powder, pressed into pellets, and then analyzed using WD-XRF. Major, minor and trace elements were analyzed, including Nb (Niobium), Y (Yttrium), Rb (Rubidium), Zr (Zirconium), Sr (Strontium), Ni (Nickel), Cr (Chromium), V (Vanadium), La (Lanthanum), Ce (Cerium), and Co (Cobalt). The diagram in Figure 3 (Sr / Rb vs Nb / Zr) shows the excellent correspondence between the absolute concentration ratios of trace elements obtained using nWD-XRF on powder and X-ray intensity ratios obtained with the non-destructive methodology on intact fragments of the obsidian sources.

Five chemical elements (Nb, Y, Zr, Rb and Sr) were used to define the various areas of origin. This choice is justified by the sensitivity of these elements that characterize the magmatic processes forming the rhyolite fluid and give origin to the obsidian when it becomes solid. The difference between the equipments used in these two labs is that the PW 1480 uses wavelength dispersive XRF while the Elva and the Bruker are energy dispersive XRF, and they do not analyze the same chemical elements.

DISCREPANCIES BETWEEN LABORATORIES IN THE CHARACTERIZATION OF LAGUNA DEL DIAMANTE-ARROYO PARAMILLOS SUB-SOURCE AND THEIR PARTIAL OVERLAPPING WITH LAS CARGAS SOURCE

A better knowledge of the regional sources allows us to improve different aspects of the obsidian landscape in central Chile and central western Argentina. At the beginning, an incomplete survey of southern and central Mendoza sources led us to misunderstand the regional obsidian use patterns, conducing to some errors in the assumed provenience of some of the analyzed archaeological samples (Durán *et al.* 2004). Further surveys in the remaining unexplored areas allowed us to fill some gaps in this regional sourcing program, as presented in a recently published paper (Giesso *et al.* 2011, see below). However, some important questions did not find definitive answers in that contribution, especially regarding the use of two sources, Las Cargas and Laguna del Diamante (Arroyo Paramillos sub-source). As we mentioned above, the information available at that time suggested that Las Cargas was represented at a larger scale than expected, appearing even in contexts located in the proximity of Laguna del Diamante source. On the contrary, Laguna del Diamante source represented an unexpectedly small area restricted to the immediate vicinity of the source, not further than 5 km (Giesso *et al.* 2011).

Research Reactor Center - University of Missouri Analysis

The studies conducted until 2009 by the MURR, that included NAA analysis for samples sources and XRF for archaeological samples, showed a high presence of Las Cargas source. In that database Arroyo Paramillos sub-source was not discriminated (Figure 2). The archaeological samples initially assigned to Las Cargas included elements that correspond to the sub-source Arroyo Paramillos, since they have very similar chemical signals. This chemical similarity had already been highlighted by De Francesco *et al.* (2006), whom on the basis of the concentrations of Nb, Y, Zr, Rb, and Sr obtained by XRF were able to discriminate both sources (Figure 3), obtaining results that show a wider archaeological distribution and more significant presence of Arroyo Paramillos obsidian in the Maipo river basin and in the eastern plains (Durán *et al.* 2012). To resolve these conflicting data, new surveys were recently extended at these sources and chemical studies by NAA at MURR were made on other samples. On this basis, archaeological samples previously assigned to Las Cargas (Giesso *et al.* 2011) have now been reassigned to Arroyo Paramillos. These recent analyses will allow us to rework previous archaeological interpretations.

Laboratory of Earth Sciences - University of Calabria Analysis

In 2006 a series of X-ray analysis were carried out in the Laboratory of Earth Sciences (Calabria University). The obtained results were not in complete agreement with the previous analysis developed by the MURR (De Francesco *et al.* 2006). The comparison between the trace elements concentrations (in ppm) acquired from obsidian in powders and the secondary X-ray intensities of the same chemical elements analyzed in entire fragments of the same samples is presented in Figure 3.

The details of the non-destructive methodology are reported in De Francesco *et al.* (2008). Because of the impossibility of correcting surface effects, the x-ray intensity ratios of only five chemical elements (Nb, Y, Zr, Rb and Sr) have been used. It is sufficient to characterize the different provenance areas. The Figure 3 is representative of the comparison between the two methodologies. The diagram indicates the presence of several groups corresponding to the different obsidian sources, and underlines the coincidence in the results obtained with both methodologies. The values of the intensity ratios never match the values of the concentration ratios exactly, as data are obtained from two different methodologies; the observed differences are not really significant for practical purposes and never influence the discriminating power of the obsidian groups. As the Figure 3 shows, the samples from Las Cargas source and Arroyo Paramillos sub-source are clearly differentiated.

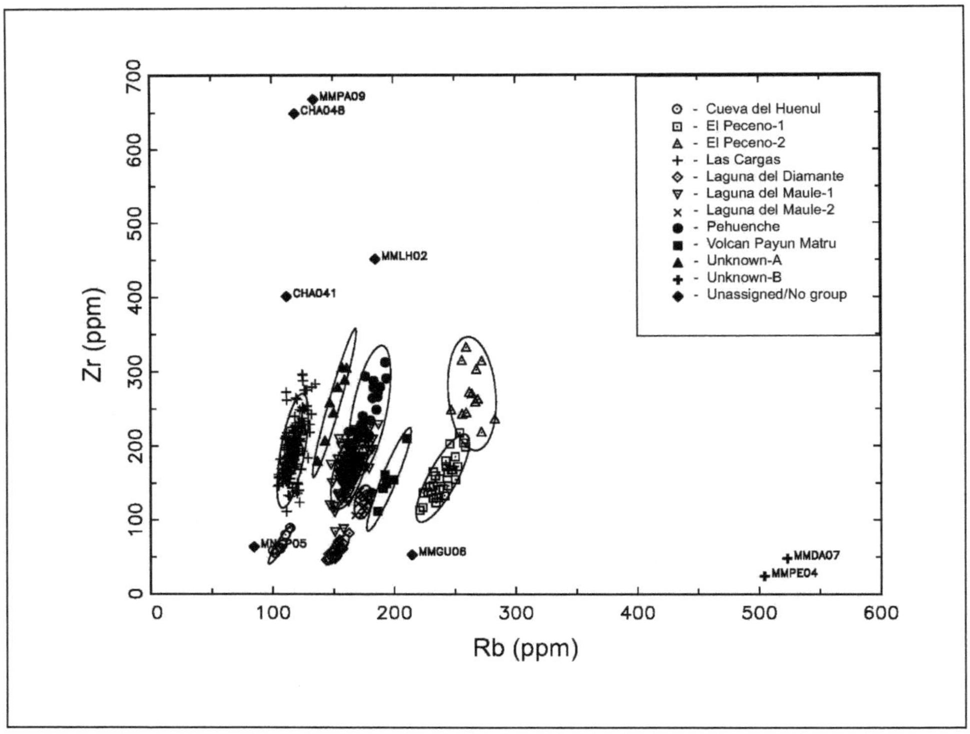

Figure 2: XRF MURR. Archaeological obsidian clusters obtained with the portable spectrometer Elva-X.

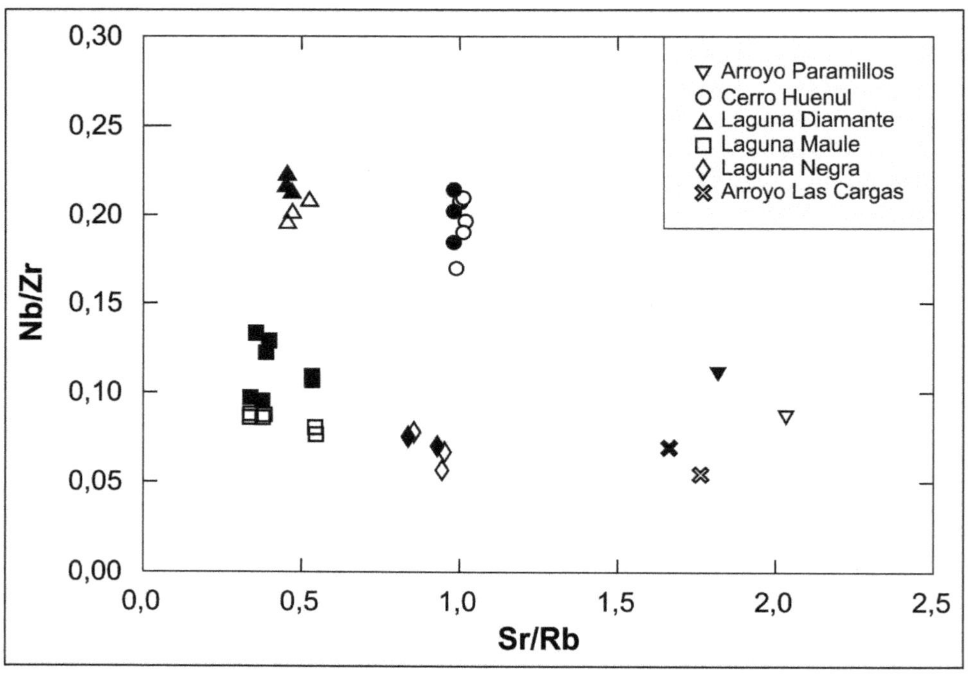

Figure 3: Chemical characterization of the sources by XRF University of Calabria: Comparison between the destructive and non-destructive XRF methodologies. Full symbols: ppm concentration ratio on powder; Empty symbols: intensity ratio on the entire fragments.

DISCUSSION OF THE RESULTS PUBLISHED IN GIESSO *ET AL.* 2011

In that previous paper, the results of XRF analysis for 428 obsidian samples from central western Argentina and central Chile were evaluated, with chronologies that extend from the early Holocene to the European contact. The results indicated differential patterns of procurement and use of obsidian through space and time. In some cases these patterns coincided with our expectations, according to the distance between sources and archaeological sites, while in other cases the patterns differed from the expected tendencies. Most of the conclusions presented in this work can be supported with the new data presented here. However, as previously mentioned, it has been necessary to review the data presented for Laguna del Diamante, taking into account the reallocation of the sub-source Arroyo Paramillos. A summary of those results showed the importance of the use of Cordilleran sources (in terms of their spatial and temporal distribution). Among the Cordilleran sources, Las Cargas and El Maule were the most heavily utilized, both on the eastern and western slopes of the Andes. The large sizes of the obsidian outcrops, the diversity and the high quality of the obsidian, are the reasons that could explain their wide spatial and temporal distribution.

The wrong assignation of the sub-source Arroyo Paramillos of Laguna del Diamante to the chemical signature of Las Cargas produced an artificially wider distribution for the latter. This affects the results for areas around Laguna del Diamante, particularly the Chilean sites of the Maipo valley, where there seemed to be a prevailing use of the far Las Cargas source. On this basis, it was suggested that Las Cargas was the most important source for Cuyo and central Chile (41.8% n=182) (Giesso *et al.* 2011; Figure 1. Areas: 2-3). Besides, Las Cargas was practically the only type of obsidian in Chilean sites, even though Laguna del Diamante is closer to most of them.

The obsidian from Laguna del Diamante had one of the most unusual patterns according to the results presented by Giesso *et al.* (2011). Its exploitation was limited to sites located in the vicinity of the source (similar results in Durán *et al.* 2006). It is difficult to explain the limited use of this source as a consequence of access problems, since the main alternatives (El Maule and Las Cargas) are also emplaced in high-altitude settings. It is possible that the lack of obsidian circulation from Laguna del Diamante was related to the very small size of the nodules available there. Surprisingly, obsidian from Laguna del Diamante source has a very restricted range. This source is not represented at sites from the western and eastern slopes of the Andes, located in the closer valleys of the area. The only two sites with obsidian from this source are located in close proximity to the source (~5 km). On the contrary, obsidian from Las Cargas is the only source present in the Maipo valley, which happens to be the closest valley to Laguna del Diamante source on the western slope.

As noted below, the results that contradicted the expected patterns for the use of this source can be explained using the reconfiguration of the sources that should be done with the new results of the characterizations obtained in different laboratories.

DISCUSSION OF THE RESULTS OBTAINED IN CALABRIA

With the results obtained in those previous works, it was difficult to explain the absence of obsidian from Laguna del Diamante source in the archaeological sites from the Maipo valley (western slope) and also in northern Mendoza sites (eastern slope). The studies showed that the Laguna del Diamante obsidian (both sub sources: Arroyo Las Numeradas and Arroyo Paramillos) did not appear in archaeological sites farther than 5 km of the sources (Durán *et al.* 2004; Giesso *et al.* 2011). In those studies it was suggested there was a late and spatially localized exploitation of this obsidian source by hunter-gatherers adapted to Andean environments. Also investigated was the possibility that the former distribution was due to a methodological problem, impeding discrimination of Laguna del Diamante-Arroyo Paramillos and Las Cargas sources (Durán *et al.* 2012). As mentioned earlier, even when these sources are far away, they have similar chemical signatures. In the first results, obtained at the University of Calabria, a separation between these two sources was highlighted (De Francesco *et al.* 2006).

Of the 101 artifacts tested in Italy, 51 are from archaeological sites located near or geographically related to the obsidian sources from Laguna del Diamante area. Results presented by Durán *et al.* (2012) allowed a further discussion of this issue. The new analysis showed the presence of obsidian from Laguna del Diamante on sites in the extra-Andean basin of the Diamante river (Alero Montiel and Potrerillos del Diamante), in the Andean Maipo river basin (Los Queltehues), and further north in the eastern piedmont (El Manzano Histórico). This allows us to recognize a much wider spatial distribution of the Laguna del Diamante source than the previously postulated. Chronological information regarding the use of different types of obsidian shows that all cordilleran sources were exploited since very early and almost continuously.

The new results obtained for several artifacts from sites in Laguna del Diamante area (LD-LD-S2 and S4) indicate a clear prevalence of the nearest obsidian sources (Laguna del Diamante sub- sources: Arroyo Las Numeradas and Arroyo Paramillos, located closer than 10 km of them) (Durán *et al.* 2012). Only 11.4% of the obsidian at these sites comes from distant sources, indicating a clear preference for local lithic resources despite the small size and low density of the nodules. The minor presence of artifacts made in obsidian from Las Cargas (about 100 km south) and Laguna Negra-Maule (both more than 220 km to the southwest) could be used to assert that the groups that occupied this mountain area had, almost during the last 2000 years, high mobility ranges. But the same results could also be used to defend acquisition by indirect means, such as exchange (De Francesco *et al.* 2006).

It is also important to take into account that studies in relation to other Cordilleran sources have corroborated some of the conclusions reached in previous publications. In particular, the chronological information confirms that all cordilleran sources were exploited since very early periods and almost continuously. It also highlights the importance of the Las Cargas and El Maule sources, which have a wide distribution that could be related to their supply of high quality obsidian and quantity of obsidian.

FINAL REMARKS

This paper emphasizes the comparison of the results and methodological discussions from the experience gained with the application of alternative techniques for the study of trace element of sources and archaeological obsidian. This research has been already performed for more than a decade, including the field survey of sources and sites, and their chemical characterization. This endeavor has proven to be fundamental for the investigation of several issues in the archaeology of the South Central Andes. The experience of working with portable spectrometers has been extremely valuable, allowing to generate a great amount of information (see also Craig *et al.* 2007; Goodale *et al.* 2012). Most of the results obtained from these tests have been confirmed with the results obtained in the laboratory at the University of Calabria.

This information has recently led to open the discussion about the exchanges systems between societies located on both sides of the Andes, the establishment of territories by hunter-gatherers, as well as changes in patterns of exploitation that occurred during the Holocene (Durán *et al.* 2012). The results allow us to affirm that all the sources within the region were exploited, but with variable spatial and temporal trends, showing a preferential use of the sources located in the mountains. In this sense, recent studies in relation to Cordilleran sources have corroborated most of the conclusions reached during the preliminary works (Giesso *et al.* 2011). The chronological information regarding the use of different types of obsidian confirms that all cordilleran sources were exploited since very early periods and almost continuously. Las Cargas, used since the early Holocene (*ca.* 9000-8000 BP), together with Maule, were the most important sources in the entire region, including eastern and western slopes of the Andes. The wide distribution of these sources could be related to their supply of high quality obsidian. At the same time, this reanalysis allows us suggest a wider spatial distribution of the Laguna del Diamante source than previously considered.

In this paper we have placed a special emphasis on clearing the discrepancy that emerged with the identification of some pieces earlier mistakenly assigned to Las Cargas source, instead of the Arroyo Paramillos sub-source from Laguna del Diamante. This problem emerged as the fingerprint of the Arroyo Paramillos sub-source had many chemical similarities with Las Cargas source. We had already indicated that these results did not match our expectations, and for this reason it was necessary to go over the results (Giesso *et al.* 2011). The new data for sources presented here from different labs allows us to improve our understanding of the supra regional obsidian landscape in the region. This paper emphasizes a concern that arose with the variation of results from different laboratories in relation to the characterization of the sources from Laguna del Diamante. To some extent the error in this source distribution was due to a striking chemical similarity of Las Cargas source with one of the sub-sources from Laguna del Diamante. In the first work done with a portable spectrometer, a large number of samples from the area linked to the Laguna del Diamante were associated with the source of Las Cargas. The use of archaeometrical techniques in this region begins to offer answers to some questions. Within this research, non-destructive X-ray Fluorescence has proved useful. The results confirm, once again, that it is a valid tool to determine the origin and dispersal of archaeological obsidian.

ACKNOWLEDGEMENTS

Two anonymous reviewers provided valuable suggestions to improve the manuscript. Programa de Investigación y Desarrollo de la Universidad Nacional de Cuyo: Perspectivas paleoecológicas para el estudio de las relaciones humano ambientales en el centro occidente argentino. Agencia Nacional de Promoción Científica y Tecnológica PICT-2006-N°421, Secretaría de Ciencia y Técnica- Universidad Nacional de Cuyo 06/G502: Holoceno medio y tardío en el Centro Oeste Argentino-Análisis del registro arqueológico de sitios cordilleranos: cambios en la subsistencia y la tecnología. Agencia Nacional de Promoción Científica y Tecnológica PICT-2006-N° 046 y PICT 2006-00046: Tendencias temporales en el uso humano del paisaje: arqueología y paleoambientes en el sur de Mendoza. PIP CONICET 114-200801-00177: Explotación de ambientes cordilleranos del centro oeste durante el Holoceno: variabilidad espacial y temporal en la organización de la tecnología lítica. PICT (IDAC-ICES): Impacto humano en el sur de Mendoza durante el Holoceno tardío-cambios en las estrategias de uso de los recursos.

REFERENCES

Barberena, R., A. Hajduk, A. Gil, G. Neme, V. Durán, M. Glascock, M. Giesso, C. Borrazzo, M. Pompei, L. Salgán, V. Cortegoso, G. Villarosa and A. Rughini
2011. Obsidian in the south-central Andes: Geological, geochemical, and archaeological assessment of north Patagonian sources (Argentina). *Quaternary International* 245: 25-36.

Biagi, P., A.M. De Francesco and M. Bocci
2007. New data on the archaeological obsidian from the middle-late Neolithic and chalcolithic sites of the Banat and Transylvania. In *The Lengyel, Polgar and Related Cultures in the Middle/Late Neolithic in Central Europe*, edited by Kozowski, J.K. and P. Raczky, pp. 309-326. Krakow, Jagiellonian University and ELTE.

Craig, N., R.J. Speakman, R.S. Popelka-Filcoff, M.D. Glascock, J.D. Robertson, M.S. Shackley and M.S. Aldenderfer
2007. Comparison of XRF and PXRF for analysis of archaeological obsidian from southern Perú. *Journal of Archaeological Science* 34(12): 2012-2024.

CRISCI, G.M., M. RICQ-DE BOUARD, U. LANZAFAME AND A.M. DE FRANCESCO
1994. Nouvelle méthode d'analyse et provenance de l'ensemble des obsidiennes néolithiques du midi de la France. *Gallia Préhistoire* 36: 299-327.

DE FRANCESCO, A.M. AND M. BOCCI
2007. Risultati dell'analisi non distruttiva in XRF sulle ossidiane dei siti archeologici di Cala Giovanna Piano (Arcipelago Toscano) e Torre Foghe. In *"Preistoria e Protostoria dell'area Tirrenica"* a cura di Carlo Tozzi and Michele Claude Weiss. Ed. Felici, Pisa, 121.

DE FRANCESCO, A.M., M. BOCCI, G.M. CRISCI AND U. LANZAFAME
2005. Caratterizzazione archeometrica delle ossidiane del Monte Arci: confronto fra metodologia tradizionale in XRF e metodologia XRF non distruttiva. *Proceeding of III° Convegno Internazionale L'ossidiana del Monte Arci nel Mediterraneo*, Ed. PTM (OR), pp. 117-128.

DE FRANCESCO, A.M., M. BOCCI, G.M. CRISCI, F. MARTINI, C. TOZZI, G. RADI, L. SARTI AND M.T. CUDA
2006b. Applicazione della metodologia analitica non distruttiva in Fluorescenza X per la determinazione della provenienza delle ossidiane archeologiche del progetto "Materie Prime" dell'I.I.P.P. *Atti della XXXIX Riunione Scientifica dell'Istituto di Storia e Protostoria* 1: 531-548.

DE FRANCESCO, A.M., G.M. CRISCI AND M. BOCCI
2008. Non-destructive analytical method by XRF for determination of provenance of archaeological obsidians from the Mediterranean Area. A comparison with traditional XRF method. *Archaeometry* 50(2): 337-350.

2011. Non destructive application of wavelength XRF in obsidian studies. In *X-Ray Fluorescence Spectrometry (XRF) in Geoarchaeology*, edited by M.S. Shackley, pp. 81-107. New York, Springer.

DE FRANCESCO, A. M., V. DURAN, A. BLOISE AND G. NEME
2006a. Caracterización y procedencia de obsidianas de sitios arqueológicos del área natural protegida Laguna del Diamante (Mendoza, Argentina) con metodología no destructiva por fluorescencia de rayos (XRF). In *Arqueología y Ambiente de Áreas Naturales Protegidas de la Provincia de Mendoza*, edited by V. Durán and V. Cortegoso, pp. 53-67. Anales de Arqueología y Etnología 61, Facultad de Filosofía y Letras, Universidad Nacional de Cuyo, Mendoza.

DURÁN, V.
2000. *Poblaciones Indígenas de Malargüe. Su Arqueología e Historia*. Serie Libros 1, CEIDER, Mendoza.

DURÁN, V., A. DE FRANCESCO, G. NEME, V. CORTEGOSO, L. CORNEJO AND M. BOCCI
2012. Caracterización y procedencia de obsidianas de sitios arqueológicos del Centro Oeste de Argentina y Centro de Chile con metodología no destructiva por Fluorescencia de Rayos (XRF). *Intersecciones en Antropología* 13(2): 423-437.

DURÁN, V., M. GIESSO, M. GLASCOCK, G. NEME, A. GIL AND L. SANHUEZA R.
2004. Estudios de redes de aprovisionamiento y redes de distribución de obsidiana durante el Holoceno Tardío en el sur de Mendoza (Argentina). *Estudios Atacameños* 28: 25-43.

DURÁN, V. AND R. MIKKAN
2009. Impacto del volcanismo holocénico sobre el poblamiento humano del sur de Mendoza (Argentina). *Intersecciones en Antropología* 10: 295-310.

DURÁN, V., G. NEME, V. CORTEGOSO AND A. F. GIL
2006. Arqueología del área natural protegida Laguna del Diamante (Mendoza, Argentina). In *Arqueología y Ambiente de Áreas Naturales Protegidas de la Provincia de Mendoza* edited by V. Durán and V. Cortegoso, pp. 81-134. Anales de Arqueología y Etnología 61, Facultad de Filosofía y Letras, Universidad Nacional de Cuyo, Mendoza.

GIESSO, M, V. DURÁN, G. NEME, M. GLASCOCK, V. CORTEGOSO, A. GIL AND L. SANHUEZA
2011. Obsidian source usage in the Central Andes of Argentina and Chile. *Archaeometry* 53: 11-21.

GIL, A.
2006. *Arqueología de La Payunia (Mendoza, Argentina): El Poblamiento Humano en los Márgenes de la Agricultura*. BAR International Series 1477, Oxford.

GOODALE, N. D.G. BAILEY, G.T. JONES, C. PRESCOTT, E. SCHOLZ, N. STAGLIANO AND C. LEWIS
2012. pXRF: A Study of Inter-Instrument Performance. *Journal of Archaeological Science* 39(4): 875-883.

LAGIGLIA, H.
1997. *Arqueología de Cazadores-Recolectores Cordilleranos de Altura*. Ediciones Ciencia y Arte, San Rafael.

MARINI, N., A.M. DE FRANCESCO, M. BOCCI, C. BRESSY AND B. GRATUZE
2007. Costa di U Monte - du Néolithique à l'âge du Fer sur la cote orientale corse: résultats de fouilles et provenance des vestiges. *Préhistoire et protohistoire de l'aire tyrrhénienne* a cura di Carlo Tozzi e Michel Claude Weiss. PROGETTO INTERREG III A Francia - Italia - Isole Toscana, Corsica,Toscana, Sardegna RICERCA 3.1. Unione Europea, Ed. Felici, Pisa, 35-42.

NEME, G.
2002. Arqueología del alto valle del río Atuel: modelos, problemas y perspectivas en el estudio de las regiones de altura del sur de Mendoza. In *Entre montañas y Desiertos: Arqueología del Sur de*

Mendoza, edited by A. Gil and G. Neme, pp. 65-84, Sociedad Argentina de Antropología, Buenos Aires.

SEELENFREUND, A., C. REES, R. BIRD, G. BAILEY, R. BÁRCENA, AND V. DURÁN
1996. Trace element analysis of obsidian sources and artifacts of central Chile (Maule River basin) and western Argentina (Colorado river). *Latin American Antiquity* 7(1): 7-20.

STERN, C.R., J. GÓMEZ OTERO AND J.B. BELARDI
2000. Características químicas, fuentes potenciales y distribución de diferentes tipos de obsidianas en el norte de la provincia del Chubut, Patagonia Argentina. *Anales del Instituto de la Patagonia* 28: 275-90.

YACOBACCIO, H.D., P.S. ESCOLA, M. LAZZARI AND F.X. PEREYRA
2002. Long-distance obsidian traffic in northwestern Argentina. In *Geochemical Evidence for Long-Distance Exchange*, edited by M. Glascock, pp. 167-203. Bergin and Garvey, Westport, CT.

APPLICATION OF PHYSICOCHEMICAL AND MACROSCOPIC METHODS TO LITHIC ARTIFACT STUDIES FROM ALERO CUEVAS SITE (SALTA, REPÚBLICA ARGENTINA): A COMPLEMENTARY APPROACH

Cecilia Mercuri[1] and Federico Restifo[1]

[1] CONICET - Instituto de Arqueología, Facultad de Filosofía y Letras, Universidad de Buenos Aires

Abstract

Nowadays, obsidian is one the rocks mostly used to understand technological organization, availability and/or accessibility of raw materials, group mobility and subsistence strategies. This rock presents a number of minor elements whose concentrations are specific to each formation or part of a geological formation, making it possible to assign a sample to a particular source. While a macroscopic characterization may be performed, not all archaeological assemblages have the same potential for distinction from particular sources. We explain and compare how a physical chemical method (XRF) and the macroscopic description works in terms of assessing the scope and limitations of its application to our lithic assemblage, which comes from two levels from the archaeological site Alero Cuevas (Santa Rosa de los Pastos Grandes, Puna of Salta), which presents a stratigraphic sequence that extends throughout the Holocene. Recovered lithic assemblages exhibit not only variety at the morphological level but also regarding the representation of different varieties of obsidian. This would suggest the existence of changes in the technological strategies and patterns of mobility (among others) over time.

Key words: *Puna of Salta - Obsidian sourcing analysis - Lithic record*

INTRODUCTION

In recent years, analyses of sources of lithic raw materials have become more relevant. One of the rocks most frequently used to study technological organization, availability and/ or accessibility of raw materials, degree of group mobility and subsistence strategy (Earle and Ericson 1977; Eerkens *et al.* 2008; Glascock 2002; Roth 2000, among others) is obsidian. On the one hand, this raw material and/ or its products have been the subject of extensive circulation networks throughout the world (cf. Earle and Ericson 1977; Hatch *et al.* 1990; Yacobaccio *et al.* 2004; Zeitlin 1982). On the other hand, it is a geologically scarce material, which occurs in concentrated locations. This volcanic extrusive igneous rock belonging to the group of silicates has a number of minor elements with concentrations below 1% (commonly referred to as trace elements), which are specific to each formation or part of a geological formation (Zumberge 1974, among others). Because of this, it is possible to assign a sample to a particular source (Glascock 2002; Hughes 1986; Shackley 1998, among others).

Usually, in a first approach, the archaeological sites are analyzed macroscopically. This is also the case for the preliminary identification of lithic raw materials, which is usually made with a reference collection, identified with the help of a geologist. This is true even when analyzing obsidian (Psota 1990). Rojas *et al.* (2004) divided the methods of obsidian analysis into three general groups: physical description, chemical and physical analysis of natural radioactivity. The first group includes macro and microscopic studies and among the chemical and physical ones, the analysis of sources.

Obsidian sources in the study area are small domes and lava domes associated with large boilers and stratovolcanoes composed mostly of andesite-dacite lavas and large deposits of ignimbrite (cf. Viramonte *et al.* 1984). These obsidians are usually related to rhyolite eruptive phases with gaps containing pumicias, Plinian eruptions, ash, and lava obsidian (Yacobaccio *et al.* 2004). There are more than fourteen obsidian lava flows in NW Argentina (NWA) (Viramonte *et al.* 1988). Three of the sources were identified; three are located in the province of Salta (Quirón, Ramadas, Alto Tocomar), six are in Catamarca (Ona-Las Cuevas, Valle Ancho, Cueros de Purulla, Chascón, Laguna Cavi and Salar del Hombre Muerto); one is located on the border of the two provinces (Archibarca) and another on the border of Jujuy, Bolivia and Chile (Laguna Blanca (Zapaleri) and Caldera Vilama 1 and 2) (Escola and Hocsman 2007; Lazzari *et al.* 2009; Seelenfreund *et al.* 2010; Yacobaccio *et al.* 2002, 2004).

It is noteworthy that for the moment, the source of Archibarca is only known from a few nodules that appeared near the volcano of the same name, on the frontier between the provinces of Catamarca and Salta and the source of the Salar del Hombre Muerto is also known from the presence of concentrations of nodules near the Salar (Hocsman, pers. comm.).

Taking into account the available published information, in this paper we explain and compare the results obtained using methods of physical chemical analysis and of macroscopic description, and show how, in our case, the complementarity of the two methods was useful for the study of different sources of obsidian. Different authors have done the same in other areas, comparing also the results obtained through different methods of chemical characterization (cf. Craig *et al.* 2007; Shackley 1988, among others). In our case, we apply these methods to the study of two archaeological levels from Alero Cuevas site, which have different dates. Finally, we briefly discuss the implications of the results of our studies considering the regional research in the highlands of South-Central Andes.

ANALITICAL METHODS FOR ROCK DETERMINATION

Analysis by X-Ray Fluorescence

X-ray fluorescence (XRF) spectroscopy is widely used for qualitative and quantitative elemental analysis of environmental, geological, biological and industrial samples, among others (Craig *et al.* 2007; Eerkens *et al.* 2002; Roux and Lennard 2006; Vázquez and Escola 1994). Compared with other techniques (AAS, ICPS, NAA), has the advantage of being non-destructive, multi-elemental, fast and positive in their cost effectiveness (Tykot and Ammerman 1997), but less accurate than other methods (Glascock and Neff 2003).

The identification of elements by methods involving X-rays is possible due to the characteristic radiation emitted by atoms from the interior of their electronic shells under certain conditions: when the beam of a high energy electron collides with a material, one of the results is the ejection of photoelectrons from the inner cortex of the atoms (Jenkins 1988). The ejected electron leaves a gap in the electronic structure of the atom. After a brief period, the atomic electrons are rearranged with an electron of higher energy, filling the vacancy. During relaxation, the atom becomes fluorescent, emitting a photon of X-rays whose energy is equal to the difference in energies of initial and final states. Detection and measurement of the photon energy allows us to determine the element and the specific electronic transition which originated (Jenkins 1988). The X-ray spectrum acquired in the process shows a number of characteristic peaks. The energy of the peaks allows the identification of elements present in the sample (qualitative analysis), while the intensity of the peaks provides the concentration of relevant or absolute elements (semi-quantitative analysis or quantitative).

Quantitative XRF analysis requires calibration of the measuring parameters, which can be done with empirical parameters that are based on analysis of standards with known elemental compositions, or fundamental with no standard key, which are based on mathematical algorithms that describe the physics of the detector response to pure elements.

There are two types of detection limits that should be taken into account: the instrumental and method. The latter is related to sample preparation and analysis (cf. Jenkins 1988). In this regard, the surface of the items under study must be clean and free of labels. If artifacts have been labeled, it is best left them the way they are than to remove the paint due sometimes residues remain and it is preferable that the location of this is clear and obvious (the presence of paint is reflected in high levels of titanium, zinc or lead). The samples selected for XRF analysis should have a size not less than 10 mm in diameter and a minimum 1.5 mm thick (Shackley 1998; and Thatcher n / d). The smaller samples tend to distort the values of the elements, but in many cases can be characterized with some degree of reliability. However, the use of small specimens is not recommended in areas or regions where the universe of obsidian sources is not well known (Potts and Webb 1992; Skinner and Thatcher n / d). The analytical results of lenticular or biconvex surfaces are comparable to those of flat surfaces or pills (made with powdered sample), so non-destructive characterization studies are viable (Hughes 1986; Tykot and Ammerman 1997, among others).

Naked Eye Macroscopic Analysis

In this case, a set of macroscopic observable attributes that are considered important for the characterization of the rocks are taken into account. It should be remarked that this is a subjective analysis so that the recorded variables and the characterization are partially subject to the needs and goals of individual research teams and especially the archaeological assemblage to analyze in addition to the knowledge of the rocks in the area.

It is important to highlight that the accuracy of this method depends not only on the knowledge of the reference collection, but also on the characteristics of the analyzed rocks. Not all assemblages have the same analytical potential. Obsidians can be macroscopically difficult to distinguish from each other, but when the group presents some variability the macroscopic method is feasible (cf. Psota 1990, among others). The time that takes the analysis is related to the size of the sample since, in principle; we analyze all the pieces of the assemblage. Specimens should be collected and examined on a table (preferably white so as not to interfere with the recording of some of the variables such as color), and with adequate lighting.

Some of the variables that can be taken into account are based on macroscopic analysis of rocks and minerals made by geologists (Clarke and Hillebrand 1897; Garrison 2003). Thus, in this case, optical and physical properties such as brightness, transparency, color, inclusions, veins, cortex, fracture, texture, homogeneity (Mercuri 2008a). Those were the characteristics that we have taken into account in our analysis.

We compare the results with the reference collection. We suggest including specimens of different sizes and thicknesses in the reference collection in order to improve comparability.

CASE STUDY: ALERO CUEVAS SITE, SANTA ROSA DE LOS PASTOS GRANDES

The cases of study are located in the Santa Rosa de los Pastos Grandes basin (SRPG, see Figure 1). It is located about 240 km to the NW of Salta Capital in the Department of Los Andes, connecting the Argentinean Puna with mesothermal valleys (northern Calchaquí Valley) and northern Chile, being also an area of transition towards South Puna (López 2008). With an approximated area of 300 km² (López 2008) and an average altitude higher than 4000 m a.s.l., the basin is formed by the waters collected from the snowy peaks of Pastos Grandes, which form a broad highland wetlands (*vega*, Vilela 1969). It participates in the Puna ecological gradient which, being a high desert, is characterized by the presence of salt deposits and a sparse and scattered vegetation consisting of mainly small shrubs (tolares) and grasses (Göbel 2002; Ruthsatz and Movía 1975; Vilela 1969). Among xerophytic vegetation, there are other plant species characteristic of the humidity of the highland wetlands. The abundance of highland grasses allows nowadays the existence of flocks of grazing llamas (López *et al.* 2004). The little soil development, coupled with the harsh conditions typical of the Puna, do not allow the establishment of stable agriculture. Regarding lithic resources, the area has good availability of resources for both the manufacturing of stone by flintknapping and by other techniques such as grinding and abrasion (cf. Mercuri 2009). However, no obsidian sources were detected within the limits of the basin (30 km around from Alero Cuevas) that we consider local in this case, the basin of Santa Rosa de los Pastos Grandes. This is the reason why, in principle, we believe that these rocks are non-local (see on this matter López 2008; Mercuri 2011).

During the archaeological fieldwork held in April 2004, a broad rock shelter of light yellowish white dacitic tuff was detected. It is located in a slope of a hill in the Quebrada de las Cuevas, a subsidiary ravine of the Quebrada de Santa Rosa, which was called Alero Cuevas (hereafter AC, see Figure 1). It is located at an altitude of 4,400 m a.s.l., in a ravine devoid of any human occupation in the present (López 2008), at a linear distance of 10 km from the village of Pastos Grandes. There are difficulties in accessing this site, given the steep slope and its altitude (going up towards the snowy peaks) and the scarcity of human traffic areas to the sides of the ravine, which is accentuated in the Quebrada de las Cuevas.

The rock shelter is located in a high visibility spot of the Quebrada, more than 5 meters from the bottom of it. This location allows a good visibility from the eaves to the ravine and its potential resources, due to its location in an area where the topography makes it an S. And moreover, it is also highly visible from nearby hills and others located farther away. The rock shelter measures 19.3 meters of frontage, and has a complex and long stratigraphic sequence of dates, ranging from 9650 ± 100 to 643 ± 35 BP (López 2008). López (2008) argues that the recurrence in occupations is related to a number of advantages such as the proximity of the rock shelter to basic resources -such as water and wood- the high visibility of most of the Quebrada and particularly, because it provides shelter in an area where there are few caves or rock shelters.

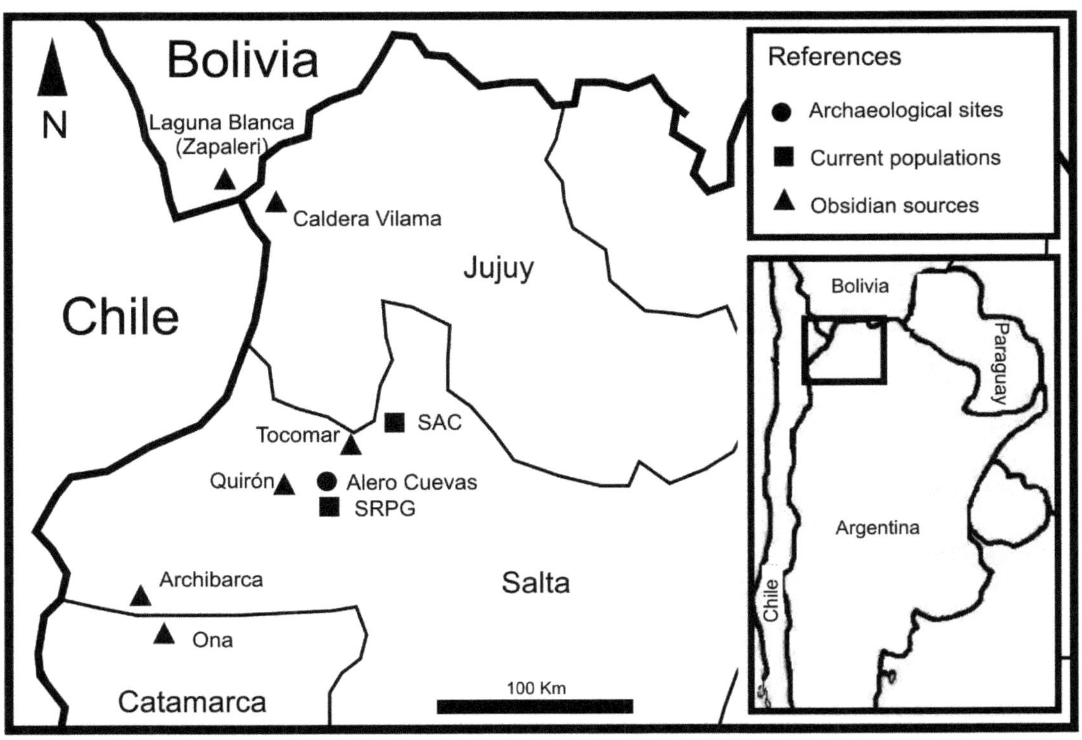

Figure 1. Map of Northwestern Argentina, showing the archaeological sites and obsidian source locations mentioned in the text. SAC= San Antonio de los Cobres, SRPG= Santa Rosa de los Pastos Grandes.

As it was mentioned, this paper presents and discusses the results of two layers with different dates. On one hand the F4 layer, dated on 8,504 ± 52 RC years BP (AA 71136 NSF, 7600-7490 cal BC), 8,838 ± 52 RC years BP (AA 71136 NSF, 8210-7750 cal BC) and 9,650 ± 100 RC years BP (LP 1736, 9300-8750 cal BC), and on the other hand layer C1, dated to 2,020 ± 60 (LP-1671 RC years B.P., 1 sigma 100 BC-60 AD).

Materials and Methods

Lithic assemblages recovered in the mentioned layers (see above) were, in the first place, analyzed macroscopically, following Aschero's techno-typological guidelines[1] (1975, 1983). In the case of obsidian identification, this analysis involved a number of variables (see above and Mercuri 2008a) that allowed us to discriminate groups, and guide the analysis. This was implemented because in other studies in the nearby area of San Antonio de los Cobres (SAC, see Figure 1) (ie. Mercuri 2006; Mercuri and Vazquez 2007; Muscio 2004) we had only detected a small variety of obsidian, and the macroscopic assignation was relatively easy for the distinctive external features of the recorded obsidians. These obsidians came from Tocomar and Zapaleri sources and to a lesser extent, Ramadas. In the case of Santa Rosa de los Pastos Grandes lithic assemblages, we found out that the variety was greater (cf. Mercuri 2008b, 2011; Mercuri and Glascock 2011), so we decided to apply this methodology (Mercuri 2008a).

A first classification allowed us to distinguish 5 groups taking into account their macroscopic characteristics (see Table 1). In order to distinguish potential sources, we compared archaeological artifacts with the reference collection of obsidians detected in the area, assigning potential sources to the each one of them. This step allowed us to select the sample we wanted to send to analyze by physicochemical methods, and, we wanted to confirm the presence of some obsidians, and on the other, we meant to clarify other we did not know (Table 1).

Group	Characteristics								Potential source	Source
	Brightness	*Colour*	*Transparency*	*Inclusions*	*Veins*	*Cortex*	*Texture, homogeneity*	*Fracture*		
1	much	translucent	yes	very small, black, small bubbles	no	no	soft but with irregularities	conchoidal	Quirón	Quirón
2	much	translucent	yes	-	parallel and white. Without influence in transparency	no	soft	conchoidal	Quirón	Quirón
3	very much	translucent black	yes, though with veins	microvesicles	parallel and black	no	very soft	very nice conchoidal fracture	Zapaleri	Zapaleri
4	soapy	gray	barely	small, golden	parallel and gray	¿?	tending to soft	conchoidal	-	Archibarca
5	little	very dark gray	barely	black, small, isolated	-	-	tending to soft	conchoidal	Cueros de Purulla	Archibarca

Table 1. Macroscopic characteristics of obsidian samples.

The obsidian samples were sent to MURR laboratory (Missouri University Research Reactor), in the U.S. in order to perform physicochemical analysis. The choice of this laboratory had to do with their large collection of reference calibrated in this lab for the area where we work (Escola 2004; Lazzari 2006; Yacobaccio *et al.* 2004, among others).

Results

In general, results obtained by XRF confirmed most of the observations made macroscopically. However, differences were recorded.

The analysis was accurate in the case of Quirón (n = 2) and Zapaleri (n = 1) (Figure 2 and Table 1). These obsidians present distinctive features such as translucency with black inclusions of mica and small bubbles (Solá 2007), and in some specimens white streaks in the first case, and in the second one, black high brightness and smooth texture (Mercuri 2011). The source of obsidian Quirón is located about 30 km from the current population of Santa Rosa de los Pastos Grandes (see Figure 1) and we were able to do some fieldwork and detect nodules of medium and small sizes, and areas with concentrations of archaeological material at different stages of reduction and shaping (cf. López 2008). In addition, hand samples were collected to constitute a reference collection. The obsidian source of Zapaleri is located along the southern and western margins of Laguna Blanca (Nielsen *et al.* 1999; Yacobaccio *et al.* 2002) about 200 km from Santa Rosa de los Pastos Grandes.

Samples of gray or dark gray colour, soapy shine, poor transparency and few inclusions and veins were assigned -through macroscopic analysis- to the source of Cueros de Purulla in Catamarca. However, XRF studies determined that the source would be Archibarca (Figures 1 and 2) on the frontier between Salta and Catamarca (see Figure 1), about 150 km SW of Santa Rosa de los Pastos Grandes.

Changes in the frequency of obsidian from different sources in the different analyzed layers could point out to changes in the technological strategies and mobility patterns (among other things) over time.

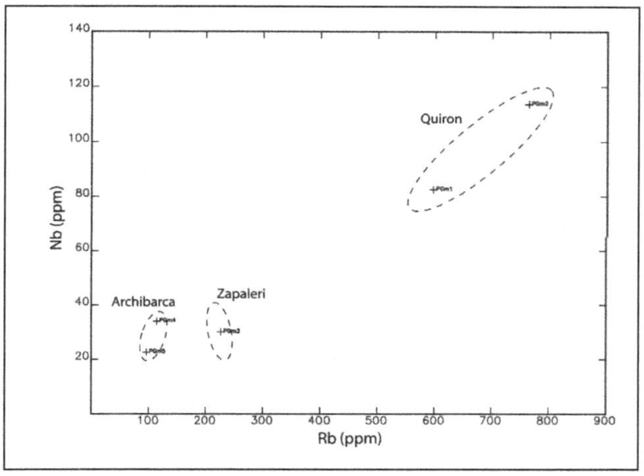

Figure 2. Bivariate plot of Rb versus Nb showing the source groups for artifacts from the site of Alero Cuevas. The source ellipses are drawn at approximately the 90% confidence level.

Layer F4 (8504 ± 52 AP, 8838 ± 52 BP and 9650 ± 100 BP)

The analyzed assemblage consists of 1,672 pieces. Undetermined flakes and flakes without platform were removed from the sample. The general fragmentation index is 69.5%, which may seem high. But in the lithic artifacts fracture rate is lower (53%). Overall, the assemblage reflects mainly the final stages of shaping artifacts. In turn, there is evidence of recycling of broken projectile points (recicled as scrappers) and tools with multiple edges (Restifo 2011). However, cases of maintenance and recycling of tools are very few (less than 5%).

In addition to local rocks such as andesite, quartzite and quartz, we determined the presence of 3 varieties of obsidian. The rock that dominates the entire lithic assemblage is obsidian from the source Quirón (n = 909, 54%), being relative frequencies of Zapaleri and Archibarca varieties significantly lower (2% and 5% respectively, Figure 3).

On the other hand, recognized artifactual classes for each variety of obsidian (Figure 4) are very different. The variety of Archibarca was identified only as debitage and as a triangular projectile point. The variety of Zapaleri occurred in debitage and as a fragment of an undifferentiated tool. Instead, Quirón obsidian was identified in most of debitage and tools, being represented in most of the artifactual classes (Restifo 2011). No cores were recovered in this layer.

Considering only the debitage, 65% of the total assemblage is composed by hyper-microflakes, followed by microflakes (32%), small flakes (2.8%) and flakes (0.2%). In the cases of Archibarca and Zapaleri only microflakes are represented (Figure 5). On the other hand, it has been detected that Quirón obsidian records debitage specimens in all sizes, though most of them are hyper-microflakes (Figure 6). In turn, flakes with sloping platforms and remnants of edges that indicate maintenance were detected, though at low frequencies. This pattern is consistent with what was emphasized in the artifact cluster.

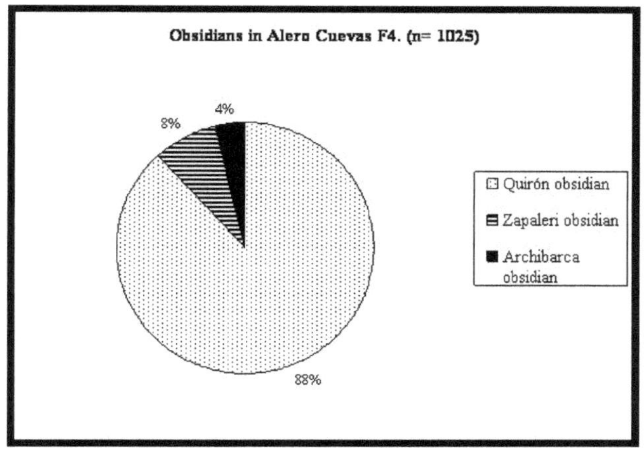

Figure 3. Obsidians in Alero Cuevas, F4.

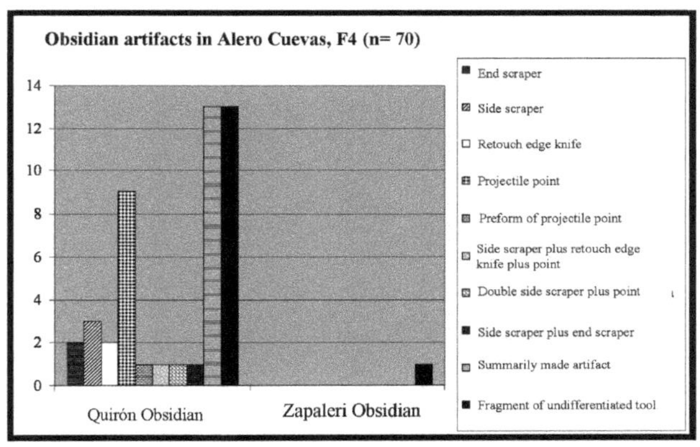

Figure 4. Obsidian artifacts in Alero Cuevas F4.

Layer C1 (2020 ± 60 BP)

The analyzed assemblage consisted of 436 pieces after removing of the sample fractured platform flakes and undifferentiated debitage. The overall fragmentation rate is 33%, which can be seen as high, however, it is noted that in the lithic artifacts the fracture rate is lower (14.3%). In general we can say that in this assemblage we observed the final stages of artifact production. We detected evidence of artifact reactivation, maintenance and reclamation practices in both lithic artifacts, such as projectile points, as well as in the debitage (cf. Mercuri 2011).

In this group a variety of raw materials, especially obsidian, was recorded. Thus, in addition to the presence of local andesite, quartzite and quartz, 5 varieties of obsidian were recorded. The rock that dominates the entire assemblage is obsidian from the source Quirón (n=

274, 62.844%), which shows a temporal continuity in the use of this rock. It is also worth noting that previous XRF analysis have shown the presence of two varieties of obsidian: Tocomar and one whose source is unknown (Unk-D), which had also been detected in Tulán 54 (Glascock, pers. comm., Mercuri and Glascok 2011). This latter rock is gray-dark gray one with very little transparency, tending to opaque. The variety Tocomar, whose source is located about 40 km W of Santa Rosa de los Pastos Grandes, is transparent and may have more or less bright white veins.

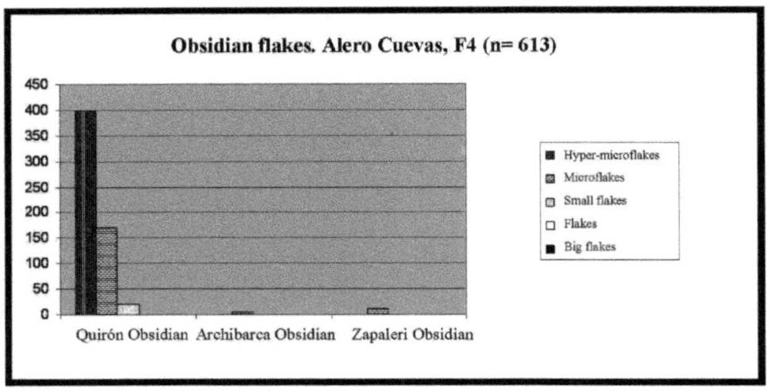

Figure 5. Obsidian flakes in Alero Cuevas F4.

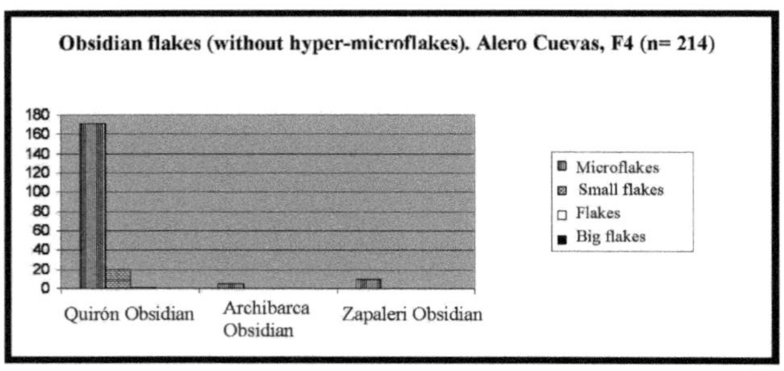

Figure 6. Obsidian flakes (without hyper-microflakes) in Alero Cuevas F4.

Varieties of Tocomar, Archibarca, Zapaleri and Unk-D obsidian are at a low frequency, with relatively similar rates, with 1%, 1%, 2% and 4% respectively (Figure 7). However significant differences regarding artifactual classes are recognized. Zapaleri and Archibarca varieties were recorded as both lithic artifacts and microflakes but not flakes or cores. The unknown source of Unk-D, was only detected in microflakes, although in more quantity than other varieties. And the Tocomar was only recorded in a few (n= 4) debitage of varying size.

With respect to lithic artifacts it is observed a relative variety of classes for obsidian of Quirón, as there were identified summarily made artifacts, one biface, projectile points, stemmed projectile points, composed artifacts of scraper plus tip between notches, and abundance of fragments of lithic artifacts (Figure 8) Some projectile points (37%), show evidence of having been rescharpened and three of the tool fragments could correspond to projectile points.

Other varieties of obsidian in which were recorded as lithic artifacts Archibarca (n= 2) and Zapaleri (n= 1). It was possible to identified only 1 stemmed projectile point and 1 side scraper for the first case, and 1 denticulate from the second (Figure 8). In the other obsidians (Tocomar and Unk-D) there were not lithic artifacts only debitage.

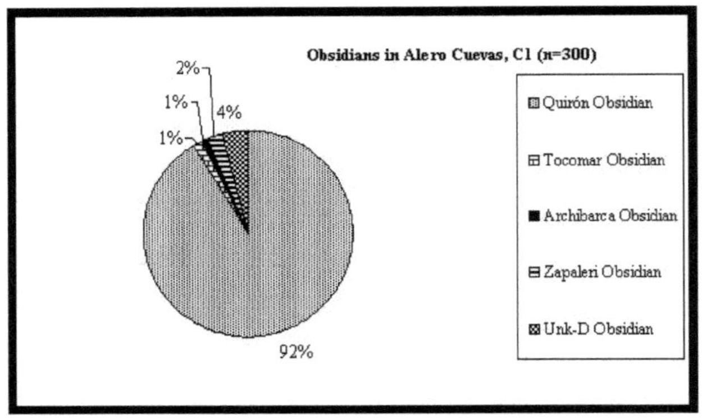

Figure 7. Obsidians in Alero Cuevas, C1.

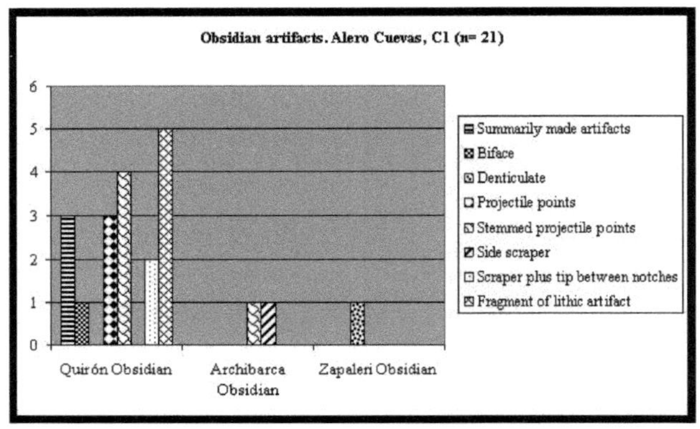

Figure 8. Obsidian artifacts in Alero Cuevas C1.

A total of 95.864% of debitage are hyper-microflakes (Figure 9), and are present in all the varieties of obsidian, although there is a majority of Quirón (88.235%, n= 225). It is noteworthy that in the case of obsidian, the smaller sizes are the most frequent ones (Figure 10). There is evidence that lead to think in stages of maintenance of the artifacts, such as bent platforms and remnants of sharpening edges.

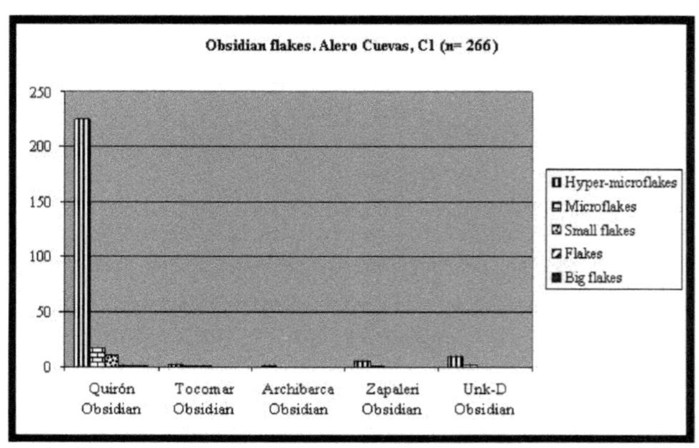

Figure 9. Obsidian flakes in Alero Cuevas C1.

DISCUSSION AND FINAL REMARKS

First, we may repeat that both methods applied in our case study were based on the knowledge of a wide reference collection of sources of obsidian, which included several samples of different sizes. A wide collection not only in a quantitative sense, but also qualitative. That is to say, a good collection, either of rock samples or core features of these is one in which it is represented the widest possible range (regardless of level of analysis).

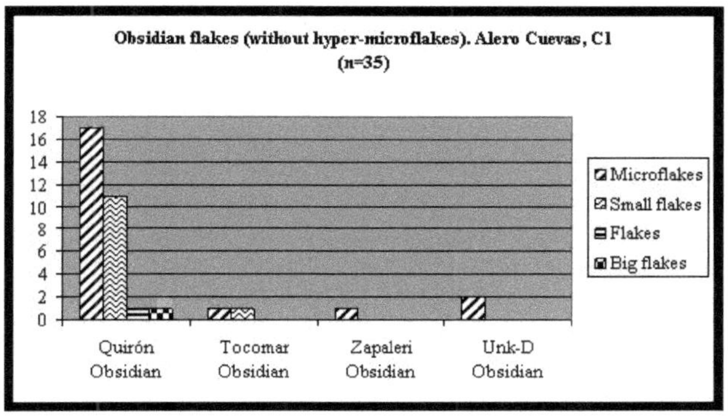

Figure 10. Obsidian flakes (without hyper-microflakes) in Alero Cuevas C1.

However, samples sent for XRF analysis were relatively few compared to the lithic assemblage recovered in the site. This information though is commonly very unclear in the literature, since it usually shows the number of items scanned, but it is not clear what percentage it represents and usually prevails the quantity over quality as even more often it is not indicated if the pieces come from surface or layer. Conversely, in macroscopic analysis, in principle, it is classified and try to identify all the artifacts, so the percentage of representation of the sample would be close to 100% (although it would be appropriate to discard the unidentified specimens).

The major advantage of XRF is that it achieves a determination with nearly 99% of reliability and is independent of the macroscopic variability of the whole rock and the region. Analysis to the naked eye is highly dependent on the macroscopic variability of the whole rocks and the region, so if the rocks have no diagnostic macroscopic features, it becomes very difficult to study and reach a determination of groups that help in the selection samples for analysis with physicochemical methods. However, in some cases macroscopic obsidian identification has reached 89% of success (see Bettinger *et al.* 1984, quoted in Psota 1990; Stern 1999, 2000, among others). However, we remark that this is very dependent on the assemblage under analysis and the characteristics of the rocks of the study region.

In our case study, two lithic assemblages with different dates from Alero Cuevas site were analyzed. The presence of obsidian varieties from Quirón (which is over 30 km of the site) and Zapaleri (300 km) was confirmed. We also discover the existence of the variety Archibarca (Table 1 and Figure 1), which is located about 150 km or so from site, and had not been previously detected in the area.

Detected patterns allow us to affirm some trends that cross the time and persist from layer F4 to layer C1. There is a predominant use of obsidian from the source of Quirón. This is recorded on both the variety of lithic artifacts and the abundant debitage, showing all stages of tool production, maintenance and recycling. As mentioned above, although the source of this material is outside of the limits of what we consider local (Santa Rosa de los Pastos Grandes basin), is located relatively close (30 km). This proximity could be a possible explanation for the continuing use and its prevalence over time.

In contrast, obsidians from farther distances are underrepresented in both analyzed assemblages. These materials appear to have entered the site as tools, which were subject to maintenance and recycling practices, economizing raw material. This is evident both in artifacts and in the debitage. With these obsidians were produced mostly projectile points.

However, there is a difference in the presence of obsidian found in the assemblages. While in the older layer only three varieties (Quirón, Zapaleri and Archibarca) were detected, in the layer dated *ca.* 2000 RC years BP we recognized five varieties (Quirón, Archibarca, Zapaleri, Tocomar and Unk-D).

It has been noted by other authors (cf. Yacobaccio *et al.* 2008) that the spatial scale at which the distribution of obsidian occurs in this area (either exchange, interaction, direct procurement), undergoes changes that accompany the changes from hunting gathering to productive economies. Thus, during the early Holocene it shows a more restricted distribution pattern, and at the middle of the Late Holocene, in moments of productive economies, the spatial distribution of obsidian becomes wider.

A plausible explanation for these patterns is that by the Late Holocene productive economies are beginning to emerge. The passage to the production of food may be consider traumatic in the sense that it is a process that involves not only economic changes (which are comparatively minor), but also changes with regard to social organization, mobility patterns, etc. (López 2008; Winterhalder and Kennett 2006, among others) and there are risks involved in this transition. That is, one way or another is subject to some type of environmental constraints in the area of study. In these contexts networks of reciprocity and social bonds that make up a reinsurance that can host populations under stress become more relevant (Mercuri 2011; Winterhalder and Goland 1993, among others).

This probably involves a series of changes related to access to different resources. For example, the ranges of action of a hunting society may include the direct exploitation of obsidian sources (Quirón obsidian variety), and to a lesser extent the achievement of other varieties, through social interactions (variety Zapaleri). In our case the assamblage from layer F4 shows a pattern that may support this hypothesis. While with the intensification of social interaction networks (cf. Tarragó 1994 among others) although it probably remained the direct access to some resources (Quirón obsidian), the variety of resources that are accessed indirectly increases (obsidians from Zapaleri, Unk-D and Tocomar) and some that may have been directly exploited during the early Holocene, see modified its accessibility to the Late Holocene (variety Archibarca). Until the moment, these are hypotheses that emerged from this work and should be tested with field and laboratory work.

Now, back to the beginning, the selection of the sample was performed macroscopically from similarities that could be seen with the naked eye, but we are aware that it should be verified with more physical-chemical analysis of obsidian from the two assemblages separately. According to macroscopical analysis (naked eye), group 5 was tentatively assigned to the variety Cueros de Purulla, as it resembles the sample of this raw material in our reference collection, but it turned out to come from Archibarca, on the frontier between the provinces of Salta and Catamarca. Because of this, we cannot rule out the differential presence of other different varieties of obsidian only distinguishable by physicochemical methods, in the two assemblages.

In short, it is important that these methods are used complementarily as other authors have already mentioned (Psota 1990; Rojas *et al.* 2004). It is best not just stick with a single one methodology and macroscopic analysis, although they do not have the resolution of the physicochemical, operate at a different level and are very important for a first classification of rocks in the assemblages. And ultimately what really matters is what is done with information obtained by these methods.

ACKNOWLEDGEMENTS

To all those who read earlier versions of this work and made comments. To those who provided samples for reference and information to carry out this work, Pato Escola, Jorge Martínez, Michael D. Glascock. To CONICET for the scholarships that allowed us to perform these studies. To the reviewers for their useful comments that helped improve the paper. Everything expressed here is our responsibility.

REFERENCES

ASCHERO, C.A.
1975. *Ensayo para una clasificación morfológica de artefactos líticos. Aplicada a estudios tipológicos comparativos*. Informe presentado a CONICET.

1983. *Ensayo para una clasificación morfológica de artefactos líticos, ficha y código descriptivo para artefactos formatizados con rastros complementarios y núcleos. Apéndice B*. Cátedra de Ergología y Tecnología, Facultad de Filosofía y Letras, UBA. Buenos Aires.

CLARKE F.W. AND W.F. HILLEBRAND
1897. *Analyses of rocks, with a chapter on analytical methods*, laboratory of the United States. *Geological Survey* 8(148): 1880-1896.

CRAIG, N., R.J. SPEAKMAN, R.S. POPELKA-FILCOFF, M.D. GLASCOCK, J.D. ROBERTSON, M.S. SHACKLEY AND M.S. ALDENDERFER
2007. Comparison of XRF and PXRF for analysis of archaeological obsidian from southern Peru. *Journal of Archaeological Science* 34(12): 2012-2024.

EARLE, T.K. AND J.E. ERICSON (EDITORS)
1977. *Exchange System in Prehistory*. Academic Press, New York.

EERKENS, J.W., H. NEFF AND M.D. GLASCOCK
2002. Ceramic production among small-scale and mobile hunters and gatherers: A case study from the western Great Basin. *Journal of Anthropological Archaeology* 21(2): 200-229.

EERKENS, J.W, K.J. VAUGHN, M.L. GRADOS AND M.J. EDWARDS
2008. La Ballena: a mining base camp in the southern Nasca region, Peru. *Antiquity* 82(315) Project Gallery.http://www.antiquity.ac.uk/projgall/eerkens315/

ESCOLA, P.S.
2004. Variabilidad en la explotación y distribución de obsidianas en la Puna Meridional argentina. *Estudios Atacameños* 28: 9-24.

ESCOLA, P. AND S. HOCSMAN
2007. Procedencia de artefactos de obsidiana de contextos arqueológicos de Antofagasta de la Sierra (*ca*. 4500-3500 AP). *Comechingonia* 10: 49-58.

GARRISON, E.G.
2003. *Techniques in Archaeological Geology*. Springer.

GLASCOCK, M.D.
2002. Obsidian provenance research in the Americas. *Accounts of Chemical Research* 35(8): 611-617.

GLASCOCK, M.D. AND H. NEFF
2003. Neutron activation analysis and provenance research in archaeology. *Measurement Science & Technology* 14(9): 1516-1526.

GÖBEL, B.
2002. La arquitectura del pastoreo: uso del espacio y sistema de asentamientos en la Puna de Atacama (Susques). *Estudios Atacameños* 23: 53-76.

HATCH J.W., J.W. MICHELS, C.M. STEVENSON, B.E. SCHEETZ AND R.A. GEIDEL
1990. Hopewell obsidian studies: Behavioral implications of recent sourcing and dating research. *American Antiquity* 55(3): 461-479.

HUGHES, R.E.
1986. *Diachronic Variability in Obsidian Procurement Patterns in Northeastern California and Southcentral Oregon*. University of California Publications in Anthropology 17, Berkeley, California.

JENKINS, R.
1988. *X-Ray Fluorescence Spectrometry*. J. Wiley & Sons, Inc., New York.

LAZZARI, M.
2006. *Traveling Things and the Production of Social Spaces: An Archaeological Study of Circulation, Value, and Material Culture in Northwestern Argentina (First millennium A.D.)*. Unpublished PhD Dissertation, Anthropology Department, Columbia University.

LAZZARI, M., L. PEREYRA DOMINGORENA, M.C. SCATTOLIN, L. CECIL, M.D. GLASCOCK AND R.J. SPEAKMAN
2009. Ancient social landscapes of northwestern Argentina: Preliminary results of an integrated approach to obsidian and ceramic provenance. *Journal of Archaeological Science* 36(9): 1955-1964.

LÓPEZ, G.E.J.
2008. *Arqueología de Pastos Grandes, Puna de Salta: Ocupaciones Humanas y Evolución a lo largo del Holoceno*. Unpublished PhD Dissertation, Facultad de Filosofía y Letras, Universidad de Buenos Aires.

LÓPEZ, G., C. MERCURI, U. CAMINO, S. FRETE AND F. RESTIFO
2004. Arqueología de Pastos Grandes, Puna de Salta: primeras aproximaciones. *Actas del XV Congreso Nacional de Antropología Argentina (CD-ROM). 1ª Ed.* Universidad Nacional de Río Cuarto, Río Cuarto.

MERCURI, C.
2006. *Diversidad en artefactos líticos de las ocupaciones del Valle de San Antonio de los Cobres, Puna de Salta, durante el Período Agro-Alfarero Temprano*. Unpublished Graduate Thesis, Facultad de Filosofía y Letras, Universidad de Buenos Aires.

2008a. *Análisis de procedencia de obsidianas en la arqueología local: pros y contras de algunas metodologías a nuestro alcance. Trabajo realizado para el Seminario de Doctorado Sobre Cosas y Tecnología*. Acercamientos arqueológicos a la Cultura Material, dictado por Dr. Luis R. González durante 2008 en el Museo Etnográfico Juan B. Ambrosetti. Facultad de Filosofía y Letras, Universidad de Buenos Aires.

2008b. El conjunto lítico de Quebrada Alta-Estructura 1: primeros pastores de Santa Rosa de los Pastos Grandes, puna de Salta. *Intersecciones en Antropología* 9: 187-196.

2009. Relevamiento de la base regional de recursos líticos en las áreas de Santa Rosa de los Pastos Grandes y San Antonio de los Cobres, Puna de Salta. *Paper presented at the 3° Congreso Nacional de Arqueometría*, Córdoba, October 2009.

2011. *Variabilidad de Estrategias Tecnológicas Líticas durante el Período Formativo (ca. 2400-1400 AP) en la Puna de Salta*. Unpublished PhD Dissertation, Facultad de Filosofía y Letras, Universidad de Buenos Aires.

MERCURI, C. AND M.D. GLASCOCK
2011. Primeros Datos sobre la Procedencia de Obsidiana de un Sitio Formativo de Santa Rosa de Los Pastos Grandes, Puna de Salta, Argentina. *Arqueología* 17: 247-258.

MERCURI, C. AND V. VÁZQUEZ
2007. Conjuntos líticos de los sitios Tempranos de Matancillas: primera aproximación. In *Arqueología Argentina en los Inicios de un Nuevo Siglo. Publicaciones del XIV Congreso Nacional de Arqueología Argentina*, compiled by F. Oliva, N. Grandis and J. Rodríguez, Tomo I, pp. 609-613. Universidad Nacional de Rosario, Facultad de Humanidades y Artes, Escuela de Antropología.

MUSCIO, H.J.
2004. *Dinámica Poblacional y Evolución durante el Período Agroalfarero Temprano en el Valle de San Antonio de los Cobres, Puna de Salta, Argentina*. Unpublished PhD Dissertation, Facultad de Filosofía y Letras, Universidad de Buenos Aires.

NIELSEN, A.E., M. VÁSQUEZ, J.C. ÁVALOS AND C.I. ANGIORAMA
1999. Prospecciones arqueológicas en la Reserva "Eduardo Avaroa" (Sud Lípez, Depto. Potosí, Bolivia). *Relaciones de la Sociedad Argentina de Antropología* XXIV: 95-124.

POTTS, P.J. AND P.C. WEBB
1992. X-Ray Fluorescence Spectrometry. *Journal of Geochemical Exploration* 44: 251-296.

PSOTA, S.
1990. Reexamining the potential to visually source western Great Basin obsidians. *Paper presented at 24° Annual Meeting of Society for California Archaeology*, Foster City.

RESTIFO, F.
2011. Tecnología lítica de cazadores-recolectores tempranos. Análisis de un conjunto del sitio Alero Cuevas, Pastos Grandes, Puna de Salta (República Argentina). *Intersecciones en Antropología* 12: 155-158.

ROJAS A., C., V.A. PEÑA RODRÍGUEZ AND M. DELGADO A.
2004. Caracterización de obsidianas arqueológicas peruanas: redes de intercambio. *Revista de Investigación de Física* 7(1-2): 51- 58.

ROTH, B.J.
2000. Obsidian source characterization and hunter-gatherer mobility: an example from the Tucson Basin. *Journal of Archaeological Science* 27(4): 305-314.

ROUX, C. AND C. LENNARD
2006. *X-Ray Fluorescence in Forensic Science. Encyclopedia of Analytical Chemistry* (online version). Willey & Sons.

RUTHSATZ, B. AND C. MOVÍA
1975. *Relevamientos de las estepas andinas del noreste de la Provincia de Jujuy, República Argentina*. FECIC, Buenos Aires.

SEELENFREUND, A., M. PINO, M.D. GLASCOCK, C. SINCLAIRE, P. MIRANDA, D. PASTEN, S. CANCINO, M.I. DINATOR AND J.R. MORALES
2010. Morphological and geochemical analysis of the Laguna Blanca/Zapaleri obsidian source in the Atacama Puna. *Geoarchaeology* 25(2): 245-263.

SHACKLEY, M.S.
1988. Sources of archaeological obsidian in the Southwest: An archaeological petrological and geochemical study. *American Antiquity* 53: 752- 772.

SHACKLEY, M.S. (EDITOR)
1998. *Archaeological Obsidian Studies. Method and Theory.* Plenum Press, New York and London.

SKINNER, C.E. AND J.J. THATCHER
N / D. *Introduction to Obsidian Characterization Studies.* http://www.obsidianlab.com/info_xrf. html (October 15 2008).

SOLÁ, P.
2007. *Informe del Análisis Petrográfico de Material Lítico*. Proyecto Doctoral Arqueología de Pastos Grandes, Puna de Salta. Buenos Aires. Ms.

STERN, C.R.
1999. Black obsidian from central-south Patagonia: Chemical characteristics, sources and regional distribution of artifacts. In *Soplando en el Viento... Actas de las III Jornadas de Arqueología de Patagonia*, pp. 221-234.

2000. Sources of obsidian artifacts from the Pali Aike, Fell´s Cave and Cañadón La Leona archaeological sites in southernmost Patagonia. In *Desde el País de los Gigantes. Perspectivas Arqueológicas en Patagonia*, pp. 43-55.

TARRAGÓ, M.N.
1994. Intercambio entre Atacama y el borde de la Puna. In *De Costa a Selva: Producción e Intercambio entre los Pueblos Agroalfareros de los Andes Centro Sur,* edited by M.E. Albeck, pp. 199-209.

Tykot, R. and A.J. Ammerman
1997. New directions in central Mediterranean obsidian studies. *Antiquity* 71(274): 1000-1006.

VÁZQUEZ, C. AND P.S. ESCOLA
1995. X- ray fluorescence analysis of obsidian objects from Catamarca, Argentina. *Journal of Radioanalytical and Nuclear Chemistry* 200(5): 373-384.

VILELA, C.
1969. *Carta geológico-económica de la República Argentina. Descripción geológica de la hoja 6c, San Antonio de los Cobres (Salta y Jujuy).* Ministerio de Economía de la Nación. Secretaría de Industria y Minería, Subsecretaría de Minería y Combustibles.

VIRAMONTE, J., H. DESTEFANIS, A. APARICIO, R. ALONSO, J. MARCUZZI, E. CINCIONI AND L. PRETINOVIC
1988. Caracterización y beneficio de las perlitas del NO Argentino. *Actas del III Congreso Nacional de Geología Económica* 1: 547-556.

VIRAMONTE, J.G., R.H. OMARINI, V. ARAÑA SAAVEDRA, A. APARICIO AND A. GARCÍA CACHO
1984. Edad, génesis y mecanismos de erupción de las riolitas granatíferas de San Antonio de los Cobres, provincia de Salta. *Actas del 9° Congreso Geológico Argentino* 3: 216-233, Bariloche.

WINTERHALDER, B. AND C. GOLAND
1993. On population, foraging efficiency, and plant domestication. *Current Anthropology* 34(5): 710-715.

WINTERHALDER, B. AND D.J. KENNETT
2006. Behavioral ecology and the transition from hunting and gathering to agriculture. In *Behavioral Ecology and the Transition to Agriculture*, edited by D.J. Kennett and B. Winterhalder, pp. 1- 21.

YACOBACCIO, H.D., M.P. CATÁ, P. SOLÁ AND M.S. ALONSO
2008. Estudio arqueológico y fisicoquímico de pinturas rupestres en Hornillos 2 (Puna de Jujuy). *Estudios Atacameños* 36: 5-28.

YACOBACCIO, H. D., P. ESCOLA, M. LAZZARI AND F. PEREYRA
2002. Long-distance obsidian traffic in northwestern Argentina. In *Geochemical Evidence for Long-Distance Exchange*, edited by M. Glascock, pp. 167-204. Bergin and Garvey, Wesport.

YACOBACCIO, H.D., P. ESCOLA, F. PEREYRA, M. LAZZARI AND M.D. GLASCOCK
2004. Quest for ancient routes: obsidian sourcing research in northwestern Argentina. *Journal of Archaeological Science* 31(2): 193- 204.

ZEITLIN, R.N.
1982. Toward a more comprehensive model of interregional commodity distribution: Political variables and Prehistoric obsidian procurement in Mesoamerica. *American Antiquity* 47(2): 260-275.

ZUMBERGE, J.H.
1974. *Geología Elemental*. C.E.C.S.A., México D.F., Third printing.

---

[1] As an exact translation was not possible, we made a conceptual approach. For flakes we discriminated by size. *Hyper-microflakes* can be placed in a 2 by 2 cm box module, *microflakes* in 4 by 4 cm, *small flakes* in 6 by 6, *flakes* in 8 by 8 and *big flakes* in 12 by 12. For tools we analyzed them mostly taking into account their sharpened edges. Some of them had knife edges (long edges with symmetrical angles of less than 50°) combined with side scrapers edges (long edges with asymmetrical angles of more than 50°) or other type of edge. For more details see Aschero 1975 and 1983.

INTRASPECIFIC VARIABILITY IN THE $\delta^{13}C$ AND $\delta^{15}N$ VALUES OF ARCHAEOLOGICAL SAMPLES OF *ZEA MAYS* COBS (NORTHEASTERN ARGENTINEAN PUNA)

Violeta A. Killian Galván[1], Nurit Oliszewski[2], Daniel E. Olivera[3] and Héctor O. Panarello[1]

[1] Instituto de Geocronología y Geología Isotópica (INGEIS) - CONICET - Universidad de Buenos Aires, [2] Instituto Superior de Estudios Sociales (ISES) - Universidad Nacional de Tucumán - CONICET, [3] Instituto Nacional de Antropología y Pensamiento Latinoamericano - Universidad de Buenos Aires - CONICET

Abstract

The aim of this article is to present the results of $\delta^{13}C$ and $\delta^{15}N$ on a sample of *Zea mays* cobs from the Río Doncellas Collection (Department of Cochinoca, Jujuy), obtained in archaeological excavations conducted between 1973 and 1975. This study includes the analysis of various races, therefore it will also contribute to the reconstruction of the isotopic ecology of the North Puna. Furthermore, it will also assess whether this diversity is accompanied by an equally variable isotopic correlation, including whether possible sources of this variation might be physiological responses to environmental factors (differential cultivation at different altitudes) or genetic influences. Likewise, we straightforwardly discuss the pertinence of the values from archaeological samples in the reconstruction of human paleodiets, taking into consideration the digenetic influences that degrade the primary isotopic signature of archaeological samples. In an attempt to generate a frame of reference for possible explanations of the *Zea mays* values obtained, we also present here the first results of present-day studies from the micro-region of Antofagasta de la Sierra (Catamarca, Southern Argentine Puna). Here we sampled three cultivated fields located at different altitudes, types of irrigation and use of fertilizers. In the assemblage (n=6) we noted a lower variability in the $\delta^{13}C$ values and a higher range in the $\delta^{15}N$ values, which indicates the importance of the use of fertilizers in arid environments and the its implications in isotopic enrichment given that the most enriched values occurred in those field with the least access to water. In an attempt to strengthen our conclusion, these values have been compared to the extant isotopic data for *Zea mays* from the larger Andean context.

Key words: *Stable isotopes - Zea mays - Museum collection - Argentine Puna*

INTRODUCTION

The production and consumption of maize (*Zea mays*) amongst the prehispanic societies that occupied the Río Doncellas basin (Province of Jujuy, North Argentine Puna) has been inferred on the basis of the high level of development that agricultural production had achieved by the Late Period (*ca.* 900 to 550 BP) and the ubiquity of *Zea mays* in the archaeological record of vegetable macro-remains (Alfaro 1988). Even so, isotopic studies on human remains to date have not shown that this cultigen was of great dietary significance in respect to other foods (Killian Galván and Olivera 2008; Killian Galván *et al.* 2012). On the other hand, paleo-dietary studies from Northwestern Argentina (known herein as NOA, using the existing Argentinian acronym for the geographical area), have shown the necessity of generating a reference database of isotopic values obtained from animal and vegetable resources from the study area, that will in turn will aid in the interpretation of the values obtained from the archaeological samples.

This is especially relevant since values obtained from the Puna cannot be attributed to purely C_3 or C_4 diets (Killian Galván and Olivera 2008; Olivera and Yacobaccio 1999), since there is the real possibility that during periods of hunting and gathering the diet might actually reflect values similar to those associated to the consumption of maize, expected for later periods (Fernández and Panarello 1999-2001). Due to these factors, it is imperative to know the factors that can introduce variability in Carbon and Nitrogen values within each species. Reflecting on the archaeological context of the NOA, the isotopic variability of Carbon in plants has been evaluated for camelid pastures showing a direct correlation between isotopic enrichment and altitude (see Fernández and Panarello 1999-2001; Samec 2011). This information has been applied to the archaeological record to evaluate the direct consumption of plants by humans; studies of present-day consumption of plants by humans still remain scarce. Similarly, a study of the isotopic relationship of ancient plant remains results in extra data given the impact that anthropogenic practices, such as irrigation and fertilizer use, have had on the isotopic relationships within cultivated plants. Given this, this article proposes a detailed isotopic study of maize whilst postulating the necessity of controlling the possible sources of isotopic variation in these cultigens.

STUDY AREA

The Archaeological Area Site of Río Doncellas

The *Zea mays* sample used in this article comes from the archaeological area site of Río Doncellas (Department of Cochinoca, Province of Jujuy), located at S 22° 49' 12,28" and W 66° 03' 54,89" forming an imaginary rectangle 35 km x 25 km (Figure 1). This area site is located in the Puna at 3900 m a.s.l., with an annual precipitation between 100 mm and 300 mm and permanent hydric resources (Alfaro 1988). The archaeological site is composed of a sector of pit houses, at the entrance of which there is a settlement conglomeration known as Poblado and a sector of cliffs, located on either side of the settlement that incorporates different funerary structures. The site also includes cultivation terraces, caves and associated shelters all within a distance of 5 km. Even though radiocarbon dates suggest a chronology of occupation that spans the period between [740 and 310] BP, it is known that the occupation of Poblado continued until the Hispanic period because of a CE 1677 Spanish coin and other indicators (Pérez de Micou 1996).

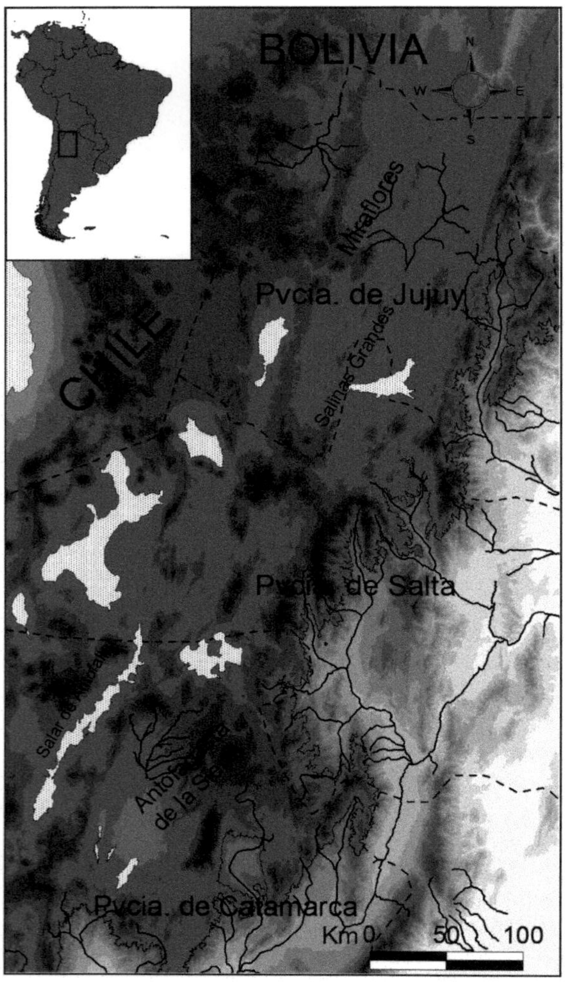

Figure 1. Map showing the study area.

The Micro-region of Antofagasta de la Sierra

The area selected for the collection of present-day samples of *Zea mays*, conducted during the summer of 2010, was the Micro-region of Antofagasta de la Sierra (Catamarca, Southern Argentine Puna). Although there are notable ecological differences between these two areas, given that the Northern Argentina Puna is wetter in respects to the Southern Argentine Puna, this micro-region is a good proxy because it is a high altitude desert whose climatic characteristics permit the successful agriculture of micro- and macro-thermic plants. Located between 3200 m and 4400 m a.s.l., this micro-region possesses a dry climate, low precipitation and large thermic amplitude (García and Rolandi 1999). On the basis of the environmental characteristics we can distinguish different sectors separated according to their agricultural potential: 1) Bottom of basin at 3400 m - 3550 m a.s.l., with availability of water and wood, which currently concentrates the largest part of the population; 2) Intermediate sectors at 3550 m - 3800 m a.s.l. that hold areas apt for agriculture and foraging; 3) High Puna at 3800-4600 m a.s.l., where the vegetation cover is sufficient only for the herding of camelids. Due to these differences, three cultivation fields where sampled at different altitudes, types of irrigation and use of fertilizers[1]. All these fields belonged to families that practiced agriculture as a supplement to their general diet since they have access to mass-produced goods.

PREVIOUS ARCHAEOLOGICAL STUDIES ON MAIZE IN DONCELLAS

Amongst the array of products consumed, maize had reached the same level as that of other cultigens (Alfaro 1988), even given that the area is more suited to the cultivation of micro-thermic plants such as the potato and quinoa. This fact is based primarily on the macro-botanical evidence, especially from the studies undertaken by Alfaro (1988). Alfaro assembled samples of well-preserved *Zea mays* currently stored in the INAPL (Doncellas Collection) that included cobs (N= 51), which was subsequently analyzed by the authors. In this assemblage we identified eight types of maize belonging to races presently known for the Quebrada de Humahuaca, emphasizing the high degree of biovariability inherent in the assemblage. The assemblage labels state that these samples came from *Cueva Tajuera* and *Cueva del Hechicero*. *Cueva Tajuera* had been excavated in quadrants by Alfaro`s team who registered at a depth of 0.6 m '...4 fragmented cobs (...) larger than those of *Puneño* maize that exists in the area. By their appearance they represent present-day types' (Alfaro 1988: 57, our translation); in the second quadrant at a depth of 0.5 m '...small cobs typical of *Puneño* maize' (Alfaro 1988: 57, our translation).

These studies also indicate that *Cueva Tajuera* was a place used by local families until recently, both as a corral and as part of ritual ceremonies linked to the fertility of animals (Alfaro 1988). Therefore in consideration of the morphological characteristics of the assemblage, and the high level of historic use of the site the author ascribes different origins to the cobs recovered there. On *Cueva del Hechicero*, located near the archaeological site, it is unfortunately not possible to see the location of the cobs recovered in the study published by Alfaro de Lanzone. Nevertheless, the importance of maize was inferred on the basis of iconographic motifs discovered at the site. Alfaro (1988) interprets, from the rock art identified in the caves, shelters and rock outcrops at the archaeological site, people tilling the earth with agricultural tools (spades with handles or digging sticks) and a 'leader' with a scepter analogous in shape, according to the author, to a maize stalk (found in the graves at Poblado). This prompted Alfaro to assign a strictly agricultural function to all the spades, adzes and ploughs found on the surface of the site[2].

ISOTOPIC VARIATION IN *ZEA MAYS* SPECIMENS

Stable Carbon Isotopes

In its natural state Carbon has two stable isotopes with different mass so that ^{12}C contains 6 protons and 6 neutrons whilst ^{13}C contains an extra neutron (Urey 1947 in Schwarcz and Schoeninger 1991). These isotopes occur in the ecosystem and in food webs since through photosynthesis terrestrial plants convert Carbon Dioxide (CO_2) from the atmosphere into glucose molecules. During photosynthesis there is a fractionationation in the species containing the heaviest isotope so that the $^{13}C/^{12}C$ isotope relation in the plant is different to that of the same isotope relation in the Carbon present in the atmosphere (Ambrose 1993). At the same time, there are three patterns of photosynthetic isotopic correlation: C_3 (arboreal species, shrubs and the majority of the wild plants of the Puna, ranging between -33 ‰ and -22 ‰), C_4 (maize, some amaranthaceae, sugar cane and sage, ranging between -16 ‰ and -9 ‰), and crasulaceans with an acidic metabolism (cacti and orchids, with values that range both in the C_3 and C_4 distribution). Therefore, an analysis of stable isotopes allows us to monitor the introduction of maize seeing as its presence is detectable in animals due to the differential values that each plant species has (Ambrose 1993). In delta notation, $\delta^{13}C$ represents the difference between the measurement of the isotope ratio that is of interest to us and the standard isotope ratio, which in this case is CO_2 extracted with 100% Phosphoric acid from Vienna-Pee Dee Belemnitella. Given that the difference between these is small, it is expressed as parts in a thousand (‰), following the equation:

$$\delta^{13}C = \left[\frac{(^{13}C/^{12}C)_{sample}}{(^{13}C/^{12}C)_{V\text{-}PDB}} - 1 \right] \cdot 1000$$

Because the level of enrichment depends on the photosynthesis pathway used by the plant, the isotopic ratio of Carbon ($\delta^{13}C$) can distinguish different sources of resources. Amongst the environmental variables that may generate variations in the $\delta^{13}C$ values of modern plants are altitude, precipitation levels, $\delta^{13}C_{CO2}$ in the atmosphere, sunlight, soil salinity, etc. Likewise, we have to consider the isotopic composition of every part of the plant; each of the different parts of a plant -roots, seed, leaf or stem- present differences in the level of $\delta^{13}C$ between 1 ‰ and 2 ‰ (Deines 1980). This might be as a consequence of the specific instant in which the Carbon fixes on that part of the plant, or because of the different biochemical composition of each of the plant tissues. For example, in the case of *Zea mays* the isotopic values of the whole cob are lower than those of the grain (in a 0.5 ‰ -1.0 ‰); because of the amount of starch and cellulose, the leaves are in turn poorer than the reproduction part (0.4 ‰). Finally, there are also variations in values due to genetic differences (Tieszen and Fagre 1993).

Stable Nitrogen Isotopes

Nitrogen possesses two stable isotopes, ^{14}N and ^{15}N. In the biosphere this element is most present as gas (N_2) in the atmosphere. The Nitrogen reserve in the atmosphere has an equivalent isotopic composition and is therefore 0‰ (Pate 1994). On the other hand in the sea this relation is of +1.0‰. Even given that Nitrogen is abundant in the atmosphere there are limiting factors to it in both aquatic and terrestrial systems. Inorganic Nitrogen in the atmosphere (N_2) is transferred into the biological sphere through the action of specialized organisms located at the root of fixing plants that convert the gaseous Nitrogen into forms that can be used by the plant (Pidwirny 2006). This process leads to little fractionation, which means that these plants have Nitrogen values similar to that of the atmosphere (0 ‰). The other plants, known as non-fixing, capture Nitrogen through the decomposition of organic material in their habitats (nitrates, ammoniac, and dissolved ammonium) that leads to a bimodal distribution of Nitrogen in these plants in respect to this isotope. Even though there is a certain superposition of values between fixing and non-fixing species, the latter possess in general, significantly more positive values (Heaton 1987); such that the median value for fixing plants is +1 ‰, with a typical range of -2 ‰ to +2 ‰, whilst in the case of the non-fixing it is +3 ‰, with a typical range of 0‰ to +6‰ (Pate 1994). In respect to modern ecosystems, it is necessary to signal out the effect of fossil fuel emissions, as well as the use of chemical fertilizers, these have, in many cases, reduced the difference between these two plant types (for example, see Hastorf and De Niro 1985).

To determine the grade and type of food consumed, the relation of Nitrogen isotopes is analyzed and expressed also as values of δ per thousand, using the following formula:

$$\delta^{15}N = \left[\frac{(^{15}N/^{14}N)_{sample}}{(^{15}N/^{14}N)_{AIR}} - 1 \right] \cdot 1000$$

Now then, the Nitrogen isotope register has to be contextualized according to its geographical origin. The availability of water, taking into consideration annual precipitation as an indicator, has been mentioned as a variable intimately related to the values obtained from the collagen of herbivore bones, with a negative relation being registered in both continents (Sealy *et al.* 1987). There are two possibilities that explain this phenomenon. Firstly, those of a metabolic slant that attributes this enrichment to adaptive response. Secondly, more Nitrogen and Ammonia is believed to occur in saline soils, such as is characteristic of arid environments (Pate 1994). For example, in coastal environments and saltpans in Namibia and South Africa plants have been discovered containing values between 4% - 10% more positive than in non-saline environments (Heaton 1987).

Indeed, it has been theories based on exogenous factors vis-à-vis the herbivores that have gained ground. Amundson *et al.* (2003) have argued that cold and humid ecosystems seem to be more efficient in the conservation and recycling of inorganic Nitrogen and, consequently, there is a greater opening of the Nitrogen cycle in drier ecosystems. Therefore the loss of Nitrogen via lixiviation and its transformation (nitrification, de-nitrification and volatility of ammonia) lead to enrichment in ^{15}N in the remaining Nitrogen in the system (Austin and Vitousek 1998). This scenario, were one has a modest, though negative, correlation between the δ^{15}N values of Nitrogen of floors and organic material, and the increase in median annual precipitation (Amudson *et al.* 2003), would go a long way in explaining the negative relation between δ^{15}N values in herbivore collagen and the availability of water, rather than the metabolic causes (Murphy *et al.* 2006).

The Use of Archaeological Plants

Plants from archaeological contexts can suffer all types of degradation, which alter the primary signal that in turn engenders a certain amount of doubt at the moment of inferring directly from the data recovered to reconstructions of past paleodiets. In the case of δ^{13}C, the values do not seem to be altered when the plants are carbonized or not (Hastorf and DeNiro 1985; Tessone *et al.* 2009). In the case of δ^{15}N values in general, and of maize in particular, evidence is even scarcer than that of Carbon values from archaeological samples; including in this, studies concerning the feasibility of using values from archaeological sources. For example, in the studies by Hastorf and DeNiro (1985) the δ^{15}N values of archaeological plants that were not carbonized were from 10 - 20% up to 35% more positive than the modern ones. Moreover, δ^{13}C and δ^{15}N values of different parts of un-carbonized archaeological plants vary between 8 ‰ and 21 ‰ respectively, whilst the same parts of modern plants have δ^{13}C or δ^{15}N values that are within 2 ‰ of each other. The cause of the formers enrichment might be as a consequence of the loss of lipids, which are isotopically lighter (Tieszen and Fagre 1993). It is important to note that the magnitude of the post-mortem ^{15}N values in archaeological samples is independent of the age of the non-carbonized specimens (in so far as Hastorf and DeNiro 1985 record).

Correction Factors

In respect to modern samples there is a systematic enrichment of ^{13}C in archaeological maize, this is probably because the modern samples have been affected by the atmospheric changes from the post-industrial period. Specifically we refer to the Suess Effect, that is, the reduction by at least 1,5 ‰ of δ^{13}C values in the atmosphere as a consequence of the use of fossil fuels during the recent industrial period. The present atmospheric CO_2 available for photosynthesis results in a systematic depletion of δ^{13}C values of modern plant samples (Friedli *et al.* 1986), it is therefore necessary to correct these samples by ~1,5 ‰ at the moment of using them to reconstruct paleodiets. Given all the factors mentioned above, we propose to explore the variability of δ^{13}C and δ^{15}N values in maize from the archaeological area of Doncellas and, on this basis, evaluate its usefulness in the reconstruction of food webs in the Argentine Puna. In monitoring these variables we should be able to evaluate possible anthropogenic effects on these same food chains, which is only possible through the joint measuring of human remains and associated plants. In the case of Doncellas, this study will have direct relevance to the discussion concerning the integration of agricultural and herding production, an integration that permitted a higher production within cultivation, thus increasing the productive potential of the area in Prehispanic periods.

MATERIALS AND METHODS

The sample includes 51 cobs, complete and fragmented, without grains, dry and in a good state of preservation. A subsample of n=22 was selected for isotopic analysis. In assigning racial denominators to the cobs we used the *Zea mays* classification of Abiusso and Cámara Hernández (1974), which is based on the external characteristics of the stem (cob with caryopsis) including aspects unique to the caryopsis and covered fourteen races: *Pisingallo, Morocho, Morocho amarillo, Chullpi, Capia, Harinoso, Harinoso amarillo, Culli, Garrapata, Azul, Marrón, Amarillo chico, Amarillo grande,* and *Bola*. The methodology employed to classify racial variability in the maize assemblage in this study then used the above, whilst adapting it to archaeological samples (Oliszewski 2008). This method implies the comparison between archaeological maize and the modern races established by Abiusso and Cámara Hernández (1974) for the Quebrada de Humahuaca (Jujuy). This procedure has been employed with very good results for the analysis of macro-remains for maize from many different archaeological sites in the NOA area (Oliszewski 2005, 2009; Oliszewski and Olivera 2009).The macroscopic description of the specimens was undertaken through simple observation and with a stereoscopic microscope (2X to 4X) taking into consideration the quantitative and qualitative characteristics of the cobs. These are described in greater detail below:

Quantitative Variables

- Length and diameter of the cob. Length may vary between 60 to 141 mm; diameter varies between 8.6 mm and 18.8 mm.
- N° of grain rows. The number of grain rows is one of the key ways in which to identify races of maize.

Although at first sight it might seem easy to count the rows, in many cases the bad state of preservation of many archaeological specimens, as well as pre- and post-depositional taphonomic processes acting on the sample, can make identification of rows difficult. The number of grain rows may vary between 8 and 20.
- N° of grains per row. This is an important measurement in the absence of the actual grains as it allows an estimation of the width of the cob. The number of grains present along a row is counted.
- Relative thickness of the grain (length of cob/number of grain rows). The thickness varies between 3.6 mm and 4.6 mm.

Qualitative Variables

- Shape of the cob. This can be: conical, oval, cylindrical or a combination of these basic forms.
- Colour of the cob. The variety of colours is very great; from light colours (yellow in the *pisincho* race) too dark ones (black in the *culli* race). To determine the exact colour a Munsell Soil Colour Chart was used.
- Quantity of glumes. It can be with: abundant glumes, with glumes, few glumes or without glumes.

Stable Isotopes

To prepare the samples for isotope measurements the maize was bathed in ultrasound, dried at 60° C in an oven for 24 hours and then ground to fragments of 1 mm. In the case of the archaeological samples it was only the cobs that could be subjected to this process, in the case of the modern examples the seeds were also available, these were measured separately. This took place in Instituto de Geocronología y Geología Isotópica (INGEIS).

Solid samples were run in Environmental Isotope Laboratory (University of Waterloo, Canada) for Nitrogen and Carbon analysis on a Carlo Erba Elemental Analyzer (CHNS-O EA1108 - Italy) coupled to a Delta Plus Continuous Flow Stable Isotope Ratio Mass Spectrometer (Thermo Finnigan / Bremen-Germany). Results were corrected to Nitrogen standards IAEA-N1 and IAEA-N2 (both Ammonium Sulphate) and Carbon standards IAEA-CH6 (sugar), EIL-72 (cellulose) and EIL-32 (graphite). The error for clean ball-milled standard material is ± 0.2 ‰ for Carbon and ± 0.3 ‰ for Nitrogen (Fry *et al.* 1992).

Once we had selected the specimens at a typological level the next step was to correlate the isotopic values with typical altitude at which these appeared and their type of maturation. A methodological strategy used to evaluate the ^{13}C and ^{15}N isotopic signal in archaeological plants is the study of these in relation to their Carbon content (% C and % N) (Araus *et al.* 1997). This information will then be compared against a database compiled from different publications (Aranibar *et al.* 2007; Falabella *et al.* 2007; Hastorf and DeNiro 1985; Oliszewski and Olivera 2009) showing the $\delta^{13}C$ and $\delta^{15}N$ values available for the Andean area.

RESULTS

The Archaeological Sample

The sample is composed of 51 dry cobs without grains both complete and fragmented. In the main the individual specimens are in a good state of preservation. The variables of the analyzed specimens are described in Table 1 and in Figure 2. In assigning race, the main characteristic considered was the number of grain rows on the cob, in this manner it was possible to identify the presence of eight races as described previously by Abiusso and Cámara Hernández (1974):

Amarillo chico cfr. (n=1). Semi-hard maize (grains with floury endosperm and hard periphery), of early maturation (approximately 60 days). Chicha is prepared from its flour (Figure 2a).

Amarillo grande cfr. (n=1). Semi-hard maize (grains with floury endosperm and hard periphery), of early maturation (approximately 60 days). Chicha is prepared from its flour (Figure 2b).

Marrón. (n=12). Semi-hard maize (grains with semi-hard endosperm), of intermediate maturation (Figure 2c).

Morocho amarillo cfr. (n=5). Semi-hard maize (grains with a semi-hard endosperm with a thick central floury layer), of intermediate maturation (Figure 2d).

Pisingallo cfr. (n=5). Hard maize (grains with a hard and crystalline endosperm), of early maturation (approximately 60 days). Used to produce cooked flour (breaking the mature grains through heat and subsequent grinding: 'pochocho') (Figure 2e).

Harinoso/harinoso amarillo. (n=4). Soft maize (grains with floury endosperm), of intermediate maturation (Figure 2f).

Culli. (n=7). Soft maize (grains with floury endosperm), of early to intermediate maturation. Used for making chicha morada (Figure 2g).

Capia. (n=3). Soft maize (grains with floury endosperm), of late maturation (Approximately 120 days). The mature grains are boiled to make mote (used in stews, dry sauce dishes - picantes, and tamales), with it flour cakes called capias are made (Figure 2h).

Finally there is a number of specimens (n=13) for which it was impossible to determine their taxonomic association given their state of fragmentation and the absence of other quantitative or qualitative diagnostic characteristics.

On the basis of these studies, we selected a minimum of two examples, when they existed, for each type for isotopic analysis. Table 1 shows the $\delta^{13}C$ and $\delta^{15}N$ results for these specimens. After consulting the available bibliography we have established a probable fractionation for the Carbon values for seeds (Tieszen and Fagre 1993). We compared the values that we obtained with the percentages of Carbon and Nitrogen in each sample (Figure 3), without discovering any correlation between the two elements (R^2= 0 for Nitrogen and R^2= 0.02 for Carbon), thus, this procedure can not be a factor to judge which of these samples are valid as primary signals.

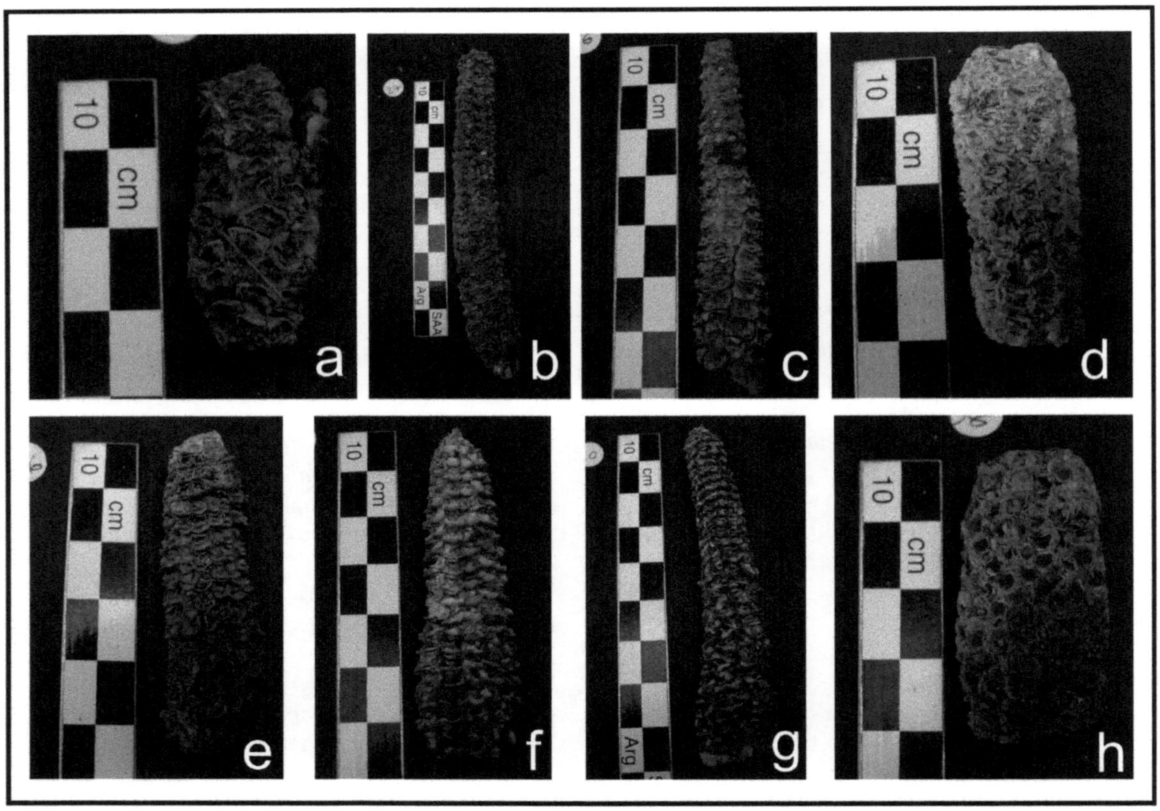

Figure 2: Types of maize identified in the Doncellas Collection.

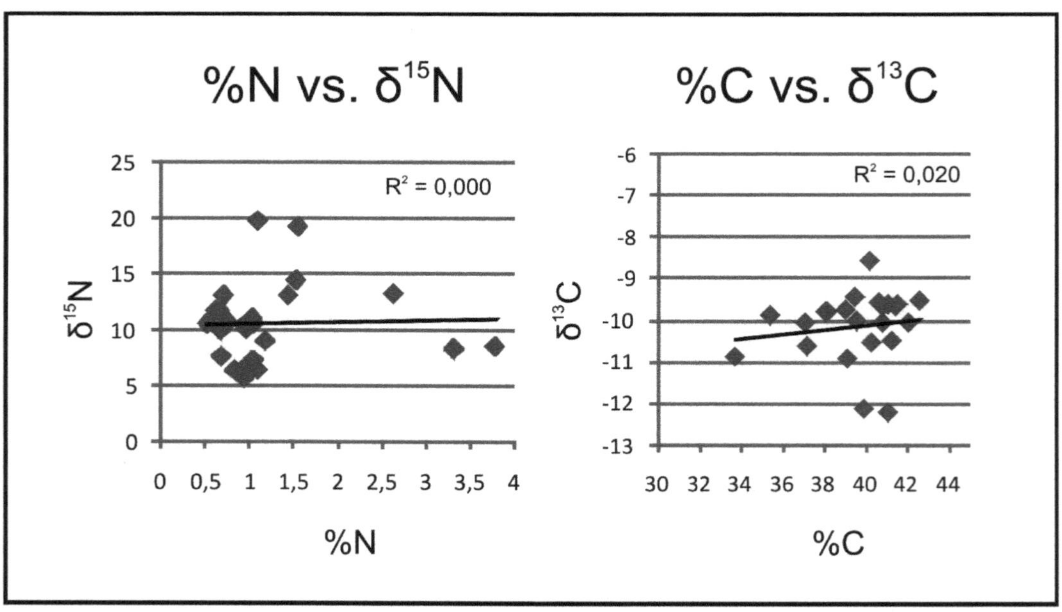

Figure 3: Relation between the $\delta^{15}N$ values and the amount of Nitrogen (% N) and relation between the $\delta^{13}C$ values and the amount of Carbon (% C)[3].

N°	Specimen	Race	$\delta^{13}C$	Δ Seed	$\delta^{15}N$	Type of maturation
1	7	Amarillo chico	-10.88	-10.13	19.76	Early
2	5	Marrón	-9.91	-9.16	13.05	Intermediate
3	17	Marrón?	-10.53	-9.78	6.4	Intermediate
4	16	Marrón	-12.19	-11.44	8.45	Intermediate
5	2	Marrón?	-9.45	-8.70	7.64	Intermediate
6	40	Culli	-9.63	-8.88	13.08	Early to intermediate
7	31	Culli	-10.50	-9.75	11.75	Early to intermediate
8	3	Culli?	-10.03	-9.28	10.67	Early to intermediate
9	12	Culli?	-9.77	-9.02	8.99	Early to intermediate
10	27	Amarillo grande?	-8.61	-7.86	10	Early
11	42	Morocho amarillo?	-10.05	-9.30	10.54	Intermediate
12	9	Morocho amarillo?	-10.09	-9.34	14.39	Intermediate
13	30	Harinoso	-9.54	-8.79	10.09	Intermediate
14	4	Harinoso amarillo?	-10.93	-10.18	10.42	Intermediate
15	26	Harinoso amarillo	-8.70	-7.95	11.20	Intermediate
16	25	Pisingallo?	-9.69	-8.94	11.03	Early
17	28	Pisingallo?	-9.59	-8.84	11.75	Early
18	36	Pisingallo?	-10.06	-9.31	19.21	Early
19	41	Pisingallo?	-10.00	-9.25	15.10	Early
20	49	Capia	-9.81	-9.06	13.22	Late
21	8	Capia	-10.64	-9.89	6.43	Late
22	22	Capia	-9.66	-8.91	7.27	Late
		Average	***-10.01***	***-9.26***	***11.38***	
		SD	***0.76***	***0.72***	***3.55***	

Table 1: $\delta^{13}C$ Values, estimated $\delta^{13}C$ seed fractionation, $\delta^{15}N$ obtained from the assemblage of *Zea mays* cobs.

We further specified the race and type of maturation inferred on the basis of phenotypic characteristics. In respect to variations within the sample of the types found, the differences are morphological without correlation at a sub-specific level. This means that we would not expect differences at the isotopic level. The only sub-sample that does show some difference within the wider assemblage is the pisingallo race (according to Poggio *et al.* 1998), which even so did not show isotopic differences within the sample studied.

On the other hand, this variation is not borne out in the $\delta^{13}C$ and $\delta^{15}N$ values (Figure 4). In respect to Carbon there is no great variation in the values, with a standard deviation of only 0.76, which is within the expected range of the same species growing in the same place, given that the normal deviation range is between ± 0,8 ‰ and ± 1,5 ‰ (Heaton 1999). Nevertheless, this is contrary to the expectation of finding a wider range of values in areas under stress, such as arid environments (*op. cit.*).

In respect to the $\delta^{15}N$ values from the Doncellas cobs, some of them were within the range expected for maize possessing a primary isotopic signal, although they are amongst the most enriched within the Andean database (between $\delta^{15}N$ 6 ‰ and 10 ‰, when the standard values from the literature are between $\delta^{15}N$ 2 and 3 ‰). Even so we should note a peculiarity of the plant matter sampled; even though greater enrichment may occur with crops cultivated with fertilizer, this enrichment occurs in the maize grain. In the case of the Doncellas sample, we could only analyze the cobs as the grains were not preserved.

If we take the maturation of the grains as a criteria for grouping specimens, we observe that the $\delta^{13}C$ do not present differences (Figure 5). Yet when we use the same criteria for $\delta^{15}N$ values, we find that there is a slight grouping tendency amongst those races that require the shortest period of maturation (Figure 5). For instance *Amarillo chico* and *Amarillo grande*, have early maturation and are located amongst the higher ranges of isotopic enrichment. We should not forget that these subspecies are the ones that do not have any cultivation problems under conditions of greater altitude and aridity. In contrast, the *Capia* type, normally associated to agricultural intensification because of its need for a longer period of maturation (Oliszewski and Olivera 2009), has a lower level of enrichment.

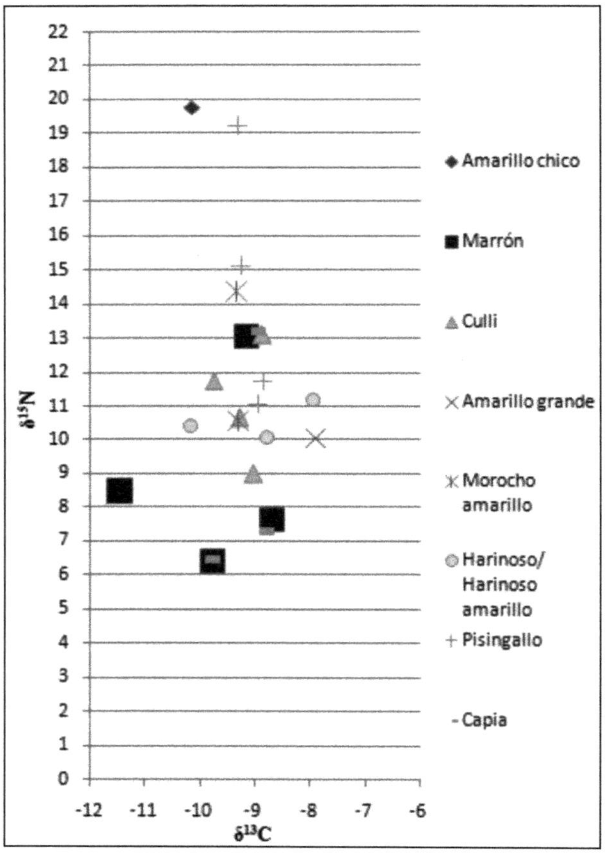

Figure 4. δ13C and δ15N values obtained when discriminating according to race.

Nevertheless, these differences cannot be explained from the perspective that these samples are from modern maize, given the digenetic consideration mentioned above. Still, if we compare the δ^{13}C values from Doncellas against the existing data (Figure 6) we find that these values behave in a fashion similar to those of modern samples, even though they are slightly more enriched. If we compare the range of Nitrogen values available for the Andes from modern, archaeological without carbonizing and carbonized archaeological (more stable chemically) samples (Figure 6), we observe that the Doncellas samples are within the expected range. This is because the Nitrogen isotopic values tend to have a wider range, even in cases where there is no material degradation. Clearly, therefore, there are values in our sample that are not completely reliable.

Now we have to ask ourselves, what is occurring to the samples that seem to preserve their primary signal? Following Hastorf and DeNiro (1985), the diagenetic changes in non-carbonized plant δ^{13}C values seem to be caused by the absorption of humic substances, whilst the changes in δ^{15}N values would not be caused by the absorption of microbes or types of organic or non-organic Nitrogen from the ground. What we therefore have to address is, by what methods can we control the source of variation in archaeological maize so as to assess the reliability of these samples?

The Modern Sample

In an attempt to generate a framework of reference as a possible explanation for the *Zea mays* values found archaeologically, we present the preliminary results from the study of present day maize from the Micro-region of Antofagasta de la Sierra (Table 2). At this location we sampled three fields located at different altitudes, types of irrigation and use of fertilizers. In the assemblage (n= 6) we noted, yet again, the lower variability in δ^{13}C values and the greater range of the δ^{15}N values, which alerts us to the importance of the use of fertilizer in arid environments and the in isotopic enrichment it produces, given that the most enriched values occurred in those fields with the least access to water.

Our results allowed us to say that the value variation, independent of digenetic processes, is always higher in Nitrogen, including modern samples (Figure 7). In line with our expectations, the lowest values in Nitrogen were from specimens at lower altitudes and with favorable hydric conditions. It is evident that we need to enlarge our samples in lieu of these variables, given that the sample that was fertilized with sheep guano, even when it does evidence an enriched value, does not surpass the higher values registered for the Peñas Coloradas sample. In this last case, we cannot discard enrichment occurring due to nitrate in a water-lixiviated soil and through a process of di-nitrification. Even so, as has already been established by other studies, we uncovered a systematic enrichment of the δ^{13}C values in the cobs and seeds, with the latter being more enriched in respect to the former. The same is the case with the δ^{15}N values.

Even though the modern samples were few, we should be aware of this differential enrichment being discovered in the future because, in such a case, the Nitrogen values for the Doncellas sample -obtained from cobs- would also have to be corrected. It is possible that the values thus produced might show even higher levels of enrichment in respect to the expected levels from modern samples. We would expect a δ^{15}N +9 ‰ value for *Zea mays* from arid contexts, in which case, at least six values from the archaeological collection would seem to be indicating reliable values. Now, how would it be possible to demonstrate that these six specimens were not subject to fertilized conditions or that they were under stressed conditions and that therefore they might have had lower primary isotopic signals that surpassed the average value of δ^{15}N +5,24 ‰ found in modern maize, or because of a digenetic influence?

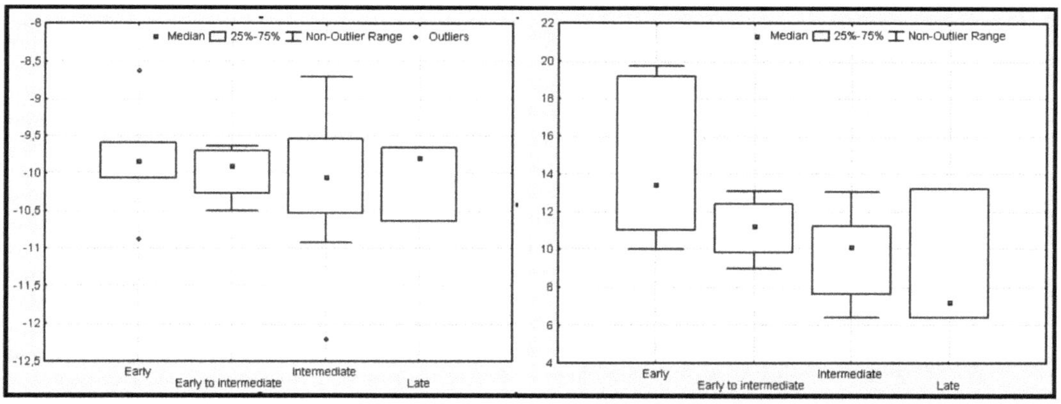

Figure 5. $\delta^{13}C$ and $\delta^{15}N$ values and maturation. Early n= 6; Early intermediate n= 4; Intermediate n= 9 and Late n=3.

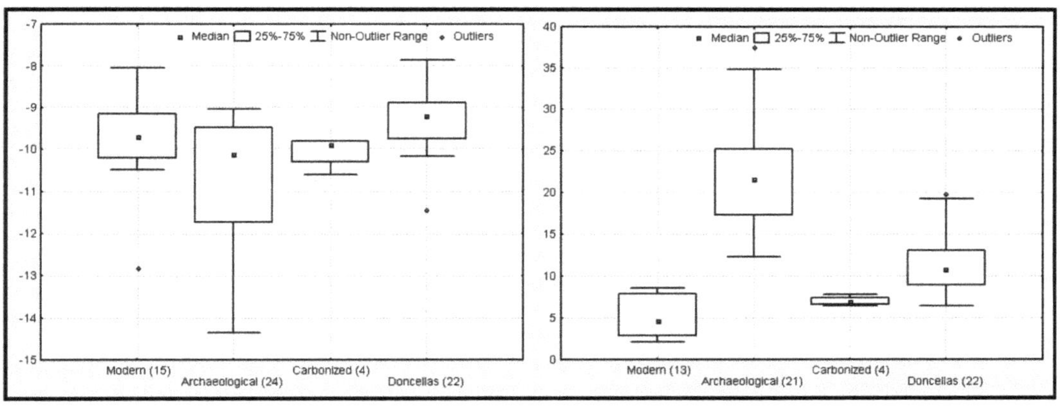

Figure 6. Comparison with the values from the Andean isotopic database.

Nº	Area	Code	Elevation m	$\delta^{13}C$‰ Grain	$\delta^{13}C$‰ Cob	$\delta^{15}N$‰ Grain	$\delta^{15}N$‰ Cob	Type of irrigation	Fertilizer
1	Villa	AS 1	3400-3550	-10.9	-11.7	6.1	5.8	Pond with pump	Water
2	Villa	AS 3	3400-3550	-11.1	-11.8	3.1	2.7	Pond with pump	Water
3	Peñas Coloradas	AS 5	3427	-9.6	-10.3	9.3	9	Major access to water	Water
4	Peñas Coloradas	AS 7	3427	-11.3	-11.5	6.8	6.5	Major access to water	Water
5	Peñas Coloradas	AS 9	3427	-10.5	-10.3	3.9	3.6	Major access to water	Water
6	Río Miriguaca	AS 11	3650	-11.2	-11.66	8.3	7.9	Every 4-8 days	Sheep guano
	Average			-10,8	-11.21	6.25	5.91		
	SD			0.63	0.71	2.41	2.42		

Table 2. $\delta^{13}C$ and $\delta^{15}N$ values obtained for the assemblage of *Zea mays* cobs and grains. Also specified are type of irrigation and fertilizer used.

DISCUSSION

The importance behind resolving these questions lies in the possibility of understanding anthropic action, as part of agro-pastoral strategies, in the isotopic values. For instance the ability to detect the use of fertilizers (as mentioned by Bogaard *et al.* 2007 and Finucane 2007) to optimize cultivation at altitude. This fact has implications for human tissues with enriched $\delta^{15}N$ values: this would not only imply the elevated consumption of animal protein, but also the consumption of plants with enriched values.

Likewise, the value of $\delta^{13}C$ in cereals, in environments impacted by anthropic action, varies depending on the water-use and irrigation economy of arid environments (Araus *et al.* 1997). Therefore, in future studies we should control for a systematic isotopic enrichment when the precipitation and humidity conditions are not favorable to agriculture.

Likewise, if our conclusions concerning the low and indirect consumption of C_4 foods by societies at this archaeological site are correct (Killian Galván and Olivera 2008), then, the discovery of *Zea mays* values with even higher levels than those present in other datasets, would distance us from the suppositions of other scholar concerning the consumption of maize by Prehispanic groups occupying the site. In this sense, we cannot ignore the typological identification utilized in this study. In the dataset we find maize races that are generally used for the production of fermented drinks, which may have little impact on the enrichment of values in human tissues (a marked difference to other types of maize processing and consumption). Following Wariner and Tuross (2009) the isotopic signal of maize does not change depending on preparation, but there are differences in assimilation of maize isotopes by living organisms that do depend on the type of preparation used - at least in the bone register (it has not been verified on dental apatite). That is to say, if maize was used to produce fermented beverage it was perhaps not processed sufficiently for it to be assimilated by an organism. Neither has the enrichment of collagen from rats fed on fermented beverages from maize husks been verified (Canal 2006).

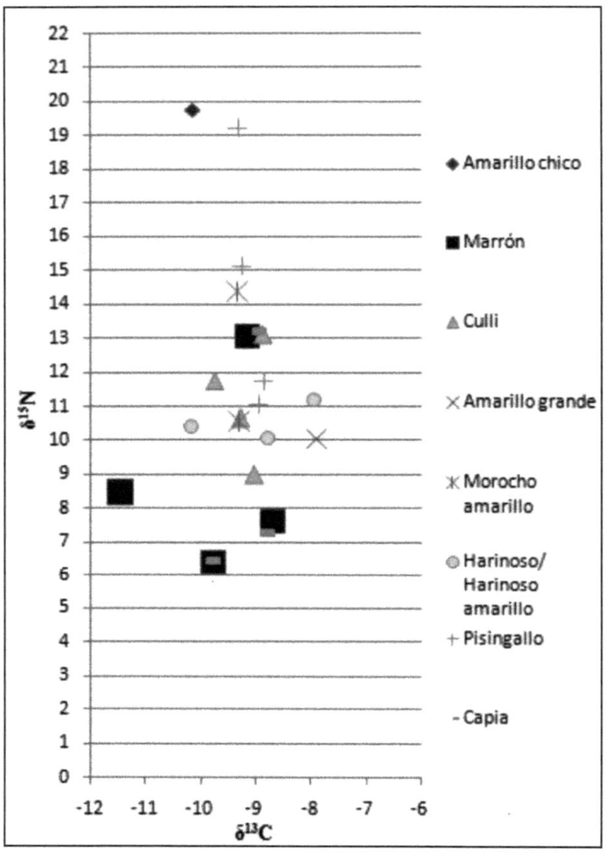

Figure 7. $\delta^{13}C$ and $\delta^{15}N$ values obtained for the assemblage of *Zea mays* grains from Antofagasta de la Sierra.

With this we do not want to affirm that maize was not an important resource for the economy of these societies. The fact that there were different varieties in a micro-region meant that the resource was fresh for the better part of the year, taking into consideration the different rates of maturation (Oliszewski and Olivera 2009), and thus was available year round.

CONCLUSION

The assemblage (n= 22) presents an ample variety of maize types associated to different uses, such as the preparation of flours, as well as those used for the production of fermented beverages (with scarce impact on the relation $\Delta^{13}C_{collagen-diet}$). On this basis, we proposed to contribute to the isotope ecology of the area, evaluating the variability within the assemblage, taking into consideration morphological differences, typical altitude of the crop and the type of maturation of each species. By mode of reference, we considered the $\delta^{13}C$ and $\delta^{15}N$ values of modern, archaeological and carbonized archaeological specimens in the Andean dataset. Our results support a low influence of the highlighted variables, although there was a tendency amongst the $\delta^{15}N$ values to group around length of maturation.

This is important given that previous to this study no substantial differences in $\delta^{13}C$ values have been registered due to altitudinal reasons for C_4 plants in natural environments, whilst we would expect increased intra-species variability in agricultural surroundings because of discretional use of water. In this investigation the studied assemblage presented variability in specimens growing at the same place (with variation oscillating between ± 0.8 ‰ and ± 1.5 ‰). This contradicts what other investigators have postulated in the past, they have emphasized the different locations of the cultivation fields to explain the racial variability within the archaeological register of the site. On the other hand, the races found are distinguished by morphological differences without there being significant genetic differences. The only group with some distance from the rest of the assemblage was the pisingallo race that does not present any differences at an isotopic level.

Similarly, we see that all the $\delta^{13}C$ are acceptable, whilst only six $\delta^{15}N$ values are within the expected range, even though these values are amongst the most enriched in the Andean dataset and are similar in enrichment value to the present-day studies conducted by us at another micro-region of the Argentine puna. Due to this, we propose that future investigations should establish a foundation for post-depositional alterations concerning the isotopic relation of non-carbonized archaeological plants. This is mainly because studies undertaken to date tend to highlight the isotopic signal of carbonized materials. Likewise, it is necessary to measure the absorption of contaminant from the earth using an electronic scanning microscope and its possible influence on isotope values. Finally, it is crucial to carry out present-day studies so as to evaluate isotopic values with pH and $\delta^{15}N$ ‰ in the earth, the use of fertilizers and pesticides. Nevertheless, the observable grouping, according to the type of crop maturation, of the Nitrogen values invites us to analyze a sample taking into consideration racial type, so as to determine whether a real link exists between the period of maturation and Nitrogen enrichment.

We consider that these results herald the need to continue our studies of the anthropogenic effects on food webs through the study of human and associated plant, prior to discarding archaeobotanical data from a consideration of paleodiets. Above all it is essential to consider this in an area were we have a crucial integration between agricultural and pastoralist production towards optimization of crop management.

ACKNOWLEDGEMENTS

We are grateful to Richard Heemskerk (Environmental Isotope Laboratory, Dept. of Earth and Environmental Sciences, University of Waterloo) who by his generous offer of lowering the cost of the isotope analysis made this study possible. We extend sincere thanks to Augusto Tessone, Estela Ducós, Pedro Salminci, Celeste Samec and Nadia Killian Galván for their valuable comments and technical assistance. Also special thanks go to Kevin Lane for translating this article. Finally, we appreciate the comments and suggestions of two anonymous reviewers, which strengthened the manuscript considerably.

REFERENCES

ABIUSSO, N. AND J. CÁMARA HERNÁNDEZ
1974. Los maíces autóctonos de la Quebrada de Humahuaca (Jujuy, Argentina), sus niveles nitrogenados y su composición en aminoácidos. *Revista de la Facultad de Agronomía*, 3° época, t. L, entrega 1-2: 1-25.

ALFARO, L.C.
1988. *Investigación Arqueológica en la Cuenca del río Doncellas. Dpto. Cochinoca, Jujuy. Reconstrucción de una cultura olvidada en la Puna Jujeña*. Imprenta del Estado de la Pcia. de Jujuy.

AMBROSE, S.H.
1993. Isotopic analysis of paleodiets: Methodological and interpretive considerations. In *Investigations of Ancient Human Tissue. Chemical Analysis in Anthropology*, edited by M.K. Sandford, pp. 59-130. Pennsylvania, Gordon and Breach Science Publishers.

AMUNDSON, R., A.T. AUSTIN, E.A.G. SCHUUR, K. YOO, V. MATZEK, C. KENDALL, A. UEBERSAX, D. BRENNER AND W.T. BAISDEN
2003. Global patterns of the isotopic composition of soil and plant nitrogen. *Global Biogeochemical Cycles* 17(1): 1031.

ARANÍBAR J., S.M.L. LÓPEZ CAMPENY, M.G. COLANERI, A.S. ROMANO, S.A. MACKO AND C.A. ASCHERO
2007. Dieta y Sociedades Agropastoriles: análisis de isótopos estables de un sitio de la Puna Meridional Argentina (Antofagasta de la Sierra, Catamarca). *Comechingonia* 10: 29-48.

ARAUS J.L., A. FEBRERO, R. BUXÓ, M.D. CAMALICH, D. MARTIN, F. MOLINA, M.O. RODRIGUEZ-ARIZA AND I. ROMAGOSA
1997. Changes in carbon isotope discrimination in grain cereals from different regions of the western Mediterranean Basin during the past seven millennia. Palaeoenvironmental evidence of a differential change in aridity during the late Holocene. *Global Change Biology* 3(2): 107-118.

AUSTIN, A.T. AND P.M. VITOUSEK
1998. Nutrient dynamics on a rainfall gradient in Haeai`i. *Oecologia* 113: 519-529.

BADECK F.W., G. TCHERKEZ, S. NOGUÉS, C. PIEL AND J. GHASHGHAIE
2005. Post-photosynthetic fractionation of stable carbon isotopes between plant organs-a widespread phenomenon. *Rapid Communications in Mass Spectrometry* 19: 1381-1391.

BOGAARD, A., T.H.E. HEATON, P. POULTON AND I. MERBACH
2007. The impact of manuring on nitrogen isotope ratios in cereals: archaeological implications for reconstruction of diet and crop management practices. *Journal of Archaeological Science* 34(3): 335-343.

CANAL, M.C.
2006. *Stable Isotope analysis and maize-stalk beer diet in rats: implications for the origins of maize.* Unpublished PhD Dissertation, The University of British Columbia, Vancouver.

DEINES, P.
1980. The isotopic composition of reduced organic carbon. In *Handbook of Environmental Isotope Geochemistry, Volume 1 "The Terrestrial Environment"*, edited by Fritz and Fontes, Chapter 8, pp. 329-406. Elsevier, Holanda.

FALABELLA, F., M.T. PLANELLA, E. ASPILLAGA, L. SANHUEZA AND R.H. TYKOT
2007. Dieta en sociedades alfareras de Chile Central: aporte de análisis de isótopos estables. *Chungara, Revista de Antropología Chilena* 39(1): 5-27.

FERNÁNDEZ, J. AND H.O. PANARELLO
1999-2001. Isótopos del carbono en la dieta de herbívoros y carnívoros de los andes jujeños. *Xama* 12-14: 71-85.

FINUCANE, B.C.
2007. Mummies, maize, and manure: multi-tissue stable isotope analysis of late prehistoric human remains from the Ayacucho Valley, Perú. *Journal of Archaeological Science* 34(12): 2115-2124.

FRIEDLI, H., H. LÖTSCHER, H. OESCHGER, U. SIEGENTHALER AND B. STAUFFE
1986. Ice core record of the 13C/12C ratio of atmospheric CO2 in the past two centuries. *Nature* 324: 237-238.

FRY B., W. BRAND, F.J. MERSCH, K. THOLKE AND R. GARRITT

1992. Automated analysis system for coupled $\delta^{13}C$ and $\delta^{15}N$ measurements. *Analytical Chemistry* 64: 288-291.

GARCÍA S. AND D. ROLANDI
1999. Viajes comerciales, intercambio y relaciones sociales en la población de Antofagasta de la Sierra, (Puna meridional Argentina). In *Propuestas para una Antropología Argentina* V, coordinated by C.E. Berbeglia, pp. 201-217. Biblos, Buenos Aires.

HASTORF, C.A. AND M.J. DE NIRO
1985. Reconstruction of prehistoric plant production and cooking practices by a new isotopic method. *Nature* 315: 489-491.

HEATON, T.H.E.
1987. The 15N/14N ratios of plants in South Africa and Namibia: relationship to climate and coastal/saline enviroments. *Oecologia* 74(2): 236-246.

1999. Spatial, species, and temporal variations in the 13C/12C ratios of C3 plants: Implications for palaeodiet studies. *Journal of Archaeological Science* 26(1): 637-649.

KILLIAN GALVÁN, V.A. AND D.E. OLIVERA
2008. First $\delta^{13}C$ for human skeletal remains from South Western Puna (Jujuy, Argentina). In *VI South American Symposium on Isotope Geology*, compiled by E. Linares, N.G. Cabaleri, M.D. Do Campo, E.I. Ducós and H.O. Panarello. Actas en CD-ROM, Resumen Extendido N° 129, Buenos Aires. ISSN 1851 6858. 4 pp.

KILLIAN GALVÁN, V.A., D.E. OLIVERA AND E. GALLEGOS
2012. Una aproximación isotópica al consumo del maíz en la Localidad Arqueológica Río Doncellas (Dpto. de Cochinota, Prov. de Jujuy). In *Las Manos en la Masa: Arqueologías, Antropologías e Historias de la Alimentación en Suramérica*, edited by M.P. Babot, M. Marschoff and F. Pazzarelli, pp. 319-338. Museo de Antropología, Universidad Nacional de Córdoba, Córdoba.

MURPHY, B.P. AND D.M.J.S. BOWMAN
2006. Kangaroo metabolism does not cause the relationship between bone collagen $\delta^{15}N$ and water availability. *Functional Ecology* 20(6): 1062-1069.

OLISZEWSKI, N.
2005. Archaeobotany of archaeological sites from Northwest Argentina (1750-1450 B.P.): ceremonial use or rubbish dumps? *Vegetation History and Archaeobotany* 14(4): 465-471.

2008. Metodología para la identificación subespecífica de maíces arqueológicos. Un caso de aplicación en el noroeste de argentina. In *Arqueobotánica y Teoría Arqueológica. Discusiones desde Suramérica*, edited by S. Archila, M. Giovannetti and V. Lema, pp. 181-202. Bogotá: Uniandes - Ceso.

OLISZEWSKI, N. AND D.E. OLIVERA
2009. Variabilidad racial de macrorestos arqueológicos de *Zea mays* (Poaceae) y sus relaciones con el proceso agropastoril en la puna Meridional Argentina (Antofagasta de la Sierra, Catamarca). *Darwiniana* 47(1): 76-9.

OLIVERA, D.E. AND H.D. YACOBACCIO
1999. Estudios de paleodieta en poblaciones humanas de los Andes del Sur a través de isótopos estables. *Paper presented at the V Congreso Nacional de Paleopatología, Alcalá La Real, Jaén.*

PATE, F.D.
1994. Bone chemistry and paleodiet. *Journal of Archaeological Method and Theory* 1(2): 161-209.

PÉREZ DE MICOU, C.B.
1996. *Los artefactos sobre materias primas vegetales flexibles de la Colección Doncellas del Museo Etnográfico (Buenos Aires) y Museo del Pucará (Tilcara).* Unpublished PhD Dissertation, Facultad de Filosofía y Letras, Universidad de Buenos Aires.

PIDWIRNY, M.
2006. *'The Nitrogen Cycle". Fundamentals of Physical Geography, 2nd Edition*. Date Viewed. http://www.physicalgeography.net/fundamentals/9s.html

POGGIO L., M. ROSATO, A.M. CHIAVARINO AND C.A. NARANJO
1998. Genome Size and Environmental Correlations in Maize (*Zea mays* ssp. mays, Poaceae). *Annals of Botany* 82(1): 107-115.

SAMEC, C.T.
2011. *Perspectiva isotópica sobre la alimentación de camélidos domésticos y silvestres de la Puna Jujeña: construyendo un marco de referencia para estudios arqueológicos.* Unpublished Graduate Thesis. Facultad de Filosofía y Letras, Universidad de Buenos Aires.

SCHWARCZ H.P. AND M.J. SCHOENINGER
1991. Stable Isotope Analyses In Human Nutritional ecology. *Yearbook of Physical Anthropology* 34: 283-321.

SEALY, J.C., N.J. VAN DER MERWE, J.A. LEE THORP AND J.L. LANHAM
1987. Nitrogen isotopic ecology in southern Africa: Implications for environmental and dietary tracing. *Geochimica et Cosmochimica Acta* 51(10): 2707-2717.

TESSONE A., C. SAMEC, V. KILLIAN GALVÁN AND H. PANARELLO
2009. Restos vegetales de origen arqueológico e isótopos estables del Carbono: su posibilidad de uso en las reconstrucciones paleodietarias y paleoclimatológicas. *Actas del III Congreso Argentino de Arqueometría y II Jornadas Nacionales*

para el estudio de Bienes Culturales. Córdoba, Argentina.

TIESZEN L. AND T. FAGRE
1993. Carbon isotopic variability in modern and archaeological maize. *Journal of Archaeological Science* 20(1): 25-40.

WARINNER, C. AND N. TUROSS
2009. Alkaline cooking and stable isotope diet-tissue discrimination in swine: archaeological implications. *Journal of Archaeological Science* 36(8): 1690-1697.

---

[1] For example, one of the Río Miriguaca (intermediate sector) neighbors used the following technic to optimize his crop. He annexed a "*bostadero*" or dungfield to his plot (each approximately 40 m^2) so that the water from the canal flows a number of times through the dungfield and his plot. In this manner, he is assured of a higher water flow. At the bottom of the basin a neighbor from Peñas Coloradas placed her crop alongside a strong water flow. This was even criticized by other neighbors who reside in Villa de Antofagasta as an overuse of the canal. In the same Villa there was a third case in which an owner watered the crops from a large water reservoir working with a pump. Conforming to expectation, this last case represented the highest production per m^2.

[2] There is currently a re-evaluation study of all the spades and adzes in the Doncellas Collection (INAPL) led by Lic. Susana Pérez with the object of interpreting their function.

[3] At the moment we only include in these figures the values for which we have Carbon and Nitrogen percentages (n=20).

ANCIENT DNA RESEARCH, SCOPE AND LIMITATIONS. A FIRST GENETIC ANALYSIS OF MUSEUM SAMPLES FROM SAN JULIÁN, SANTA CRUZ, ARGENTINA

Cristina B. Dejean[1 2 3], Cristian M. Crespo[2 3 6], Francisco R. Carnese[1 2], José L. Lanata[4 5 6]

[1] Sección Antropología Biológica, ICA, Facultad de Filosofía y Letras, UBA, [2] Centro de Estudios Biomédicos, Biotecnológicos, Ambientales y Diagnóstico (CEBBAD) Universidad Maimónides, [3] Fundación de Historia Natural Félix Azara. Dpto. de Cs. Naturales y Antropológicas, Universidad Maimónides, [4] Universidad Nacional de Río Negro, [5] IIDyPCa, [6] CONICET

Abstract

Thirty years ago the first ancient DNA studies have began. New methodologies have facilitated the task and ancient genomes can now be analyzed. The knowledge has accumulated in the field of the so-called paleogenetic that enable us to better understand the genetic relation between *H. sapiens* and *H. neanderthalensis*. Archaeogenetic in American samples has confirmed the Asian origin of Native Americans, establishing five founder maternal lineages for North America (A, B, C, D and X). For South America, the first four lineages are present A, B, C and D. Analyses of samples from Argentina shows a particular distribution, high B haplogroup frequencies in North west, and decreasing presence A and B towards the South, with a concomitant increase of C and D haplogroups, which are the only described so far in Tierra del Fuego. The analysis of specimens from Rosa Novak Museum from San Julián, in Santa Cruz Province, also demonstrated C and D haplogroups in Southern Patagonia.

Key words: *Ancient DNA - Paleogenetic - Mitochondrial DNA - Native Americans - Maternal haplogroups*

INTRODUCTION

The aim of this work is to introduce the methodology employed to analyze ancient DNA, including a brief review about its applications in paleoanthropology and archaeology, presenting also the first results obtained with samples from the Rosa Novak Museum, San Julián, Santa Cruz, in the Argentinian Patagonia.

Paleogenetics is a field of research in which many groups are working nowadays in several countries. It comprises research involving ancient DNA (Hummel 2003). Some authors prefer the term Archaeogenetics when it refers specifically to their applications in Archaeology (Renfrew 2000). But, what is ancient DNA? Herrmann and Hummel (1994) define "...the term ancient DNA covers any bulk of trace DNA from a dead organism or part of it, as well as extracorporeally encountered DNA of a living organism. Therefore, any DNA that has undergone autolytic or diagenetic processes or any kind of fixation is considered to be ancient DNA". No matter its antiquity, ancient DNA would be that which may be present in any kind of archaeological or historical remains. DNA extraction has been essayed and done from different tissues and remains. Bone and teeth are most frequently employed but hair, skin, tissues and coprolites can be also a DNA source for investigations (Hummel 2003; Kaestle and Horsburgh 2002).

As in any study involving human remains ethical, legal and social implications should always be considered. Because these samples are unique, we agree with other authors (Kaestle and Horsburgh 2002), that analysis should be made only in cases where data will contribute to understand o answer a valuable question or elucidate a debate. In some cases, ancient DNA may not be very informative because of the own limitations of this type of analysis. So it is very important to evaluate, before starting, the specimen preservation state to be studied. Chemical processes can affect the DNA molecule: oxidation, hydrolysis, alkylation, condensation may introduce changes in the original sequence and must be considered in the analysis (Lindahl 1993; Willerslev and Cooper 2005).

Sample contamination is a huge problem and preventing its occurrence is always a priority for researchers. External contamination must be eliminated and several methods have been reported with this aim in the literature: abrasion, treatment with hypochlorite, irradiation with UV, etc. (Kaestle and Horsburgh 2002; Kemp and Glenn Smith 2005; O'Rourke *et al.* 2000). Extreme measures are usually reported during the procedures of extraction and amplification. The working area must be exclusively employed for ancient specimens and separated from the one where modern DNA is manipulated to avoid that any trace may contaminate samples. UV irradiation of this area is also recommended as well as the employ of sterile materials, solutions, water, etc. and free from enzymes, which may destroy DNA. Masks, gloves and lab coats must be employed by operators during the manipulations involved in the different processes. No matter how much care has been taken there is still the possibility of no amplifiable DNA may be present. Amplification inhibitors may also be co-purified in the extracts obtained and its presence produces negative results. Many protocols have been developed and several commercial kits are available to eliminate substances that may act as inhibitors, so this situation must always be considered and essayed when negative results are systematically obtained (Kemp *et al.* 2006).

Finally reproducibility of the data obtained is always pursued. Several protocols are followed to confirm that data obtained came from the specimen studied and not from any possible contaminant: independent sample study in different laboratories, cloning amplified DNA, processing of two samples from the same individual, or several extractions and amplifications (Cooper and Poinar 2000; Gilbert *et al.* 2005; Poinar 2003).

APPLICATIONS IN PALEOANTHROPOLOGY AND ARCHAEOLOGY

The first reports involving ancient DNA came from China Hunan Medical School (Hunan Medical School 1980), about 30 years ago, where the presence of preserved DNA was demonstrated in a Chinese mummy, from Mawanghtui. DNA was extracted from the extinct species quagga (Higuchi *et al.* 1984) and mammoth (Johnson *et al.* 1985). In 1984 DNA presence was demonstrated in an Egyptian infant mummy (Pääbo *et al.* 1984). Five years later, Hagelberg and co-workers reported the first amplification from an archaeological material (Hagelberg *et al.* 1989). From then on, successful amplifications have been achieved; one of the most known are those that involve the Russian Tsar family, whose remains were found in Ekaterinenburg in 1991 (Gill *et al.* 1994; Ivanov *et al.* 1996).

Methods and new sequencing techniques have been applied in paleoanthropology, including the significant findings about Neanderthals genetics in relation to that of Homo sapiens (Green *et al.* 2010; Krings *et al.* 2000). Mitochondrial data obtained showed a great distance between DNA from both species, but the publication of the first Neanderthal genome seemed to indicate the possibility of a closer genetic distance in nuclear genes (Green *et al.* 2010). The presence of FOX P2 suggests *Homo neanderthalensis* could produce language (Krause *et al.* 2007). These new techniques, were applied to the study of the so called Denisovian man, buried in a cave from Altai, a phalange and a molar found together have made it possible to analyze its mitochondrial and nuclear genome (Krause *et al.* 2010; Reich *et al.* 2010). The authors have postulated that the samples belong to a different species from *H. sapiens* and *H. neanderthalensis* and found that part of their genetic characteristics can still be found in extant Melanesians.

It is also remarkable that many technical developments for forensic sciences are rapidly applied to ancient DNA analysis, for example phenotypic characteristics were assigned to Neanderthals through certain allelic variants determining the possibility of them having light color eyes and clear skin (Lalueza-Fox *et al.* 2007). Recently the phenotype for en ancient human from Greenland was inferred, extracting his DNA from a hair sample and based in the study of these genetic markers, by the Danish team headed by Eske Willesev (Rasmussen *et al.* 2010).

The most frequently genetic marker studied in Archaeogenetics research is mitochondrial DNA. Not only because it is an excellent marker for maternal lineages, due to fact that it is inherited from mother to child, but also because its presence in numerous copies in each cell, so it is more probably found in ancient material. Usually native populations inhabiting a geographical area exhibit a limited number of mitochondrial lineages, the so-called haplogroups, which are by convention named with letters. Haplogroups may be sub classified in haplotypes, which share some basic mutations among them but present additional changes in their DNA sequences. Native American's haplogroups are found in Asia and constitute part of its original diversity. In the case of America, a particular distribution is generally accepted. In North America five are present: A, B, C, D and X. In South America, four are found A, B, C, and D (Dornelles *et al.* 2005) and in the very south extreme of Patagonia only C and D are generally detected. When the presence of X haplogroup in North American natives was detected, it was considered the result of mixture with European populations, but later it was demonstrated that it presented particular mutations only seen in this region and different from those coming from Eurasia. Its presence in North America pre-Columbian samples was an additional finding to be accepted as one of the founder maternal lineages coming from Asia (Mahli and Glenn-Smith 2002). So far no X Amerindian haplogroup have been detected neither in extant nor in extinct South American natives.

Concerning South America, the first studies with ancient DNA have been carried out in different archeological sites from the extreme north to Tierra del Fuego in the south. Eight pre-Columbian mummies from Colombia were analyzed by Monsalve *et al.* (1996), who detected three out of four Amerindian mitochondrial haplogroups (A, B and C). At the same time, Ribeiro dos Santos *et al.* (1996) amplified and sequenced the hipervariable control region of mitochondrial DNA in 18 Pre-Colombian skeleton remains (500-4000 BP) from the Amazonia Region. Ten of them presented typical mutations for haplogroups A, C and D. However, a more ancient sample, 4000 year BP, was assigned to haplogroup B, which is characterized by a deletion of 9 base pairs in the V region of the codifying mitochondrial DNA; the remaining samples could not be assigned to Amerindian mitochondrial lineages although they exhibited mutations usually detected in Asian populations. Previously in mummies from northern Chile, Horai *et al.* (1991), also observed a low prevalence of the deletion of 9bp, which was only detected in one of the eleven remains tested.

It should be noted that the haplogroup B could not be detected in soft tissue, bone and teeth of 56 pre-Hispanic remains from northern Chile and the north, center and south of Argentina, so it seems to be a typical characteristic in the South American prehistoric groups, particularly, in those that inhabited the extreme south of the continent (Demarchi *et al.* 2001; Merriwether *et al.* 1994; Ramos *et al.* 1995; Rogan and Salvo 1990). These results showed low frequency or absence of haplogroup B in Prehispanic-South Amerindian remains. However, Moraga *et al.* (2001) in mummies of Azapa (2990-6000

BP), and Rothhammer *et al.* (2003) in skeletal remains of Tiwanaku, detected a relatively high frequency of B (15% and 22%, respectively). Later, a high prevalence of B was also observed in prehistoric groups coming from the Lluta, Azapa and Camarones valleys of Northern Chile. In this region, haplogroup B varies from 35% in the Late Archaic Period (6000-3900 BP) to 53% in the Late Intermediate Period (1000-500 BP) (Moraga *et al.* 2005). Besides, Shinoda *et al.* (2006) also detected high frequencies of haplogroup B (66%) in Peruvian remains from Paucarcancha, Patallacta and Huata sites, which show strong similarities with those of modern Quechua and Aymara Peruvian populations. Feheren-Schimtz and coworkers (2010 and 2011) have also studied several Peruvian ancient samples and determined also B haplogroup in different frequencies. For Colombian samples, Silva *et al.* 2008 and Casas-Vargas *et al.* 2011 also published high frequencies of haplogroup B in samples from the Herrera period and the Guane territories, respectively.

In specimens from the North Argentina, Carnese *et al.* (2010) have also detected B in high frequencies (47%) in the Pampa Grande population (1310 ± 40 B.P.) some of the B haplotypes detected have never been described neither in ancient nor in modern samples. A high frequency of D (42%) and low A haplogroup (11%) presence were also found. Absence of C haplogroup was probably due to the small number of samples studied. Different frequencies were found when Los Amarillos site was studied (1100-1500 AD), in this opportunity 65.2% of the individuals were carriers of the A haplogroup (Mendisco *et al.* 2011).

Nores and coworkers (2011) for 48 samples from the center region of our country determined the presence of the four haplogroups, but with a prevalence of C in the plains (40.9%) and hills (38.5%) where haplogroup frequencies seem to change with the location: higher B prevalence in the hills area (38.5%) and lower in plains (0.046%). Meanwhile A haplogroup seem to be absent in ancient samples coming from the plains compared to the 31.3% in more recent ones.

On the other hand, in the southern extreme of the continent, in the Fueguian-Patagonian Region, Lalueza *et al.* (1997) did not detect the haplogroups A and B in 60 remains of Fueguian aboriginals from Chilean and Argentine collections (100-200 BP), including two samples from skeletal remains of 4030 and 5000 years of age). Later, through the sequenciation of 24 samples, the same team (Garcia-Bour *et al.* 2004) detected the mutations characteristic of the haplogroups C and D in all cases, confirming their previous RFLP results. Two Yamana individuals from the Beagle Channel exhibited the same both haplogroups (Dejean *et al.* 2008).

In summary, regarding the series analyzed so far in South America, we have found a relatively elevated prevalence of the haplogroups A and C, in the Amazonia Region, high frequencies of B, especially in the Andean Region, and decreasing A and B and high frequencies of C and D in Patagonian samples.

Although several Argentine samples have been analyzed (Carnese *et al.* 2010; Dejean *et al.* 2008; Lalueza *et al.* 1997; Mendisco *et al.* 2011) studies have been carried out in laboratories abroad. The first results obtained in laboratories from our country have been only recently published (Crespo 2011; Nores and Demarchi 2011; Nores *et al.* 2011). As one of the groups working in this area, our aim is to contribute to the understanding the genetic mitochondrial diversity among the groups that inhabited the Patagonian region in Argentina studying samples from pre and post-Columbian times. In this work we present the first results obtained when analyzing museum samples from this area.

GENETIC ANALYSES OF MUSEUM SPECIMENS FROM "ROSA NOVAK" MUSEUM MATERIALS AND METHODS

Samples Studied

This research was carried out as a part of the project *"Paisajes Arqueológicos en el paralelo 49°"*, directed by J.L. Lanata. Samples analyzed came from the collections of Regional Museum "Rosa Novak", Puerto San Julián, Santa Cruz Province. Studies involving human remains should be carried with the maximum respect and care, the AABA (*Asociación de Antropología Biológica de Argentina*) has issued recommendations for their treatment, research and conservation (AABA 2007). This study was conducted taking these considerations and authorizations from the province and the curator were obtained. The ethical committee from Maimónides University approved the working protocol.

Materials deposited in the museum were donated by private collectors for their study and conservation. Some of them, called Hoffman collection, consist of several skulls and jaws and also several long bones and vertebrae. The museum has no record of the context in which the samples were found, and the only data available is that they came from the surroundings of the village.

The other collection was donated by Edward Walker, comes from San Julián neighborhood area and was collected while performing works on a ranch. To our knowledge, these remains were taken to the museum at the time of their finding in 1987. It is also composed by several skulls and jaws, and some long bones. Both collections present a good state of preservation. No datation was done to the samples, no associated objects, natural or cultural, that may have accompanied the skeletal remains, were found or recorded. So far it is not possible to estimate if samples are from pre or post Columbian times.

Samples from five different individuals were taken, three premolars, one molar, and one left femur (Table 1) for genetic analysis. They were selected based on their preservation status: absence of caries in teeth, fractures and fissures in the pieces. Data about age and sexual assignment was not available.

Anatomical classification	Collection	Codification
Second left upper pre-molar	Hoffman	Da1
Second right upper pre-molar	Walker	Da2
Third right upper molar	Hoffman	Da3
First right upper pre-molar	Walker	Da4
Left Femur	Hoffman	Fsj

Table 1. Samples studied.

Precautions Taken to Avoid Contamination

Numerous protocols for ancient DNA extraction from different sources try to ensure the elimination of foreign DNA. The authors have not only emphasized on sampling methods, but also on the extraction process and the possible presence of modern DNA contamination before or during the manipulations to be sure that results obtained come from the specimen studied (Gilbert *et al.* 2005; Hofreiter *et al.* 2001; Hummel 2003; Kaestle and Horsburg 2002; Montiel *et al.* 2001; Mulligan 2006; O'Rourke *et al.* 2000; Pääbo *et al.* 1989, 2004; Willerslev and Cooper 2005; Yao and Zhang 2003). Due to the fact that samples included in this research were not initially collected and preserved under optimal conditions for ancient DNA analysis, strict measures were taken to eliminate and avoid contamination with modern DNA and to authenticate the results obtained during all steps for DNA extraction and amplification.

Manipulations were performed in a physically separate laboratory, dedicated only to ancient DNA analysis. Operators involved in research (CBD and CMC) used gloves, masks, caps and lab coats. All sterile disposable plastics used were irradiated for 45 minutes before use. ART tips were employed. Surfaces, instruments and tools employed in laboratory procedures were clean employing 10% bleach, 70% ethanol and rinsed in DNase and RNase free water and irradiated with UV for 45 minutes before and after each stage of work. The experimental areas of pre and post-PCR were strictly separated.

A blank extraction was performed in parallel to each sample in order to evaluate contamination during the procedure. For each sample at least two extractions were completed and three amplifications from each extract were done to confirm the results obtained. All personnel involved in the sample processing were tested using the same techniques to compare the DNA profiles obtained to those of the sample. Both CBD and CMC belong to H haplogroup. We used this information to evaluate the authenticity of the ancient DNA obtained in this analysis.

DNA Extraction and Purification

In order to preserve the morphology of the dental specimens analyzed we proceed to obtain dentine from the interior of the teeth, which afterwards were reconstituted with a self-curing monomer commonly used in dental prosthesis. The procedure can be resume as follows: to eliminate contaminants in the outer surfaces each dental piece was placed in a 15 ml sterile conical tube with cap with sodium hypochlorite 5.5% for 15 minutes, rinsed four times, air dried and UV irradiated 45 minutes from each side. A vice was used to hold the tooth during cutting and dentine removal, the tooth root was removed using a cutting wheel to access to the piece pulp. Afterwards, the dentine was removed by grinding using a milling cutter and collected in another 1.5 ml sterile tube with cap. In the case of the femur a small incision was done in the intercondylar fossa, and bone powder was obtained from the inside (Figure 1). The amount of powder obtained was about 500 mg in all cases.

We employed between 60-70 mg of dentine or bone powder in each extraction performed as suggested in literature (Kuch *et al.* 2007). The powder was transferred to a new sterile conical tube and 1ml of 0.5 M EDTA at pH 8 was added and was left overnight under rotation at room temperature; a blank tube was processed in parallel.

Samples were centrifuged for 10 minutes at 3000 rpm and supernatant was discarded and the pellet was resuspended in 1 ml of digestion solution (0.8 ml of extraction buffer, each containing 250 ml of H_2O, 1.25 ml of 2M Tris Cl, 2.5 ml 0.5 M EDTA, 10.2 g of sodium acetate - 0.2 ml of 10% SDS-dodecyl sodium-and 50µl of proteinase K (20 mg / ml). The sample and its blank were incubated at 55 °C overnight, on an orbital incubator. The solution was extracted in 1 ml phenol: chloroform: isoamilic alcohol (25:24:1) (Hummel 2003). The aqueous supernatant containing the DNA in solution was transferred to a sterile tube. As a final step the samples were purified and concentrated employing the commercial kit Wizard SV Gel and PCR Clean-Up System (Promega), according to the manufacturer's technical recommendations[1].

Figure 1. Femur showing incision done to obtain bone powder.

Sex Determination

In order to determine the sex of the individuals studied a dimorphism in the amelogenin gene was analyzed. This gene encodes a protein involved in the formation of dental enamel; it is a single copy gene located at position Yp11.2 of Y chromosome with its counterpart in the region Xp22.3-p.22.1 the X chromosome. In the X chromosome a 6bp deletion is present. This marker is used in forensic studies and it is of great importance in archaeology, allowing sexing individuals even when not all the skeletal parts are available to establish it by their morphological characteristics. In this case, we amplified a small region where the dimorphism is located, in the case of Y chromosome amplicon size is 112 base pairs and it is 106 base pairs on chromosome X. The sizes of the fragments obtained were compared in 2.5% agarose electrophoresis.

mtDNA Analysis

Amerindian mitochondrial haplogroups were studied by RFLP (restriction fragment length polymorphisms). Primers, PCR conditions and restriction enzymes were described in Stone and Stoneking 1993 and Carnese *et al.* 2010. Through the RFLP technique, the absence or presence of the characteristic polymorphism were visualized on 2% agarose gels, stained with ethidium bromide and visualized through a UV transilluminator. This procedure was carried out for haplogroups A, C and D employing a restriction enzyme to reveal the point mutations. For B haplogroup the presence of a typical deletion of 9 bp in the region V, between the lysine tRNA and COII genes, was analyzed directly with the DNA fragments obtained with a 3% agarose gel.

RESULTS AND DISCUSSION

Results so far achieved through the analysis of ancient genomes, as those for Neanderthals and Denisovian remains demonstrate the relevance that these kinds of studies can acquire. Genetic diversity in pre-Columbian Amerindians contributes to our knowledge about the origin of the first Americans, to the comprehension of their demographic distribution through space and time. The development of studies involving ancient DNA in Argentina will provide new data about the peopling of our territory in the South American context.

After thirty years, it is well known that working with ancient DNA is not an easy task. Difficulties with the extraction and amplification are diverse, the need to corroborate results is generally expensive and time consuming. Contamination and PCR inhibitors presence cannot always be controlled and excluded, but when consistent results are obtained they are always of great value. Results from mitochondrial diversity obtained from ancient South America samples begin to appear and will make it possible to map geographically and temporarily the distribution of maternal lineages in the region. Argentine samples from north and south have demonstrated good percentage of ancient DNA recovery. This seems to be the case for Patagonian samples here reported. The genetic study of Museum Rosa Novak samples has been performed with both aims analyzing and trying to cause the minimal damage to the material employed. Reconstruction of the dental pieces to return them to the collection has been done, so that external dental morphology can still be studied. It is also possible to extract DNA from long bones with a minimal damage (Gibbon *et al.* 2009), as it was done in the present work.

Ethical and legal aspects must not be neglected. Currently, in Argentina the national law 1 25.517/01 states that the remains of the aborigines that are deposited in museums and public or private collections should be made available to people and / or communities who claim them. In turn, states that those remains not claimed by their communities could remain available to the institutions that house them, having to be all scientific endeavor aimed consented with the communities. In 2010, the 701/2010 decree has been issued to rule the 25.517 law, establishing the INAI (*Instituto Nacional de Asuntos Indígenas*) as the institution in charge of coordinate, develop and assist in the monitoring and study of compliance with the policies and actions established by this law, with the additional support of the INAPL (*Instituto Nacional de Antropología y Pensamiento Latinoamericano*). We believe that all actors involved when human remains are analyzed (scientists, indigenous communities, museum curators, institutions), should discuss in the frame of this law the scopes of the studies, and establish the relevance of the results that may be obtained.

Concerning the studies done in samples from the Rosa Novak Museum, they were extracted and their results compared to the personnel profiles involved in this study to demonstrate absence of contamination. Additionally, no positive amplification was obtained from each extraction blank and PCR reaction blank.

Analysis of the amelogenin gene determined that Da2 and Da3 samples had profiles corresponding to female individuals. DNA extracted from Da1, Da4 and Fsj could not be amplified in any of the PCRs performed. Results were replicated twice from two different DNA extractions, always obtaining a negative amplification result. In spite of the sex marker negative results in 3/5 samples, amelogenin demonstrated to be an excellent tool for sexing individuals, and it will be probably very useful in the case of sub-adults or when sexual dimorphism is diminished in adults. Neither could be mitochondrial DNA amplified in the same sample. We believe that in San Julián specimens the presence of PCR inhibitors or extremely degraded DNA may be responsible for both negative results. Amplifying smaller portions of DNA may be useful if the second hypothesis is right.

Mitochondrial lineage could be determined in two of the five samples analyzed (Da2 and Da3) the presence of the cleavage site characteristic of Native American haplogroup D (Crespo *et al.* 2009). However, subsequent re-analysis of the samples allowed us to verify the D haplogroup in Da2 and in the case of Da3 C haplogroup was determined (Table 2). Samples Dal, Da4 and Fsj could not be amplified in any of the PCRs performed. Presence of C and D haplogroups in the samples was expected according to previous findings of these two maternal Amerindian lineages in other Patagonian samples, no other haplogroup has been determined in extinct and historical specimens studied so far, no matter this results the presence of another founders lineages cannot be excluded for the region (Dejean *et al.* 2008; García-Bour *et al.* 2004; Lalueza *et al.* 1997).

Sample	Mitochondrial haplogroup	Amelogenin
Da1	N/A*	N/A
Da2	Haplogroup D	Female
Da3	Haplogroup C	Female
Da4	N/A	N/A
Fsj	N/A	N/A

Table 2. Mitochondrial and amelogenin amplifications results.
*N/A: No amplification obtained.

The results obtained in this first approach to Patagonian samples indicate the possibility to perform this kind of research to contribute to the history of the populations in the South of our country. We continue to study the material obtained. Sequences of the mitochondrial HVRI control region for Da2 and Da3 have confirmed the haplogropus obtained by RFLP (unpublished results). Da2, for example, has a D4h3 haplotype, described in extant native Amerindians following a coastal route towards South America (Perego *et al.* 2009) and was also found by Fehren-Smicht and co-workers (2011) in ancient samples from North Perú coast. The amplification of the control region of the mitochondrial DNA to establish the haplotype in each sample seems to offer a better chance to describe this region genetic variability due to the fact that less DNA is consumed. New samples are being included as part of the project and will contribute to increase data available to understand the relationships among South American past populations. Radiocarbon dates should be done in samples studied in this work, as we consider that temporal dimension is necessary to describe both, time and diversity for evaluating *tempo* and *modo* of the peopling of the Argentine territory in the South American context.

CONCLUSIONS

We demonstrated the presence of two females with C and D maternal mitochondrial lineages in Patagonian samples. For both individuals enough DNA was obtained with the techniques employed, but new extraction protocols to eliminate inhibitors and improve results must be developed. Each collection must be tested to achieve the best performance for extraction and amplification.

Working with ancient DNA is laborious, and time consuming but the data obtained are valuable and it is worth the effort. Dating and increasing number of samples analyzed will contribute to a better understanding of the America peopling process and also to evaluate how much of the original genetic diversity was lost with the European conquest.

ACKNOWLEDGEMENTS

We acknowledge the financial support of UBACyT, CONICET, Fundación Felix de Azara, Universidad Maimónides and Universidad de Río Negro.

REFERENCES

AABA (Asociación de Antropología Biológica de Argentina)
2007. *Declaración de la asociación de Antropología Biológica Argentina en relación con la ética del estudio de restos humanos.* Available in: www.fcnym.unlp.edu.ar/aabra.

Carnese, F.R., F. Mendisco, C. Keyser, C.B. Dejean, J.M. Dugoujon, C.M. Bravi, B. Ludes and E. Crubezy
2010. Paleogenetical study of pre-Columbian samples from Pampa Grande (Salta, Argentina). *American Journal of Physical Anthropology* 141: 452-462.

Casas-Vargas, A., A. Gómez, I. Briceño, M. Díaz-Matallana, J.E. Bernal and J.V. Rodríguez
2011. High genetic diversity on a sample of pre-Columbian bone remains from Guane territories in northwestern Colombia. *American Journal of Physical Anthropology* 146: 637-649.

Cooper, A. and H. Poinar
2000. Ancient DNA: do it right or not at all. *Science* 289: 1139.

Crespo, C.M.
2011. *Viajando con genes: una metodología molecular aplicada a interrogantes arqueológicos.* IIDyPCa - UNRN - CONICET, Colección Tesis. Bariloche. ISBN 978-987-26198-3-1.

Crespo, C., C. Dejean, F. Carnese and J.L. Lanata
2009. Extracción y tipificación de ADN a partir de piezas dentales de muestras arqueológicas de la Prov. de Santa Cruz, Argentina. Paper *presented at the Novenas Jornadas Nacionales de Antropología Biológica*. Puerto Madryn, Argentina.

Dejean, C.B., C. Keyser, S. Avena, J.M. Dugoujon, B. Ludes, E. Crubezy and F.R. Carnese
2008. Análisis genético de muestras esqueletales provenientes del Canal de Beagle. In *Actas del X Congreso de la Asociación Latinoamericana de Antropología Biológica*, La Plata, October, 2008. Published in CD-ROM.

DEMARCHI, D., G. PANZETTA-DUTARI, S. COLANTONIO AND A. MARCELINO
2001. Absence of the 9-bp deletion of mitochondrial DNA in pre-historic inhabitants of Argentina. *Human Biology* 73: 575-582.

DORNELLES, C., S. BONATTO, L. DE FREITAS AND F. SALZANO
2005. Is haplogroup X present in extant South American Indians? *American Journal of Physical Anthropology* 127: 439-448.

FEHEREN-SCHMITZ, L., M. REINDEL, E.T. CACIGAO, S. HUMMEL AND B. HERRMANN
2010. Pre-columbian population dynamics in coastal Southern Peru: a diachronic investigation of mtDNA patterns in the Palpa region by ancient DNA analysis. *American Journal of Physical Anthropology* 141: 208-221.

FEHEREN-SCHMITZ, L., O. WARNBERG, M. REINDEL, V. SEIDENBERG, E. TOMASTO-CAGIGAO, J. ISLA-CUADRADO, S. HUMMEL AND B. HERRMANN
2011. Diachronic investigations of mitochondrial and Y-Chromosomal genetic markers in pre-columbian Andean highlanders from South Peru. *Annals of Human Genetics* 75: 266-283.

GARCIA-BOUR, J., A. PERÉZ-PÉREZ, S. ÁLVAREZ, E. FERNÁNDEZ, A.M. LÓPEZ-PARRA, E. ARROYO-PARDO AND D. TURBÓN
2004. Early population differentiation in extinct aborigines from Tierra del Fuego-Patagonia: ancient mtDNA sequence and Y-chromosome STR characterization. *American Journal of Physical Anthropology* 123: 361-370.

GIBBON, V., C. PENNY, G. STRAKLJ AND P. RUFF
2009. Brief communication: minimally invasive bone sampling method for DNA analysis. *American Journal of Physical Anthropology* 139: 596-599.

GILBERT, M., H. BANDELT, M. HOFREITER AND I. BARNER
2005. Assessing ancient DNA studies. *Trends in Ecology and Evolution* 20: 541-544.

GILL, P., P. IVANOV, C. KIMPTON, R. PIERCY, N. BENSON, G. TULLY, I. EVETT, E. HAGELBERG AND K. SULLIVAN
1994. Identification of the remains of the Romanov family by DNA analysis. *Nature Genetics* 6: 130-135.

GREEN, R., J. KRAUSE, A. BRIGGS, T. MARICIC, U. STENZEL, M. KIECHER, N. PATTERSON, H. LI, W. ZHAI, M. FRITZ, N. HANSEN, E. DURAND, A. MALASPINAS, J. JENSEN, T. MARQUES-BONET, C. ALKAN, K. PRUFER, M. MEYER, H. BURBANO, J. GOOD, J. SCHULTZ, A. PETRI, A. BUTTHOF, B. HOBER, B. HOFFNER, M. SIEGEMUND, A. WEIHMANN, C. NUSBAUM, E. LANDER, C. RUSS, N. NOVOD, J. AFFOURTIT, M. EGHOLM, C. VERNA, P. RUDAN, D. BRAJKOVIC, Z. KUCAN, I. GUSIC, V. DORONICHEV, L. GOLOVANOVA, C. LALUEZA-FOX, M. DE LA RASILLA, J. FORTEA, A. ROSAS, R. SCHMITZ, P. JOHNSON, E. EICHLER, D. FALUSH, E. BIRNEY, J. MILLIKIN, M. SLATKIN, R. NIELSEN, J. KELSO, M. LACHMANN, D. REICH AND S. PÄÄBO
2010. A draft sequence of the Neanderthal genome. *Science* 328: 710-722.

HAGELBERG, E., B. SYKES AND R. HEDGES
1989. Ancient bone DNA amplified. *Nature* 342: 485-485.

HERRMANN, B. AND S. HUMMEL
1994. Introduction. In *Ancient DNA: Recovery and Analysis of Genetic Material from Paleontological, Archaeological, Museum, Medical and Forensic Specimen*, pp. 1-12. Springer-Verlag, New York.

HIGUCHI, R., B. BOWMAN, M. FREIBERGER, O. RYDER AND A. WILSON
1984. DNA sequences from the quagga, an extinct member of the horse family. *Nature* 312: 282-284.

HOFREITER, M., D. SERRE, H. POINAR, M. KUCH AND S. PÄÄBO
2001. Ancient DNA. *Genetics* 2: 353-359.

HORAI, S., R. KONDO, K. MURAYAMA, S. HAYASHI, H. KOIKE AND N. NAKAI
1991. Phylogenetic affiliation of ancient and contemporary humans inferred from mitochondrial DNA. *Philosophical Transactions: Biological Sciences* 333: 409-417.

HUMMEL, S.
2003. *Ancient DNA Typing. Methods, Strategies and Applications*. Springer-Verlag, Berlin Heidelberg.

HUNAN MEDICAL COLLEGE
1980. *Study of an Ancient Cadaver in Mawangtui Tomb No. 1 of the Han Dynasty in Changsha*. Ancient Memorial Press, Beijing, pp. 184-187.

IVANOV, P., M. WADHAMS, R. BOBY, M. HOLLAND, V. WEEDN AND T. PARSONS
1996. Mitochondrial DNA sequence heteroplasmy in the Grand Duke of Russia Georgij Romanov establishes the authenticity of the remains of Tsar Nicholas II. *Nature Genetics* 12: 417-420.

JOHNSON, P., C. OLSON AND M. GOODMAN
1985. Isolation and characterization of deoxyribolucleic acid from tissue of the woolly mammoth, *Mammuthus primigenius*. *Comparative Biochemistry and Physiology* 81: 1045-1051.

KAESTLE, F. AND K. HORSBURGH
2002. Ancient DNA in Anthropology: Methods, Apliccations and Ethics. *Yearbook of Physical Anthropology* 45: 92-130.

KEMP, B. AND D. GLENN SMITH
2005. Use of bleach to eliminate contaminating DNA from the surface of bones and teeth. *Forensic Science International* 154: 53-61.

KEMP, B., C. MONROE AND D. GLENN SMITH
2006. Repeat silica extraction: a simple technique for the removal of PCR inhibitors from DNA extracts. *Journal of Archaeological Science* 33(12): 1680-1689.

KRAUSE, J., Q. FU, J. GOOD, B. VIOLA, M. SHUNKOV, A. DEREVIANKO AND S. PÄÄBO
2010. The complete mitochondrial DNA genome of an unknown hominin from southern Siberia. *Nature* 464: 894-897.

KRAUSE, J., C. LALUEZA-FOX, L. ORLANDO, W. ENARD, R. GREEN, H. BURBANO, J. HUBLIN, C. HANNI, J. FORTEA, M. DE LA RASILLA, J. BERTRANTPETIT, A. ROSAS AND S. PÄÄBO
2007. The derived FOXP2 variant of modern human was shared whit Neandertals. *Current Biology* 17: 1908-1912.

KRINGS, M., C. CAPELLI, F. TSCHENTSCHER, H. GEISERT, S. MEYER, A. VON HAESELER, K. GHOSSSCHMIDT, G. POSSNERT, M. PAUNOVIC AND S. PÄÄBO
2000. A view of Neanderthal genetic diversity. *Nature* 26: 144-146.

KUCH, M., D. GROCKE, M. KNYF, M. GILBERT, B. YOUNGHUSBAND, T. YOUNG, I. MARSHALL, E. WILLERSLEV, M. STONEKING AND H. POINAR
2007. A preliminary analysis of the DNA and diet of the extinct beothuk: a systematic approach to ancient human DNA. *American Journal of Physical Anthropology* 132: 594-604.

LALUEZA, C., A. PÉREZ, E. PRATS, L. CORNUDELLA AND D. TURBON
1997. Lack of founding Amerindian mtDNA lineages in extinct Aborigines from Tierra del Fuego-Patagonia. *Human Molecular Genetics* 6: 41-46.

LALUEZA-FOX, C., H. RÖMPLER, D. CARAMELLI, C. STÄUBER, G. CATALANO, D. HUGHES, N. ROHLAND, E. PILLI, L. LONGO, S. CONDEMI, M. DE LA RASILLA, J. FORTEA, A. ROSAS, M. STONEKING, T. SCHÖNEBERG, J. BERTRANTPETIT AND M. HOFREITER
2007. A melacortin 1 receptor alleles suggest varying pigmentation among Neanderthals. *Science* 318: 1453-1455.

LINDAHL, T.
1993. Instability and decay of the primary structure of DNA. *Nature* 362: 709-715.

MALHI, R. AND D. GLENN-SMITH
2002. Brief communication: haplogroup X confirmed in prehistoric North America. *American Journal of Physical Anthropology* 119: 84-86.

MENDISCO, F., C. KEYSER, C. HOLLARD, V. SELDES, A. NIELSEN, E. CRUBEZY AND B. LUDES
2011. Application of the iPLEX Gold SNP genotyping method for the analysis of amerindian ancient DNA samples: benefits for ancient population studies. *Electrophoresis* 32: 1-8.

MERRIWETHER, D., F. ROTHHAMMER AND R. FERRELL
1994. Genetic variation in the New World: ancient teeth, bone and tissue as sources of DNA. *Experientia* 50: 592-601.

MONSALVE, M., F. CARDENAS, F. GUHL, D. DELANEY AND D. DEVINE
1996. Phylogenetic analysis of mtDNA lineages in South American Mummies. *Annals of Human Genetics* 60: 293-303.

MONTIEL, R., A. MALGOSA AND P. FRANCALACCI
2001. Authenticating Ancient Human Mitochondrial DNA. *Human Biology* 73: 689-713.

MORAGA, M., E. ASPILLAGA, C. SANTORO, V. STANDEN, P. CARVALLO AND F. ROTHHAMMER
2001. Análisis de ADN mitocondrial en momias del Norte de Chile avala la hipótesis de origen amazónico de poblaciones Andinas. *Revista Chilena de Historia Natural* 74: 719-726.

MORAGA, M., C. SANTORO, V. STANDEN, P. CARVALLO AND F. ROTHHAMMER
2005. Microevolution in prehistoric Andean populations: chronology mtDNA variation in the desert valleys of Northern Chile. *American Journal of Physical Anthropology* 127: 170-181.

MULLIGAN, C.
2006. Anthropological applications of ancient DNA: problems and prospects. *American Antiquity* 71: 386-380.

NORES, R. AND D. DEMARCHI
2011. Análisis de haplogrupos mitocondriales en restos humanos de sitios arqueológicos de la provincia de Córdoba. *Revista Argentina de Antropología Biológica*. In press.

NORES, R., M. FABRA AND D. DEMARCHI
2011. Variación temporal and espacial en poblaciones prehispánicas de Córdoba. Análisis de ADN antiguo. *Revista del Museo de Antropología* 4: 187-194.

O´ROURKE, D., M. HAYES AND S. CARLYLE
2000. Ancient DNA Studies in Physical Anthropology. *Annual Review of Anthropology* 29: 217-242.

PÄÄBO, S.
1984. Über den Nachweis von DNA in altägyptischen Mumien. *Das Altertum* 30: 213-218.

PÄÄBO, S., R. HIGUCHI AND A. WILSON
1989. Ancient DNA and the Polimerase Chain Reaction. *The Journal of Biological Chimestry* 264: 9709-9712.

PÄÄBO, S., H. POINAR, D. SERRE, V. DESPRES, J. HEBLER, N. RHOLAND, M. KUCH, J. KRAUSE, L. VIGILANT AND M. HOFREITER
2004. Genetic Analyses of Ancient DNA. *Annual Review of Genetics* 38: 645-679.

PEREGO, U.A., A ACHILLI, N. ANGERHOFER, M. ACCETTURO, M. PALA, A. OLIVIERI, B. HOOSHIAR KASHANI, K.H. RITCHIE, R. SCOZZARI, Q.P. KONG, N.M. MYRES, A. SALAS, O. SEMINO, H.J. BANDELT, S.R. WOODWARD AND A. TORRONI
2009. Distinctive Paleo-Indian Migration Routes from Beringia Marked by Two Rare mtDNA Haplogroups. *Current Biology* 19: 1-8.

POINAR, H.
2003. The top 10 list: criteria of authenticity for DNA from ancient and forensic samples. *International Congress Series* 1239: 575-579.

RAMOS M., C. LALUEZA, E. GIRBAN, A. PÉREZ-PÉREZ, S. QUEVEDO, D. TURBÓN AND D. ESTIVILL
1995. Amplyfing dinucleotide microsatellite loci from bone and tooth samples of up to 5000 years of age: more inconsistency than usefulness. *Human Genetics* 96: 205-212.

RASMUSSEN, M., Y. LI, S. LINDGREEN, J. PEDERSEN, A. ALBRECHTSEN, I. MOLTKE, M. METSPALU, E. METSPALU, T. KIVISILD, R. GUPTA, M. BERTALAN, K. NIELSEN, T. GILBET, Y. WANG, M. RAGHAVAN, P. CAMPOS, H. KAMP, A. WILSON, A. GLEDHILL, S. TRIDICO, M. BUNCE, E. LORENZEN, J. BINLADEN, X. GUO, J. ZHAO, X. ZHANG, H. ZHANG, Z. LI, M. CHEN, L. ORLANDO, K. KRISTIANSE, M. BAK, N. TOMMERUP, C. BENDIXEN, T. PIERRE, B. GRONNOW, M. MELDGAARD, C. ANDREASEN, S. FEDOROVA, L. OSIPOVA, T. HIGHAM, C. RAMSEY, T. HANSEN, F. NIELSEN, M. CRAWFORD, S. BRUNAK, T. PONTÉN, R. VILLEMS, R. NIELSEN, A. KROGH, J. WANG AND E. WILLERSLEV
2010. Ancient human genome sequence of an extinct Paleo-Eskimo. *Nature* 463: 757-762.

REICH, D., R. GREEN, M. KIRCHER, J. KRAUSE, N. PATTERSON, E. DUREN, B. VIOLA, A. BRIGGS, U. STENZEL, P. JOHNSON, T. MARICIC, T. MARQUES-BONET, C. ALKAN, Q. FU, S. MALLICK, H. LI, M. MEYER, E. EICHLER, M. STONEKING, M. RICHARDS, S. TALAMO, M. SHUNKOV, A. DEREVIANKO, J. HUBLIN, J. KELSO, M. SLATKIN AND S. PAABO
2010. Genetic history of an archaic hominin group from Denisova cave in Siberia. *Nature* 468: 1053-1060.

RENFREW, C.
2000. *Archaeogenetics: Towards a Population Prehistory of Europe*. Cambridge, UK, Mc Donald Institute Monographs.

RIBEIRO DOS SANTOS, A., S. SANTOS, L. MACHADO, V GUAPINDAIA AND M. ZAGO
1996. Heterogeneity of mitochondrial DNA Haplotypes in pre-columbian natives of the Amazon región. *American Journal of Physical Anthropology* 101: 29-37.

ROGAN, P. AND J. SALVO
1990. Molecular genetics of pre-columbian South American mummies. *UCLA Symposium in Molecular Evolution* 122: 223-234.

ROTHHAMMER, F., M. MORAGA, M. RIVERA, C. SANTORO, V. STANDEN, F. GARCIA AND P. CARVALLO
2003. Análisis de ADNmt de restos esqueletales del sitio arqueológico de Tiwanaku y su relación con el origen de sus constructores. *Chungara* 35: 269-274.

SHINODA, K., N. ADACHI, S. GUILLENS AND I. SHIMADA
2006. Mitochondrial DNA analysis of ancient Peruvian highlanders. *American Journal of Physical Anthropology* 131: 98-107.

SILVA, A., I. BRICEÑO, J. BURGOS, D. TORRES, V. VILLEGAS, A. GÓMEZ AND J.E. BERNAL
2008. Análisis de ADN mitocondrial en una muestra de restos óseos arcaicos del periodo Herrera en la sabana de Bogotá. *Biomédica* 28: 569-567.

STONE, A. AND M. STONEKING
1993. Ancient DNA from a pre-columbian Amerindian population. *American Journal of Physical Anthropology* 92: 463-471.

WILLERSLEV, E. AND A. COOPER
2005. Ancient DNA. *Proceedings of the Royal Society B*, 272: 3-16.

YAO, Y. AND Y. ZHANG
2003. Pitfalls in the analysis of ancient human mtDNA. *Chinese Science Bulletin* 48: 826-830.

---

[1](http://www.promega.com/resources/protocols/technical-bulletins/101/wizard-sv-gel-and-pcr-cleanup-system-protocol/).

DETERMINATION OF AGE AND SEX ON DENTAL PIECES OF *LAMA GUANICOE*: A METHODOLOGICAL APPROACH

Vanesa Parmigiani[1]

[1] Facultad de Ciencias Naturales y Museo, UNLP; CONICET-CADIC

Abstract

Archaeozoological studies in central Tierra del Fuego are limited by the possibilities of bone conservation. This region is characterized by the development of the sub-Antarctic forest, where survival of organic material is very low, due to factors such as the acidity produced by leaves decomposition, humidity, roots action, ground freezing, etc. In order to study the archaeofauna and to discuss models of resource utilization, selection and management by past hunting-gatherer societies, it is essential to have data regarding sex and age of death of individuals. Therefore, we decided to undertake an intensive study of one type of archaeozoological remains: dental pieces of guanaco (*Lama guanicoe*), the main animal resource exploited by the Selk´nam and pre-Selk´nam societies. Because of their chemical and physical properties, especially their hardness (due to the inorganic hydroxiapatite matrix), teeth are expected to survive the acidic conditions of forest soils. In order to determine age and to infer seasonality from teeth samples, we use cementocronology, a histological observation technique used to count incremental lines by means of the analysis of cross sections of incisors and canines. As for sex determination, we use the amplification of the PCR, method designed to maximize the recovery of DNA in ancient samples, since it has been shown that the genetic material is better preserved in hard tissues such as bone or tooth. In this paper we present some preliminary results from the experimental stage, where present day samples of *Lama guanicoe* are used to adjust the different variables involved in the application of these techniques, previous to the analysis of archaeological samples from different environments of the Island. This paper is intended to show how techniques that are widely developed in different areas of biology, from population ecology to molecular biology, could contribute to solve problems of archaeological contexts.

Key words: *DNA - Incremental lines - Teeth*

INTRODUCTION

The intensive and systematic archaeological researches in Tierra del Fuego were truly initiated during the 1970's, resuming previous sporadic studies carried out by different researchers since the beginnings of the twentieth century. However, despite the short time passed since then, they already allow to glimpse an almost complete panorama of the human occupation since the first settlements in the Isla Grande, which took place around 10 millennia ago. In the beginning of these researches, fieldworks were concentrated on the coastal areas or relatively close to the coasts, particularly in the southern part of the archipelago (northern coast of Beagle channel and islands extending to the south) and in the coasts of the Atlantic Ocean and the Strait of Magellan. In the mid 1990's, archaeological studies were initiated in the inner territories of the Isla Grande, in the Sub Antarctic forest area (Mansur 2002).

Until the late nineteenth century, all the northern and central portions of the Island, including coasts and inner territories, were occupied by a hunter-gatherer society: the Selk´nam society. The most exploited faunal resource was the guanaco (*Lama guanicoe*), performing an essential role in the economy of the Selk´nam and pre-Selk´nam societies, according to ethnographic writings (Bridges 1951; Chapman 1986; Gallardo 1910; Gusinde 1937) and the archaeological investigations in the region.

As a result of the archaeological investigations developed in the northern and central parts of Tierra del Fuego, different publications refer to the exploitation of guanaco (*Lama guanicoe*) by hunter-gatherer societies, and some of them concern isotopic and taphonomic studies on guanaco remains; this research results are very important to understand archaezoological assemblages (Barberena *et al.* 2009; Bogdanovic *et al.* 2008; Borrero 1979, 1986, 1990, 1994-95; Borrero *et al.* 1985; Camaros *et al.* 2008; Favier Dubois and Borrero 2005; Laming-Emperaire *et al.* 1972; Mansur 2002; Mansur and Piqué 2009; Massone 1987, 1996, 2004; Massone and Morello 2007; Morello *et al.* 1999; Muñoz 2002; Parmigiani *et al.* 2010; Prieto *et al.* 2007; Salemme *et al.* 2007; Santiago 2007, 2009; Santiago and Salemme 2008).

Nevertheless, these archaeozoological studies especially concern the steppe region of the Island. In sites of the forest region, archaeozoological studies are limited by the possibilities of bone conservation, which depend on soil composition as well as on site formation and taphonomic processes. In fact, the archaeological record of most sites in forest soils of the central strip of the island is very poor, presenting scarce amounts of bone remains and generally, in a very bad state of conservation (Mansur 2002).

In this context, it is difficult to discuss the strategies of hunter-gatherer groups who inhabited this area, because zooarchaeological assemblages are mainly formed by bone specimens on which there are few chances to use accepted techniques for sex and age determination. Up to now, age determination is generally done by estimations from epiphysal fusion, tooth eruption and wear (Davis 1987; De Lamo 1990; Herrera 1988; Kaufmann 2004, 2008; Mengoni Goñalons 1999; O'Connor 2000; Oporto *et al.* 1979; Puig and Monge 1983; Raedecke 1978).

Sex determination is possible in function of certain differences in pelvis form and structure; it is also possible by means of analysis of canines teeth, in this case metric variables are studied on the tooth itself as on the alveoli of lower canines (Kaufmann 2009; Kaufmann and L´Heureux 2009; Lefèvre *et al.* 2003; Raedecke 1978; Santiago 2009).

These techniques can be used and may give good results when the archaeozoological assemblages are in good conservation and integrity conditions.

Our research seeks to study archaeofauna of different sites as a means to discuss models of use, selection and management of animal resources by the hunter-gatherer societies in the region; for this reason, we need to obtain data regarding sex, age and eventually death seasonality of the individuals hunted, specially the guanaco. Therefore, knowing the characteristics of the soils as well as the bad possibilities of conservation of bone remains in the archaeological sites of this part of the island, we decided to carry out an intensive research on a special type of remain: the dental pieces of *Lama guanicoe*. Due to their physical and chemical characteristics, particularly their strength based on the presence of an inorganic hydroxiapatite matrix, dental pieces are expected to survive in the acidic conditions of the forest soil. Besides, they present a large information potential, since they constitute a first order taxonomic element, providing information about age of death, allowing to infer seasonality and to determine the sex of the animal.

In this work we present the first results of the proposed research, developed within the frame of the archaeological project Heart of the Island[1].

AREA OF STUDY

The Fueguian archipelago, located in the southern extreme of South America (between 52° S and 56° S), is formed by a main island, called Isla Grande, and a series of islands and islets extending South until Cape Horn. It is surrounded by the Atlantic Ocean, the Pacific Ocean and the Strait of Magellan (Figure 1). Its extension is 400 km East-West, and about 300 km North-south, with a total surface of 66.000 km^2 (Tuhkanen 1992).

The area of study covers the central strip of the Isla Grande de Tierra del Fuego (Argentinean territory). It is a mountain environment. Mountain ranges are formed by sub-parallel chains oriented in west-east direction. This zone has many lakes, lagoons, rivers and streams. Among them, the Fagnano lake stands out due to its extension, 100 km length and 40 km width. The landscape is dominated by the Sub Antarctic forest, which in sectors is interrupted by plains and bogs.

On the contrary, the northern part of the island corresponds to plateaus that gradually ascend, from the coasts of the Strait of Magellan and the Atlantic Ocean, to the mountain range. In it, the environment is dominated by steppe vegetation, practically lacking all trees.

The forest in the central strip of the island is characterized by three species of Antarctic beech: the *Nothofagus pumilio* ("Lenga") is the predominant species, with fleeting leaves. The *Nothofagus betuloides* ("Guindo" or "Coihue") is the only one of these three species to have perennial foliage and appears associated with *N. pumilio*, forming mixed forests in the most humid areas. It is common to find it within the wood masses towards the southern coast of Fagnano Lake. Finally, the *Nothofagus antarctica* (Ñire) dominates especially the ecotone between the forest and the steppe in the north of the island (Frangi *et al.* 2004).

Tierra del Fuego was inhabited by hunter-gatherer societies since the end of the last glacial episode until the European colonization, in the beginning of the twentieth century (cf. Mansur and Piqué 2009). Upon arrival of the first colonizers, practically all the Isla Grande was the territory of the Selk'nam, a hunter-gatherer society oriented towards the exploitation of the biggest terrestrial mammal in Tierra del Fuego: the guanaco (*Lama guanicoe*). Due to their own mobility dynamics and the differences in resources between the environments of the north of the island and our own area of study, it is possible to establish that the Selk´nam and pre-Selk´nam hunter-gatherer societies adapted their animal resource management strategies to each one of these environments.

However, until now, only the archaeological sites in the northern zone of the island could be analyzed from the archaezoological point of view. Although bone specimens present variable qualities of conservation, they are more abundant and better preserved than the bone record from the forest area. In the latter, the archaeozoological studies are seriously limited because of the possibility of preservation and conservation of the bone material, which depend on both taphonomic and site formation processes, especially the soil composition. These woody formations characterize by thin acid soils, called "Acid Brown ground of the Forest" (Tuhkanen 1992), where the survival of bone remains from archaeological sites is very low, due to different factors. Among them, the acidity produced by the decomposition of withered leaves, humus, humidity, the action of roots, but also the permanent process of ground freezing and melting.

Most of the archaeological sites excavated in the Sub Antarctic forest region show scarce bone material. It is generally in a very bad state of preservation. Besides hindering the archaezoological study, these characteristics make impossible to compare the data from the forestry environment with that from the island's northern steppe, where preservation conditions are very different. Therefore, as previously stated, we decided to carry out an intensive study on dental pieces of *Lama guanicoe*. In fact, due to their physical and chemical attributes, it is expected that these dental remains could survive within the archaeological context of the forest.

Figure 1: Geographic location of Tierra del Fuego.

MATERIALS AND METHODS

The differential preservation of the archaeofaunistic remains depends, among other factors, on the hardness, thickness and size of the bone material, along with the taphonomic processes that affect bone survival (Chaix and Méniel 2005; Gifford 1981; Lyman 1987; O´Connor 2000). In this aspect, the soil pH has a big influence in the conservation probabilities of the bone remains. These tend to be good in alkaline grounds and bad in slightly acid to acid grounds. There are also other factors that take part in the conservation process, as humidity and ventilation of the sedimentary deposits.

Just like remain preservation, the conservation does not affect all vestiges equally; here also, it depends on the characteristics of each type of bone. It is possible to emphasize that the corrosion caused by the acidity of the ground permits to establish a scale of resistance, being the dental enamel the last bastion before complete dissolution (Chaix and Méniel 2005).

Due to its particular hardness, the dental pieces could constitute an essential indicating element for studying the use of faunal resources and particularly in the case of the *Lama guanicoe*. The characteristic hardness of the teeth is based fundamentally in the presence of an inorganic hydroxiapatite matrix [$Ca_{10}(PO_4)_6(OH)_2)$] (Mones 1979). Due to its biological origin, some authors have proposed to call it bioapatite (Belasse 2002, 2003; Belasse *et al.* 2002; Krueger 1991; Metcalfe *et al.* 2009; Richards *et al.* 2009).

To approach the faunistic studies in archaeological sites where conservation of bone remains is bad, we considered that it would be interesting to apply and to explore the potentiality of diverse techniques used in biological studies of conservation and handling of current *Lama guanicoe* populations.

Within this context, we wondered what kind of data would be useful to discuss models of use, selection and management of animal resources by the Selk´nam and pre-Selk´nam hunter-gatherer societies that inhabited this region. We thought it would be relevant to be able to determine sex and age of death, and if possible, to infer seasonality of the guanaco (*Lama guanicoe*).

Age Determination

The determination of the age of individuals hunted is essential to understand the management of animal resources by hunter-gatherer societies.

To be able to determine age, we propose to use the cementochronology method, concept that is part of the term "skeletochronology" (Castanet *et al.* 1977), proposed to designate every methodological tool applied to estimate the time parameter, including the age of the recorded events until the accretion by the elements of the skeleton (bone, dentin and cement). Cementochronology is based on the analysis of the growth lines that can be observed in the cement that covers the teeth roots in mammals. The study of cement deposits have demonstrated that they are annual and that their count allows us to estimate the age of the individuals (Klein *et al.* 1981; Klevezal 1996, 2001), since the cement deposits continue throughout their life. These growth marks are recognized in both bones and teeth of vertebrates; based on previous studies, it is known that the cement covering the teeth roots reflects better the animal's growth, since it is not subject to modifications that might alter its reading (Castanet *et al.* 1977; Chaix and Méniel 2005).

The dental pieces selected to carry out this study were incisives and canines of *Lama guanicoe* (Figure 2). This selection is based on the positive results published by Klevezal (1996) with ungulates in general, and specifically for the guanaco by Raedecke (1978), Sasco (2000) and Cévoli Romeo (2005).

Sex Determination

Many animals present a pronounced sexual dimorphism. Thanks to this, sex can be determined through morphologic or morphometric criteria. Being the males generally bigger than the females, some animals present a dimorphism that allows estimating the sexual index in a bone sample.

Distribution by sex can tell us about the economy of a society, and this is why it is an important matter to establish which parts of the skeleton allow us or not to determine the sex. Usually sex can be determined using complete artiodactyls crania, with horns or antlers (Davis 1987) or through canine morphology. Its determination is also possible in function of other parts of the postcranial skeleton, such as the pelvic girdle, the presence of the penile bone, the morphometry of long bones, among others. The South American camelids are characterized by the fact that they do not present clear external sexual differences, with the exception of the sexual organs. In spite of this, it is possible to reach a determination through certain appreciable differences in the canine teeth and in the pelvic girdle (Kaufmann and L´Heureux 2009; Raedecke 1978).

Unfortunately, it is very unlikely to find whole pelvises in the archaeological sites, and even less so in the area of the center of the island, for the reasons explained before. Anyhow, sex determination based on these skeletal elements is not always possible. Consequently, here also, we emphasize in the interest of using dental material. In order to approach the problem of sex determination, we decided to undertake sex determination through studies of ancient DNA (Svensson *et al.* 2008).

DNA is the molecule in which all the genetic information of organisms is encoded. Therefore, by analyzing their DNA, it is possible to determine the species to which they belong and moreover, some of the individual traits, like sex. Modern methods of DNA extraction and multiplication in present day and fossil individuals provide us information with a minimal margin of error (Adler *et al.* 2011; Chaix and Méniel 2005; Weinstock *et al.* 2009).

Therefore, since it is possible to recover DNA from ancient specimens, we proposed ourselves to explore the determination of the sex through the amplification of a segment of the Y chromosome, the Sry locus, and another of the X chromosome, the locus ATP7A.

It has been demonstrated that the genetic material is better conserved in hard tissues, such as bone or tooth (Hagelberg *et al.* 1989; Haynes *et al.* 2002). The cause could be the low water and enzyme content, although it has also been suggested that DNA could be trapped by the hydroxiapatite in the tissues, which would protect it from the syntrophic enzymes (Valverde 2006).

Then to do this, we proposed to use the PCR (polymerase chain reaction), which is a molecular biology technique used to obtain a large number of copies of a DNA specific fragment, starting from a minimum of that fragment or mold. Through the application of nuclear and mitochondrial DNA with the use of primers of exonic origin and PCR, it is tested on a region determinant of sex (Unzaga *et al.* 2008).

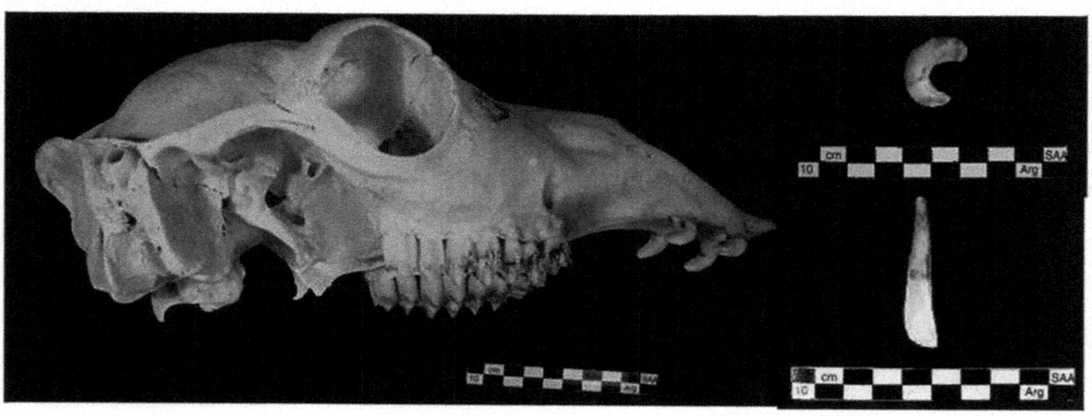

Figure 2: Dental pieces selected incisives and canines.

RESULTS

In the scope or our research project, we started by applying these techniques in an experimental phase, in order to build a frame of reference with samples of current *Lama guanicoe*. For this, we started with the skull collection belonging to the Anthropology laboratory of CADIC and with fresh samples obtained through specific field trips.

Preparation for the Technique of Dental Cementum Layer Counting (DCL)

In order to apply the DCL technique, we start by cutting each tooth transversally between the root and the crown. Next, the root is selected to undergo a demineralization process with nitric acid (Figure 3). This is an important step in which the time factor need to be calibrated, since the time that the sample must be submerged in acid varies in relation to tooth size and type (incisive or canine). For this, we leave the sample for continuous 24 hours and then verify the process by reviewing the state of the piece every four hours. The decalcification time of the piece must be adequate, so that the following steps are successful. Once decalcified, the sample is immediately submerged in successive washes with water, to neutralize the action of the nitric acid. This step is being evaluated, since we found two different ways of carrying it out. The first one is doing consecutive washes with tap water; the second is leaving the piece to soak 24 hours in a salt solution. Once this time passed, we proceed to cut the sample with a freezing microtome, separating the best slices to dye them with hematoxiline.

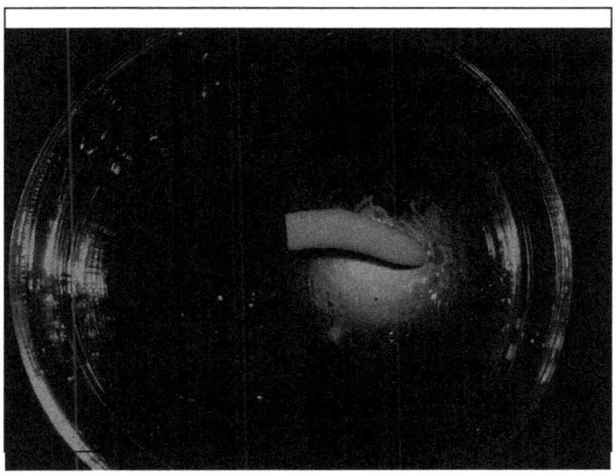

Figure 3: Decalcified tooth piece with transversal cut.

Here, it is also important to calibrate time, since if it is excessive, the thin sheet will be over dyed and the growth lines will not be distinguishable. Once this step is over, we treat the sheets with two consecutive washes in glycerin with different concentrations and then we mount them on the glasses. Finally, they are ready for the microscopy stage to evaluate the lines and carry out the counting.

PCR Technique Set-up

The PCR technique is used to amplify regions of the DNA, which allows obtaining several copies from minimal amounts of genetic material. By the use of this process, millions of copies from the selected segment are produced, from variable DNA regions, usually short fragments. This technique allows to amplify DNA more effectively from very few original molecules.

In order to carry out the sex determination through DNA, it is necessary to go through an experimental stage to calibrate the sample extraction protocol. To do so, we started extracting DNA samples from fresh tissue (muscle) of *Lama guanicoe*. Once the experimentation with fresh samples is over, we proceed with the DNA extraction from dental pieces, from the skull collection at CADIC.

Until now, we have only worked with nuclear DNA. We did not use segments with more than 200 pb, which is recommended to avoid including contamination. We already amplified a segment of the Y chromosome, the Sry locus, designing primers of 180 pb, and amplified a segment of the X chromosome, ATP7A locus, of 130 pb.

To design the Primers, we utilized the published sequences and the suggested program of the NCBI (National Center for Biotechnology Information). Once the technique is calibrated, current teeth samples will be processed. For this, we will evaluate application of the technique published by Nadin Rohland and Michael Hofreiter 2007 in the Protocol Ancient DNA extraction from bones and teeth.

DISCUSSION

The general characteristics of Tierra del Fuego and, particularly, of our study area, that we presented at the beginning of this paper show the difficulties to carry out archaeological investigations that allow to study the dynamics of past occupations in an intensive and complete way. Due to this, one of the main points of the researches carried out in the past years has been trying to develop methodological tools that allow us to improve field data recovery techniques, as well as analysis of the several types of materials (Mansur and Piqué 2010; Mansur *et al.* 2007).

Within this frame, the research we have faced will allow to advance in the zooarchaeological study. This investigation tries to show how techniques widely developed in the different areas of Biology, from population ecology to molecular biology, can contribute to solve problems of archaeological contexts, taking the appropriate measures for sample collection and processing.

We have presented here some of the first results obtained in the construction of the frame of reference (experimental stage) with *Lama guanicoe* specimens from current and recent collections. In fact, since the moment when the first communication was presented in the XVII Congreso Nacional de Arqueología Argentina up to now, we could make important advances in the application of this methodology. Once the different variables to control are adjusted, we will be able to start the analysis of the archaeological samples from the sites on the central zone and other environments within the island, using a hard frame of reference which considers the specificity of archaeological materials of Tierra del Fuego.

ACKNOWLEDGEMENTS

I would like to thank specially to Lic. Ceballos and the Ecophysiology Lab at CADIC for their great help and support related to DNA; to Dr. Dellabianca for transmitting her experience and support related to the DCR technique, to Hernán De Angelis for his time, to Dr. Mansur for her critical reading, Gastón Delgado for the time dedicated to translation of this paper, and finally to two anonymous reviewers whose comments helped to improve this text.

REFERENCES

ADLER, C.J, W. HAAK, D. DONLON, A. COOPER AND THE GENOGRAPHIC CONSORTIUM
2011. Survival and recovery of DNA from ancient teeth and bones. *Journal of Archaeological Science* 38(5): 956-964.

BALASSE, M.
2002. Reconstructing dietary and environmental history from enamel isotopic analysis: time resolution of intra-tooth sequential sampling. *International Journal of Osteoarchaeology* 12: 155-165.

BALASSE, M., S.H. AMBROSE, A.B. SMITH AND T.D. PRICE
2002. The seasonal mobility model for prehistoric herders in the south-western Cape of South Africa assessed by isotopic analysis of sheep tooth enamel. *Journal of Archaeological Science* 29(9): 917-932.

BALASSE, M., A. B. SMITH, S.H. AMBROSE AND S.R. LEIGH
2003. Determining sheep birth seasonality by analysis of tooth enamel oxygen isotope ratios: The late stone age site of Kasteelberg (South Africa). *Journal of Archaeological Science* 30(2): 205-215.

BARBERENA, R., A.F. ZANGRANDO, A.F. GIL, G.A. MARTÍNEZ, G.G. POLITIS, L.A. BORRERO AND G.A. NEME
2009. Guanaco (*Lama guanicoe*) isotopic ecology in southern South America: spatial and temporal tendencies, and archaeological implications. *Journal of Archaeological Science* 36(12): 2666-2675.

BOGDANOVIC, I., E. CAMAROS, H. DE ANGELIS, A. LASA, M.E. MANSUR, A. MAXIMIANO, V. PARMIGIANI, R. PIQUE HUERTA AND O. VICENTE
2009. El paraje de Ewan, un lugar de reunión selknam en el centro de la Isla (Tierra del Fuego, Argentina). In *Arqueología de la Patagonia. Una Mirada desde el Último Confín*, compiled by M. Salemme *et al.*, pp. 941-956. Ed. Utopías. T. 2, ISBN 978-987-1529-11-7.

BORRERO L.A.
1979. Excavación en el Alero Cabeza de León (Tierra del Fuego). *Relaciones, Revista de la Sociedad Argentina de Antropología* 13: 255-271.

1986. *La economía prehistórica de los habitantes del Norte de la Isla Grande de Tierra del Fuego*. Unpublished PhD Dissertation, Facultad de Filosofía y Letras, Universidad de Buenos Aires.

1990. Taphonomy of guanaco bones in Tierra del Fuego. *Quaternary Research* 34: 361-371.

1994-95. Arqueología de la Patagonia. Palimpsesto. *Revista de Arqueología* 4: 9-69.

BORRERO L., M. CASIRAGHI AND H. YACOBACCIO
1985. First guanaco-processing site in southern South America. *Current Anthropology* 26: 273-276.

BRIDGES, L.
1978 [1951]. *El Último Confín de la Tierra*. Buenos Aires, Marymar.

CAMARÓS, E., V. PARMIGIANI AND E. VERDÚM
2008. Espacio ritual, espacio doméstico: diferencias en el uso del recurso faunístico en la sociedad Selknam. In *Arqueología de la Patagonia. Una Mirada desde el Último Confín*, compiled by M. Salemme *et al.*, pp. 707-716. Ed. Utopías. T. 2, ISBN 978-987-1529-11-7.

CASTANET, J., F. MEUNIER AND A. RICQLES
1977. L'enregistrement de la croissance cyclique par le tissu osseux chez les Vertébrés poikilothermes: données comparatives et essai de synthèse. *Bulletin Biologique de la France et de la Belgique* 111: 183-202.

CÉVOLI ROMEO, S.
2005. *Dinámica de la población de guanacos (Lama guanicoe, Muller) de la reserva Cabo dos Bahias, Chubut.* Unpublished Graduate Thesis, Facultad de Ciencias Biológicas, Universidad Nacional de la Patagonia San Juan Bosco.

CHAIX, L. AND P. MÉNIEL
2005. *Manual de Arqueozoología*. Barcelona, Ariel Prehistoria.

CHAPMAN, A.
1986. *Los Selk'nam. La Vida de los Onas*. Emecé Editores, Buenos Aires.

DAVIS, S.J.M.
1987. *The Archaeology of Animals*. London, Battsford Ltd.

FAVIER DUBOIS, C. AND L. BORRERO
2005. Playas de acreción: cronología y procesos de formación del registro arqueológico en la costa central de la bahía san Sebastián, Tierra del Fuego (Argentina). *Magallania* 33(2): 93-108.

FRANGI, J., M. BARRERA, J. PUIGDEFÁBREGAS, P. YAPURA, A. ARAMBARRI AND L. RICHTER
2004. Ecología de los bosques de Tierra del Fuego. In *Ecología y Manejo de los Bosques de Argentina*. edited by M.F. Arturi, J.L. Frangi and J.F. Goya, Chapter 20. Laboratorio de Investigación de Sistemas Ecológicos y Ambientales, Facultad de Ciencias Agrarias y Forestales, Facultad de Ciencias Naturales y Museo, Universidad Nacional de La Plata.

GALLARDO, C.
1910. *Los Onas*. Buenos Aires, Cabaut & Cia.

GIFFORD, D.P.
1981. Taphonomy and Paleoecology: a critical rereview of Archeology´s sister disciplines. In *Advances in Archeological Method and Theory 4*, edited by M.B Schiffer, pp. 365-438. New York, Academic Press.

GUSINDE, M.
1982 [1937]. *Los indios de Tierra del Fuego. Tomo 1: Los Selk'nam*. 2 vols. Buenos Aires, Centro Argentino de Etnología Americana.

HAGELBERG, E. B., B. SYKES AND R. HEDGES
1989. Ancient bone DNA amplified. *Nature* 342 (6249): 485.

HAYNES, S., J.B. SEARLE, A. BRETMAN AND K.M. DOBNEY
2002. Bone preservation and ancient DNA: The application of screening methods for predicting DNA survival. *Journal of Archaeological Science* 29 (6): 585-592.

HERRERA, O.N.
1988. Los camélidos y sus indicadores óseos de estacionalidad: apuntes para la discusión. In *De Procesos, Contextos y otros Huesos,* edited by N. Ratto and A. Haber, pp. 101-110. FFyL,UBA, Buenos Aires.

HILLSON, S.
1986. *Teeth*. Cambridge Manuals in Archeology, Cambridge University Press.

KAUFMANN, C.A.
2004. La fusión ósea como indicador de edad y estacionalidad en guanaco (*Lama guanicoe*). In *Contra Viento y Marea: Arqueología de Patagonia*, edited by M.T. Civalero, P. Fernández and G. Guraieb, pp. 477-487.

2008. *Metodologías para la construcción de perfiles de mortalidad y determinación de la estacionalidad en restos óseos de Guanaco (*Lama guanicoe*): su aplicación a sitios arqueológicos de la región pampeana*. Unpublished PhD Dissertation, Facultad de Ciencias Naturales y Museo, Universidad Nacional de La Plata.

2009. *Estructura de edad y sexo en guanaco: estudios actualísticos y arqueológicos en Pampa y Patagonia*. Sociedad Argentina de Antropología, Buenos Aires.

KAUFMANN, C. AND G.L. L`HEUREUX
2009. El dimorfismo sexual en guanacos (*Lama guanicoe*). Una evaluación osteométrica de elementos poscraneales. *Revista del Museo de Antropología de Córdoba* 2: 181-198.

KLEIN, R., C. WOLF, L.G. FREEMAN AND K. ALLWARDEN
1981. The use of dental crown heights for constructing age profiles of red deer and similar species in archaeological samples. *Journal of Archaeological Science* 8(1): 1-31.

KLEVEZAL, G.A.
1996. *Recording Structures of Mammals. Determination of Age and Reconstruction of Life History*. A.A. Balkema, Rotterdam, Brookfield.

2001. Assessment of the season of death of ancient human from cementum annual layers. *Journal of Archaeological Science* 28(5): 481-486.

KRUEGER, H.
1991. Exchange of carbon with biological apatite. *Journal of Archaeological Science* 18(3): 355-361.

LAMING-EMPERAIRE, A., D. LAVALLÉE AND R. HUMBERT
1972. Le site de Marazzi en Terre de Feu. *Objets et Mondes* 12: 225-244.

LEFÈVRE, C., S. LEPETZ AND D. LEGOUPIL
2003. ¿Cazadores terrestres, cazadores marítimos? Explotación de los recursos animales en el Locus 1. Cazadores-recolectores de Ponsonby (Patagonia austral) y su paleoambiente desde VI al III milenio A.C. *Magallania* (bilingual edition) 31: 63-116.

LYMAN, R.L.
1987. Zooarcheology and taphonomy: a general consideration. *Journal of Ethnobilology* 7(1): 93-117.

MANSUR, M.E.
2002. El Corazón de la Isla. Arqueología de la zona central de Tierra del Fuego. In *Mundos Fueguinos. Doce Miradas. Sobre Selk`nam, Yaganes y Kawesqar*, edited by C. Odone and P. Mason, pp. 148-166. Taller Experimental Cuerpos Pintados, LTD, Santiago de Chile.

MANSUR, M.E., A. MAXIMIANO, R. PIQUÉ AND O. VICENTE
2007. Arqueología de rituales en sociedades cazadoras-recolectoras. Una aproximación desde el análisis del espacio socialmente producido. In *Arqueología de Fuego-Patagonia. Levantando Piedras, Desenterrando Huesos... y Develando Arcanos*, edited by F. Morello, A. Prieto, M. Martinic and G. Bahamondes, pp. 741-754. Centro de Estudios del Cuaternario Antártico (CEQUA), Punta Arenas, Chile.

MANSUR, M.E. AND R. PIQUÉ
2009. Between the forest and the sea: hunter-gatherer occupations in the subantarctic forests in Tierra del Fuego, Argentina. *Arctic Anthropology* 46(1-2): 144-157.

2010. An Ethnoarchaeological approach to the Selknam ceremony of Hain. A discussion of the impact of ritual on social organization in hunter-gatherer societies. Archaeological invisibility and *forgotten knowledge. Conference Proceedings, University of Łódź, Poland, 5th-7th September 2007*, edited by E.K. Hardy *et al.*, p. 180-192. Oxford. BAR - S2183.

MASSONE, M.
1987. Los cazadores paleoindios de Tres Arroyos (Tierra del Fuego). *Anales del Instituto de la Patagonia* 17: 47-60.

1996. Hombre temprano y paleoambiente en la región de Magallanes: evaluación critica y perspectivas. *Anales del Instituto de la Patagonia* 24: 81-97.

2004. *Los cazadores después del hielo*. Centro de Investigaciones Diego Barros Arana, Col. Antropología, Santiago.

MASSONE, M. AND F. MORELLO
2007. Los cetáceos en el mundo Selknam: una evaluación arqueológica. In *Arqueología de Fuego-Patagonia. Levantando Piedras, Desenterrando Huesos... y Develando Arcanos*, edited by F. Morello, A. Prieto, M. Martinic y G. Bahamondes, pp. 709-721. Centro de Estudios del Cuaternario Antártico (CEQUA), Punta Arenas, Chile.

MENGONI GOÑALONS, G.L.
1999. *Cazadores de Guanaco de la Estepa Patagónica*. Colección de Tesis Doctorales, Sociedad Argentina de Antropología, Buenos Aires.

METCALFE, J.Z., F.J. LONGSTAFFE AND C.D. WHITE
2009. Method-dependent variations in stable isotope results for structural carbonate in bone bioapatite. *Journal of Archaeological Science* 36(1): 110-121.

MONES, A.
1979. *Los dientes de los vertebrados. Una introducción a su estudio*. pp. 1-97, láms. 1-10. Facultad de Humanidades y Ciencias, Montevideo.

MORELLO, F., L. CONTRERAS AND M. SAN ROMÁN
1999. La localidad de Marazzi y el sitio arqueológico Marazzi 1, una reevaluación. *Anales del Instituto de la Patagonia* 27: 183-197.

MUÑOZ, A.S.
2002. *La explotación de mamíferos por cazadores-recolectores terrestres en Tierra del Fuego*. Unpublished PhD Dissertation, Facultad de Filosofía y Letras, Universidad de Buenos Aires.

O'CONNOR, T.
2000. *The Archaeology of Animal Bones*. Sutton Publishing.

OPORTO, N., R. BIGATTI AND E. LARRIEU
1979. Determinación de edades en guanaco (*Lama guanicoe*) en base a su dentición. *Revista Argentina de Producción Animal* 4(9): 965-983.

PARMIGIANI, V., H. DE ANGELIS AND M.E. MANSUR
2010. ¿Reemplazo o ampliación?: Cambios en la economía de los grupos cazadores recolectores Selknam del centro de la Isla Grande de Tierra del Fuego. In *Circumpolaridad como Fenómeno Sociocultural. Pasado, Presente, Futuro*, pp. 199-216.

PRIETO, A., E. CALÁS, F. MORELLO AND J. TORRES
2007. El sitio arqueológico Myren 2, Tierra del Fuego, Chile. *Magallania* 35(2): 89-103.

PUIG, S. AND S. MONGE
1983. Determinación de la edad en *Lama guanicoe* (Müller). Deserta 7: 246-270, Mendoza.

RAEDECKE, K.
1978. *El guanaco de Magallanes, Chile. Su distribución y biología*. Ministerio de Agricultura de Chile, CONAF. Publicación Técnica 4. MS.

RICHARDS, M., V. GRIMES, C. SMITH, T. SMITH, K. HARVATI, J.J. HUBLIN, P. KARKANAS AND E. PANAGOPOULOU
2009. Response to Nowell and Horstwood. *Journal of Archaeological Science* 36(7): 1657-1658.

ROHLAND, N. AND M. HOFREITER
2007. Ancient DNA extraction from bones and teeth. *Nature* 2(7): 1756-1762.

SALEMME, M., G. BUJALESKY AND F. SANTIAGO
2007. La Arcillosa 2: La ocupación humana durante el Holoceno Medio en el Río Chico, Tierra del Fuego, Argentina. In *Arqueología de Fuego-Patagonia. Levantando Piedras, Desenterrando Huesos... y Develando Arcanos*, edited by F. Morello, M. Matinic, A. Prieto and G. Bahamonde, pp. 723-740. Centro de Estudios del Cuaternario Antártico (CEQUA), Punta Arenas, Chile.

SANTIAGO, F.
2007. Análisis tafonómico en La Arcillosa 2. Tierra del Fuego. Argentina. *Intersecciones en Antropología* 8: 15-26.

2009. *La ocupación humana en el norte de Tierra del Fuego durante el Holoceno medio y tardío. Su vinculación con el paisaje*. Unpublished PhD Dissertation, Facultad de Ciencias Sociales, Universidad Nacional del Centro de la Provincia de Buenos Aires.

SANTIAGO, F. AND M. SALEMME
2008. Las vueltas 1: un sitio de matanza de guanacos del Holoceno tardío en el norte de Tierra del Fuego (Argentina). En: *Arqueología de la Patagonia. Una mirada desde el último confín*. M. Salemme et al. (comps). Ed. Utopías. T.2, pp. 825-844.

SASCO, H.G.A.
2000. *Estudio preliminar de líneas incrementales de cemento en las piezas dentarias de guanaco (Lama guanicoe) para determinación de edad.*

Unpublished PhD Dissertation, Facultad de Medicina Veterinaria, Universidad de Concepción, Chile.

SVENSSON, E.M., A. GÖTHERSTRÖM AND M. VRETEMARK
2008. A DNA test for sex identification in cattle confirms osteometric results. *Journal of Archaeological Science* 35(4): 942-946.

TUHKANEN, S.
1992. The Climate of Tierra del Fuego from a vegetation geographical point of view and ecoclimatic counterparts elsewhere. *Acta Botnica Fennica 145*, editor in chief: Saara Bäck, pp. 1-64. Departament of Botany, University of Helsinki, Finland.

UNZAGA, R., C. CATNESI, F. DI ROCCO AND L. VIDAL RIOJA
2008. Marcadores del cromosoma Y de camélidos sudamericanos. *Libro de Resúmenes del 13° Congreso Latinoamericano de Genética y VI Congreso Peruano de Genética, Lima, Perú.*

VALVERDE, G.
2006. *Aplicación y perspectiva del análisis molecular de ADN antiguo (ADNa) y ADN mitocondrial (ADNmt) en las ciencias forenses y la antropología molecular*. Trabajo para optar al título de especialidad en ciencias bioquímicas forenses. Universidad Mayor de San Andrés. Facultad de Ciencias Farmacéuticas y Bioquímicas, La Paz, Bolivia.

WEINSTOCK, J., B. SHAPIRO, A. PRIETO, J.C. MARÍN, B. A. GONZÁLEZ, M. GILBERT, P. THOMAS AND E. WILLERSLEV
2009. The Late Pleistocene distribution of vicuñas (*Vicugna vicugna*) and the ''extinction'' of the gracile llama (''*Lama gracilis*''): New molecular data. *Quaternary Science Reviews* 28: 1369-1373.

---

[1] The Archaeological Project Heart of the Island is a large project approved by the province of Tierra del Fuego and the CONICET, developed in CADIC under direction of Dr. M.E. Mansur. It includes different lines of research on exploitation of biotic and abiotic resources in the Sub Antarctic forest region. Within this scope, the results presented here correspond to a doctoral research project in course at La Plata University.

METHODOLOGICAL PROPOSAL TO IDENTIFY IRRIGATION CANALS USING DIATOMS AS BIOMARKERS: PEÑAS COLORADAS (ANTOFAGASTA DE LA SIERRA, SOUTHERN PUNA OF ARGENTINA)

Lorena Grana[1 2], Ma. Lorena Cohen[1 3] and Nora I. Maidana[1 2]

[1] CONICET, [2] Dpto. de Biodiversidad y Biología Experimental, FCEyN, UBA - Buenos Aires, [3] Inst. Superior de Estudios Sociales - Inst. de Arqueología y Museo de la UNT - San Miguel de Tucumán

Abstract

Studies about prehispanic irrigation canals have an important role in the agricultural landscape reconstruction in archaeological investigations. However, these types of evidence are characterized in many areas by their low visibility and poor preservation, rendering difficult their recognition and subsequent study. For this reason, it is necessary to generate diverse independent methodologies. Therefore the study aims to propose a new methodological tool suitable for the use of biomarkers, such as diatom assemblages, for the identification and characterization of ancient irrigation canals. We have analyzed a furrow-like feature lying in proximity to the archaeological site of Peñas Coloradas (Antofagasta de la Sierra, Argentina), to evaluate which aspects of the diatom analysis (specific composition, absolute abundances, preservation stages of valves) are relevant to achieving this objective. Our results suggest that the analysis of the composition and abundance of diatom assemblages and their comparison with modern analogues was found to be useful for identifying the existence of ancient canals and thus, of archaeological water management. Besides, we have statistically proven that the degree of valve preservation was not a suitable tool for this type of investigation. This study has presented the first evidence for this type of archaeological hydraulic technology ever found in the locality and the sub-basin Las Pitas.

Key words: *Irrigation canals - Diatoms - Puna*

INTRODUCTION

Agricultural societies living in semi-desert environments are dependent upon hydraulic technology to deal with water constraints. Hydraulic features are complex and heterogeneous, ranging from irrigation canals to dikes and terraces, thus creating a particular agricultural landscape of high archaeological potential. However, these types of evidence are characterized in many areas by their low visibility and poor preservation, rendering difficult their identification and subsequent study. For this reason, it is necessary to generate diverse independent methodologies that can help us to identify and characterize these features. In this way we will succeed in developing a more accurate interpretation of this technology and will therefore provide a faithful reconstruction of past agrarian landscapes.

The increased interest in hydraulic technology has led researchers to develop diverse methodologies in several regions, such as the micromorphological approach (consisting in the analysis of sedimentological, pedological and climatic-pedo facies; summarized in Purdue *et al.* 2010), infrared aerial photographs studies (Nichols 1988), planimetric analyses and irrigation networks designs - analyses of size, flow directions, and spatial distributions (Díaz 2009; Quesada 2006); the mapping of soils based on their chemical analysis (Tchilinguirian and Olivera 2000), and the study of irrigation water quality though infiltration tests (Salminci 2011), among others.

In relation to biomarkers and irrigation canals, the proxies that are most frequently studied for investigating and characterizing canals or water reservoir, are ostracodes and pollen (Bayman *et al.* 2004; Palacio Fest *et al.* 2001), as well as phytoliths, the latter used for identifying past irrigation infrastructures (Rosen and Weiner 1994). However, we have not found references for the use of diatoms as biomarkers aimed at the identification and analysis of irrigation canals. It is therefore the goal of this paper to evaluate the viability of the use of diatoms analysis to identify and characterize past irrigation canals. We have analyzed a furrow-like feature lying in proximity to the archaeological site of Peñas Coloradas (Antofagasta de la Sierra, Argentina), to evaluate which aspects of the research (specific composition, absolute abundances, preservation stages of valves) are relevant to achieving this objective.

DIATOMS AND ARCHAEOLOGY

Diatoms are unicellular algae placed in the class Bacillariophyceae. Their cell walls are made of silica, a resistant material which is the main responsible for their preservation within the sediments when they die and subsequently deposit. These algae have a high survival capacity in virtually every aquatic environment, although the physicochemical characteristics of the water determine the set of species living at any given time and in a certain area (Round *et al.* 1990). In the scientific literature, diatoms have been the focus of intensive research (including systematic and taxonomic studies), paying special attention to their ecological relationships and their palaeoecological significance (Mannion 1987). In the last three decades, the relationship between diatoms and certain archaeological issues has become increasingly evident (Battarbee 1988; Cameron 2007; Fernández and Maidana 2008; Juggins and Cameron 1999; Mannion 1987; Martínez Macchiavello *et al.* 1999). These analyses can be applied at a range of spatial and temporal scales to place archaeological materials in their broader environmental and cultural context, and they have been proved to be a useful tool in archaeological discussions.

In summary, diatom analysis can focus on three major subjects: a) palaeoenvironmental reconstructions, b) site formation processes, and c) archaeological records, the first one being the most common approach found in archaeological studies (Cameron 2007; Ognjanova-Rumenova 2008). Palaeoenvironmental reconstructions have been applied elsewhere at local (e.g. Fernández and Salemme 2010) and regional scales (e.g. Grana 2007; Morales 2010) to contextualize and better understand the cultural processes within a well-known archaeological landscape. They can also be found in evaluations of the human cultural impact on past environments. These reconstructions are most frequently aimed at understanding prehistoric agricultural impacts on lakes (e.g. Bradshaw *et al.* 2006) and, to a less extent, at analyzing the impact of hunter-gatherer settlements on aquatic ecosystems, e.g. the activity of prehistoric Inuit whalers in Arctic Sea (e.g. Hadley *et al.* 2010).

A different approach consists in the application of diatom analysis to the study of archaeological sediments in order to determine site formation processes or to identify differentiated areas of activity within the sites (e.g. Kligmann 2003; Mannion 1987).

In the last decades, diatom analysis has been increasingly used by archaeologists in clay provenance studies and the development of pottery typologies. The identification of the diatom assemblages contained in the clay and the water used in ceramic manufacturing, has made it possible for researchers to identify local clay sources (e.g. Jansma 1990), and to calculate the distance between these clay sources and the archaeological sites (e.g. Solá and Morales 2007). In this line of research, Cameron (2007) has proposed the use of diatoms for identifying the material sources and the manufacturing techniques used in ancient building construction. But these analyses have been criticized by some authors, who postulate that they present important limitations. As an example, Kligmann and Calderari (2012) have stressed the fact that there are different moments in the production chain of ceramic, in which the diatoms could be incorporated, so they could interfere in our interpretation of procedence of clay.

The study of diatoms has also shed light into the analysis of coprolites (both human and nonhuman) from prehistoric settlements. It has normally been used together with other microfossil analysis (such as phytoliths or pollen) to provide information relating to seasonal occupation and the palaeoenvironmental conditions of the sites (e.g. Horrocks *et al.* 2003). Diatom investigation has also proven to be efficient in elucidating issues related to agricultural societies. For instance, diatoms found in mortars have been used to identify wet milling (e.g. Babot 2009), while those found in fields have aided researchers in the reconnaissance and the classification of agricultural practices (e.g. Korstanje and Cuenya 2010).

We have found further examples of diatom studies applied to the localization and characterization of hydric features associated with archaeological sites, such as the turf-based structures from the Viking Age (e.g. Barthust *et al.* 2006), the water wells (e.g. Neely *et al.* 1995), and the agricultural terraces in prehispanic Mesoamerica (e.g. Trombold and Israde-Alcantara 2005). However, we have not found much evidence for the use of diatoms related to the investigation of earlier canals. An exception to this trend is Cameron (2007) who, in his paper, cites previous diatom studies of early Romano-British ditches aimed at the exploration of past water flows, tidal influences, and maintenance activities on the infrastructures. Within this line, the research presented by Beneš *et al.* (2002) focuses on the drainage ditches of a moat, which surrounded the old fortifications in Prague, stressing the evolution of medieval ditches and pits and the increment of water pollution through time. No further information has been found in the revised literature that concerns diatom analysis and the identification of past irrigation canals.

PEÑAS COLORADAS ARCHAEOLOGICAL SITE AS A CASE STUDY

Taking into account the scarcity of research involving diatoms and past drainage features, we have chosen an archaeological locality named Peñas Coloradas, in Antofagasta de la Sierra (26° 02' 26.92'' S and 67° 22' 00.86'' W, Figure 1), as a case study to critically test the applicability of this approach for investigating these past agricultural features.

Antofagasta de la Sierra is a basin located above 3400 m a.s.l. in Catamarca province (Northwestern Argentina). This is an extremely dry environment, which is part of the Salt Puna, characterized by its extreme conditions of aridity and environmental instability. Today, water in this area is a limiting factor. Thus, we can postulate that irrigation features have played an important role in the agricultural and pastoral activities of past human groups living in this area.

Olivera *et al.* (2004) have suggested that by *ca.* 1000 years BP the extremely dry environment of this area (average annual precipitation < 200 mm) could have favored the incorporation, at that time, of new agricultural strategies focused on artificial irrigation on a steeper terrain. However, it seems that the effects of this drought were not so severe in the sub-basin of Las Pitas River (Figure 1), since wetlands located upstream showed less reduction and changes of moisture

(Tchilinguirian and Olivera 2009). In the southern part of Punilla River, the most important in this geological basin, there is evidence of prehispanic farming fields (Bajo del Coypar I) covering almost 800 ha. (Olivera and Vigliani 2000-2002; Vigliani 2005). It has been proposed that this extensive agricultural production system began at the end of the Late Regional Process (*ca.* 1000 BP) at the same time as the development of the main regional urban center, named La Alumbrera (Olivera and Vigliani 2000-2002).

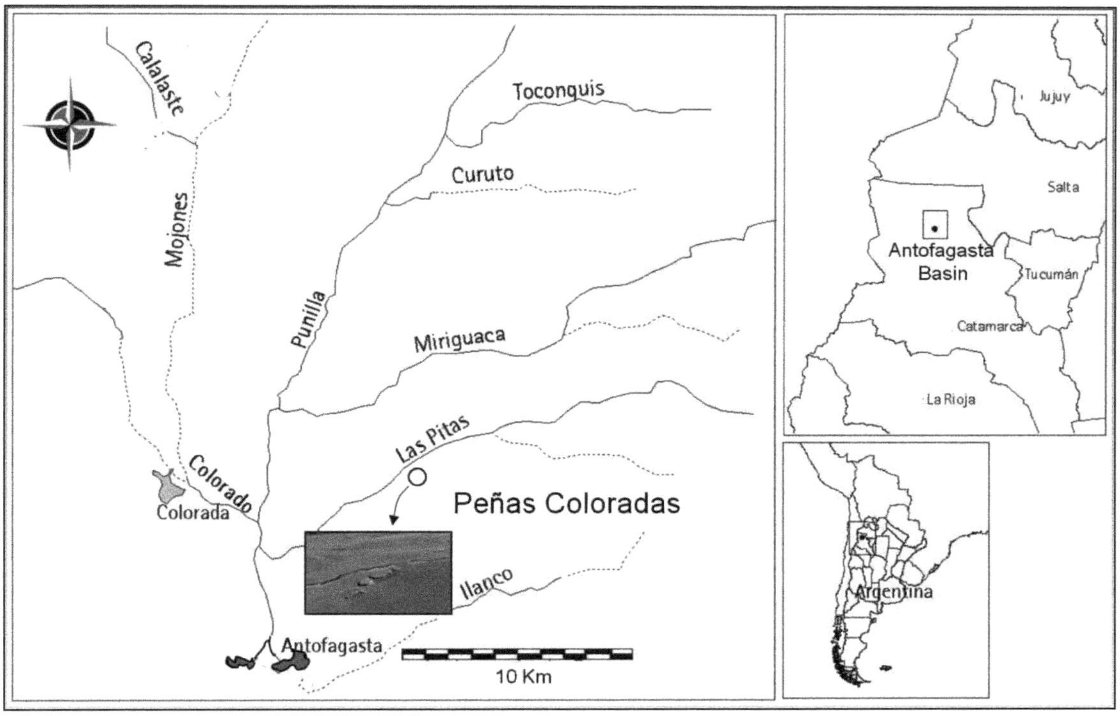

Figure 1: Map of the region Antofagasta de la Sierra and the location of Peñas Colorada site.

The archaeological site of Peñas Coloradas lies in the eastern mid-lower basin of the subsidiary river Las Pitas. The surrounding hills are formed of up to four different types of ignimbrite rocks called Peñas Coloradas 1, 2, 3, and 4. This middle course of Las Pitas River is well known by locals for its topography and productivity, which enables herding and agricultural activities in a small scale. It was during the morphological study of this archaeological locality using satellite images that we discovered a furrow-like structure, which is the object of analysis in this paper (Figure 2).

Peñas Coloradas has been mentioned in the archaeological literature mainly for its high concentration of rock art (Aschero 1999, 2000; Aschero *et al.* 2006; Martel 2009, 2010; Martel and Aschero 2007), although other archaeological aspects and areas of the site are currently under study, such as Peña Colorada 3 (Cohen 2010). Its peak, called Peña Colorada 3 cumbre (PC3c; Figure 2), has been dated in 850 ± 60 years BP (Later Period) following reoccupation (Cohen 2009b). This place consists of 22 architectural structures of various shapes. Based on their localization and visual relation, Cohen (2010) has proposed that this was a place with a restricted access and where one could have a visual control over the basin without being seen by those standing below. This particular visual characteristic of the landscape has been related to the visual control of its resources, such as pastures, crops and water, and also the traffic along the basin.

The existence of a productive landscape near PC3c can be inferred from a number of features identified in its close periphery: a) the presence of several big mortars, b) the agricultural shovels found on its surface (Elias 2010), and c) a single row of stones, probably an ancient agro-pastoral square structure. We have photographic evidence for the latter, obtained by Weisser in 1923, where a complete squared structure in a sector at the foot of the Peña Colorada, near the furrow, is visible (Figure 3) (Cohen 2009b). In addition, we have identified another row of stones, which follows the direction of the furrow. We consider here that they could be contemporary and also, that they could have been part of the exploited landscape of Peña Colorada 3 site, basing our hypothesis on the visual relation of features mentioned above (see more in Cohen 2009a, 2009b, 2010)[1].

The furrow feature stretches along the terraced plain on a flat surface, which shows a minimal negative slope towards the river, located 10 m below. Its outline could be come off the actual active canal, which lies in front of site Peñas Coloradas 1 and 2. However, today it is impossible to see the complete furrow because it is interrupted by a local road. In addition, the owner of the land, who lives a short distance from the site, affirms that he built the modern canal which is used today. However, he does not know the origin of the ancient furrow, so we can therefore assumed that it has an earlier origin. Nevertheless, we have not succeeded in estimating the age of this feature. Thus, its chronology remains hypothetical, and is based only on the already mentioned archaeological context.

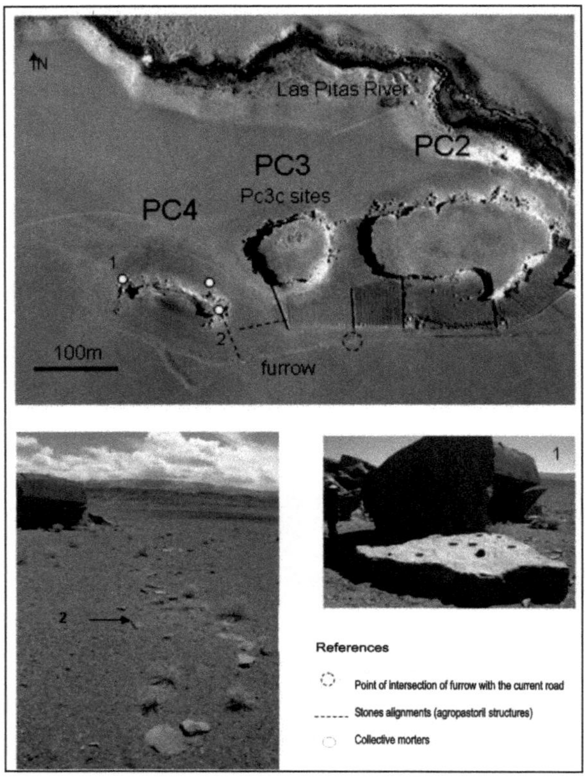

Figure 2: Satellite image of Peñas Coloradas site and location of the furrow.

Two different sections are evident in the furrow: section A (SA, 3.10 m wide) and section B (SB), which is shallower and wider (14.20 m wide) and was highly modified by eolian erosion, especially in its eastern side (Figure 4).

Up to present, researchers have only found indirect indicators of irrigation technology, such as models of hydraulic systems sculpted on rocks in the sites of Punta de la Peña 9 and 4, near Peñas Coloradas (Aschero *et al.* 2009). Thus, the find of this type of feature for the first time in sub-basin Las Pitas, is of great importance. Therefore, we are particularly interested in testing the hypothesis of the use of the furrow as an irrigation canal through the analysis of the diatom assemblages recovered from several places of this feature.

MATERIALS AND METHODS

To obtain the sediment samples, we traced two transects perpendicular to the main axis of the furrow in sections A and B (Figure 5). We determined initially a number of sampling points along each transects. Considering every section of the furrow as if they were part of true water bodies, we designed sampling strategy looking for different environmental situations (without influences of water or sporadic inundation, marginal environment, and permanent influence of water). After this, we chose several sampling points: two external points (E-1 and E-2) located one meter out of each margin of the furrow; two in the upper part of each bank (BU-1 and BU-2); two in the lower part of each bank (BL-1 and BL-2); and finally one more point in the middle of the furrow (C). In sampling points E and C, we took two samples, one in the surface (EU and CU, 0-5 cm) and another in the subsurface (EL and CU, 5-10 cm) with the aim of exploring the possible effects of wind erosion and to evaluate changes of minimum and maximum flood (Figure 5). In total, we have analyzed 16 sediments samples of the furrow.

In order to improve our interpretations on the evolution and characteristics of this particular feature, additional samples, coming from two modern irrigation canals that are currently under different use, were taken using the same methodology. We have compared the diatom assemblages found in the two sections of the furrow with those of the two irrigation canals (Figure 6). The irrigation canal, named here Canal 1 (C1, 3.35 m wide), is used all year long according to communal regulations. It was sampled five days after the last time it was used. The other sampled irrigation canal (C2, 18 m wide) is only used in winter to redirect the river, mainly to prevent water freezing in the surrounding meadow (*vega*) and to allow for livestock grazing (*Ovis* sp and *Lama glama*) to continue. When consulted, local inhabitants said that the large amplitude of this canal is natural, and it is due to the combined action of water and ice. All modern samples were fixed *in situ* with formaline (4%).

In the laboratory, 3g of each sediment sample were processed following the standard protocol of Battarbee (1986), using H_2O_2 and heating; clay particles were partially removed by decantation. Permanent slides were mounted with Naphrax®. Diatom valves and fragments were examined under a Polivar Reichert Jung light microscope using 100x oil immersion objective. Relative abundances were calculated counting all the valves and fragments along three selected transects in each slide. Valve densities were calculated using the aliquot method (Battarbee 1986).

In order to undertake the taxonomic identification, a number of morphological and morphometrical features of the valves have been considered: the length of the apical and transapical axes, striae density, the presence or absence of raphe, stigmata, and apical pore fields, among others. Specific taxonomic and ecological references have been consulted (Krammer and Lange-Bertalot 1986, 1988, 1991a, 1991b; Lange-Bertalot 2000; Lowe 1974; Round *et al.* 1990; Straub 1990; Van Dam *et al.* 1994, etc.), together with the local floras (Díaz and Maidana 2006; Maidana and Seeligmann 2006; Rumrich *et al.* 2000; Seeligmann and Maidana 2003; Seeligmann *et al.* 2008).

Two cluster analyses have been performed to compare the diatom composition of the two modern canals with each of the two sections of the furrow (Square Euclidean distance index, minimum distance grouping method; MVSP 3 - Kovach Computing Services). Only samples

References
→ Quadrangular structures photographed by Weisser in 1923, at the foot of the site Peña Colorada 2.

Figure 3: Squared structure near the furrow; taken by Weisser in 1923 (photographic record of Natural Science Museum of La Plata, Argentina).

Figure 4: Sectors of the furrow: Sector A (SA) and Sector B (SB).

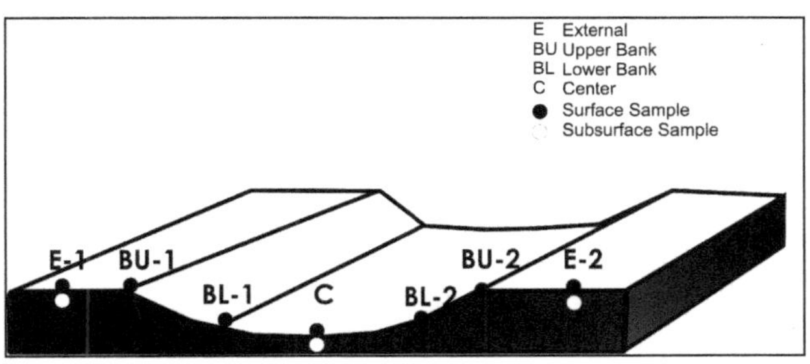

Figure 5: Transects laid perpendicular to the main axis of the furrow showing the different sampling points.

coming from the center and the exterior parts of the canals have been used.

Like in many other archaeological contexts, the interpretation of the information obtained from the diatoms in this feature has not only been based on the species composition but also on the degree of preservation of the frustules. Therefore, we have analyzed the percentage of non-fragmented and fragmented valves as well as the variations in diatom diversity in the different sampled sites in the furrow.

For evaluating the relationship between the number of broken and unbroken valves and their localization in the canal, we have plotted the percentages of the two types of siliceous elements mentioned and performed a Pearson's X^2 Test ($X^2_{0,95}$) to check the statistical significance of the results (Zar 1974). Our null hypothesis was that the number of broken and unbroken valves is independent of the type of sampling site considered. The alternative hypothesis was that there exist a relationship between broken and unbroken valves and the sampling points.

RESULTS AND DISCUSSION

Diatom Analysis in Two Sections of the Furrow

In the 16 studied samples we have identified 65 diatom taxa in SA and 59 in SB. The majority of the identified taxa were cosmopolitan and characteristic of shallow aquatic environments (Table 1, Figures 7 and 8).

In SA samples, *Fragilaria capucina* var. *vaucheriae*, an opportunistic planktonic species, dominate in the center of the furrow, together with some scarce epiphytic taxa.

In the edges and the exterior of the furrow, *Planothidium lanceolatum* and *Pseudostaurosira brevistriata* were the dominant species. These are commonly found in epipsammic communities living in rivers, with no territorial distinction. Other co-dominant species found are the aerophilic diatoms, such as *Denticula elegans*. This species can survive out of a water body, for instance in wet soils. Several littoral species have also been identified, such as *Staurosira venter* and *S. construens*.

Figure 6: Modern irrigation canals: Canal 1 (C1) and Canal 2 (C2)

As we had expected, the maximum peak in valve abundance was found in the two samples taken from the center of the furrow (SA-CU and SA-CL) (Figure 8).

In SB samples, *Planothidium lanceolatum* and *Pseudostaurosira brevistriata* were dominant or co-dominant (Table 1, Figures 7 and 9). As opposed to the SA findings, *Fragilaria capucina* var. *vaucheriae* was only present in the bank samples of SB, and in low percentages. A number of different epiphytic and epipsammic species, such as *Planothiidum dubium* and *Cocconeis placentula*, have only been found in the center of the furrow. Although facultative aerophilic species were present in all sampling sites, *D. elegans* and *Luticola muticopsis* have been identified only in the center, while other species, such as *Hantzschia amphioxys* and *L. mollis*, were exclusive to the exterior and bank of furrow.

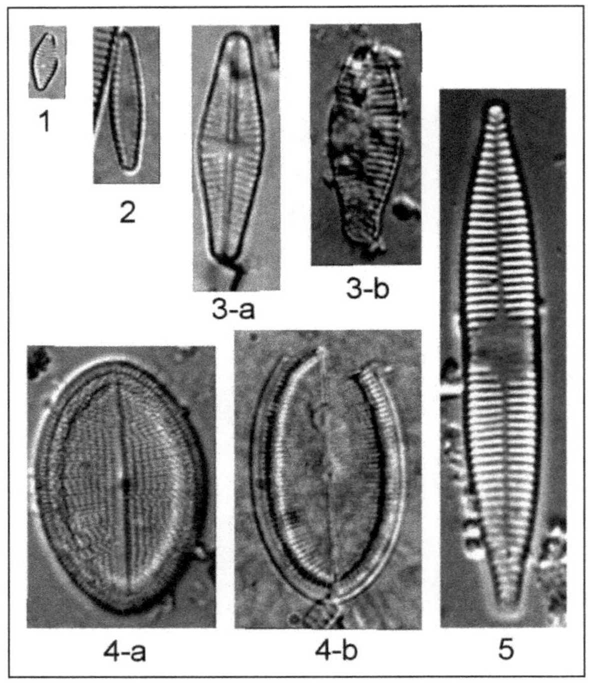

Figure 7: Examples of the identified diatoms. 1. *Staurosira venter*; 2. *Pseudostaurosira brevistriata*; 3-a-b. *Planothidium lanceolatum* (a. unbroken valve; b. broken valve); 4a-b. *Cocconeis placentula* (a. unbroken valve; b. broken valve); 5- *Ulnaria ulna* (LM: x1500).

The highest peak in valve abundance has been found in the superficial sample of the furrow center, but contrary to what we expected, the same result was not obtained in the subsurface samples, where the highest abundance was found in the upper banks of furrow (Figure 9).

The species abundance in the diatom assemblages, together with the highest abundance encountered in the two SA central samples, let us to conclude that the furrow acted in the past as a water channel, whereas the diatom composition of the assemblages recovered from the SB samples has not given a clear signal. We should take into account that the SB samples have been more exposed to wind action than the SA ones. Since diatoms can be easily transported by wind action, it can be assumed that the record of the depositional history of the site SB was probably modified by wind activity. Nevertheless, taking into account that the dominant species in SB were characteristic of rivers, we can assume the fluvial origin of this feature.

Comparison with Modern Canals

In the 14 studied samples of the two modern canals (C1 and C2), we have identified 67 and 69 diatom taxa, respectively (Table 1).

The cluster analysis performed to compare the two modern canals with each of the two sections of the furrow shows different groupings. When compared to SA (Figure 10) a relation between all the central samples is noticed. This confirms our assumption of the use of the SA furrow as an irrigation canal. As it was the case in the previous analysis of diatom assemblages in the two sections of the furrow, when comparing the modern canals with the SB samples (Figure 11), there is no clear differentiation between the central and the external samples. The two major groups are made up of samples coming from both the exterior and the center of the furrow. This supports the already mentioned hypothesis of the wind influence in SB diatom assemblage composition.

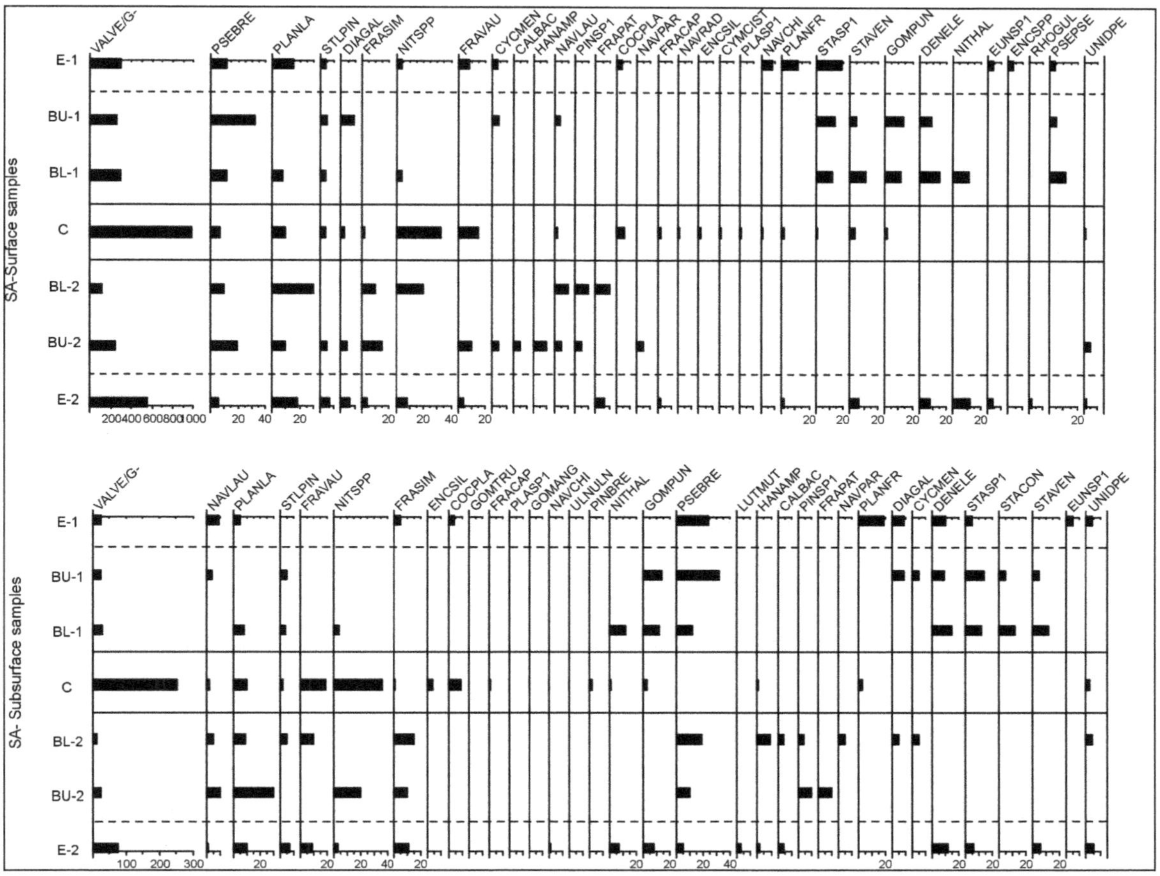

Figure 8: Relative abundance (%) of species in SA (UNIDPE: unidentified diatoms).

Unbroken and Broken Valves in Relation to Sampling Points

We assume that, when the water flow is over, the center of the shallow channel would be more protected than both its edges and the external sections from the physical damages caused by wind activity and water movements. Taking this into account, we would expect the unbroken valves to be more abundant than the broken ones in all the central samples.

In fact, independently of the current use of canals C1 and C2, the percentages of unbroken valves in the center were higher than in other locations within the same canals (Figure 12a), an observation which was statistically confirmed by the results of the Pearson's X^2 Test (Table 2). Therefore, we have accepted the assumption that such a distribution is a direct consequence of running waters in that canals.

Surface and subsurface SA samples show a similar distribution of broken and unbroken valves percentages (Figure 12b), but the results of the Pearson's X^2 Test are different: in surface samples the distribution is independent, whereas in the subsurface ones it varies according to the sampling site (Table 2).

Surface and subsurface SB samples show a completely different pattern of distribution of unbroken and broken valves (Figure 12c). As opposed to the situation seen in the SA samples, here the results of Pearson's X^2 Test indicate that the distribution is independent of the sampling site in the surface samples but dependent in the sub surficial samples (Table 2).

In sum, while the signal of the presence of running waters is clear in the modern canals we sampled, we have not obtained statistically significant results from the two sections of the ancient furrow. The data obtained from the samples coming from two sectors of the furrow suggest that the determination of the percentage of broken and unbroken valves might not be the most suitable method for identifying ancient canals.

DIATOM TAXA	TAXA CODE	SITES SAMPLING			
		SA	SB	C1	C2
Achnanthes coartacta (Brébisson) Grunow	ACHCOA	-	-	x	-
Achnanthidium exiguum (Grunow) Czarnecki	ACHEXI	-	-	-	x
Amphora coffeaeformis (Agardh) Kützing	AMPCOF	-	-	-	x
Amphora sp 1	AMPSP1	-	-	x	-
Amphora veneta Kützing	AMPVEN	-	x	x	x
Caloneis aff. *bacillum* (Grunow) Cleve	CALBAC	x	x	-	x
Cocconeis placentula Ehrenberg	COCPLA	x	x	x	x
Cyclotella meneghiniana Kützing	CYCMEN	x	-	-	-
Cymbella cistula (Ehr.) Kirchner	CYMCIST	x	-	-	x
Denticula elegans Kutzing	DENELE	x	x	x	x
Denticula kuetzingi Grunow	DENKUE	-	x	-	-
Denticula sp 1	DENSP1	-	x	-	-
Diadesmis gallica W. Smith	DIAGAL	x	x	x	-
Encyonema silesiacum (Bleisch In Rabenhorst) D. G. Mann	ENCSIL	x	x	x	x
Encyonema aff. *schneideri* Krammer	ENCSCH	-	x	-	-
Encyonema aff. *lange-bertalotti* Krammer	ENCLBR	-	-	-	x
Encyonema aff. *minutiforme* Krammer	ENCMIN	-	-	x	x
Encyonema sp 1	ENCSP1	-	-	x	-
Encyonema spp	ENCSPP	x	x	-	-
Eunotia aff. implicata Nörpel et al.	EUNIMPL	-	-	x	x
Eunotia aff. muscicola var. tridentula Nörpel & Lange-Bertalot	EUNMUS	-	x	-	-
Eunotia sp 1	EUNSP1	x	-	-	-
Fragilaria capucina Desmaziéres var. *capucina*	FRACAP	x	x	x	x
Fragilaria capucina var. *vaucheriae* (Kützing) L.-Bertalot	FRAVAU	x	x	x	x
Fragilaria aff. *patagonica* (Cleve) De Toni	FRAPAT	x	-	x	x
Fragilaria spp	FRASPP	-	x	-	x
Frankophila similioides Lange Bertalot & Rumrich	FRASIM	x	x	x	x
Gomphonema parvulum (Kützing) Kützing	GOMPAR	-	-	-	x
Gomphonema punae Lange-Bertalot & Rumrich	GOMPUN	x	x	x	x
Gomphonema truncatum Ehrenberg	GOMTRU	x	-	-	x
Gomphonema aff. *acuminatum* Ehrenberg	GOMACU	-	-	x	-
Gomphonema aff. *angustum* (Kützing) Rabenhorst	GOMANG	x	-	-	x
Gomphonema sp 1	GOMSP1	-	-	x	-
Haloroundia speciosa (Hustedt) Diaz & Maidana	HALSPE	-	x	-	-
Hantzschia amphioxis (Ehrenberg) Grunow	HANAMP	x	x	x	x
Luticola mollis Lange-Bertalot & Rumrich	LUTMOL	-	x	x	x
Luticola muticopsis (Van Heurck) D. G. Mann	LUTMUT	x	x	-	x
Luticola nivalis (Ehrenberg) D.G. Mann	LUTNIV	-	x	-	-
Navicula gregaria Donkin	NAVGRE	-	-	-	x
Navicula lauca Rumrich & Lange-Bertalot	NAVLAU	x	x	x	x
Navicula parinacota Rumrich & Lange-Bertalot	NAVPAR	x	-	x	x
Navicula radiosa Kützing	NAVRAD	x	x	x	-
Navicula sp 1	NAVSP1	-	-	x	-
Navicula spp	NAVSPP	-	x	x	x
"Naviculadicta" chilensis Rumrich & Lange-Bertalot	NAVCHI	x	x	x	x
Nitzschia amphibia Grunow	NITAMP	-	x	x	x
Nitzschia communis Rabenhorst	NITCOM	-	-	x	x
Nitzschia conmutata Grunow	NITCON	-	x	-	-
Nitzschia halloyi Maidana & Herbst	NITHAL	x	x	x	x
Nitzschia microcephala Grunow	NITMIC	-	-	x	-
Nitzschia palea (Kützing) W. Smith	NITPAL	-	x	x	-
Nitzschia aff. *fonticola* Grunow	NITFON	-	-	x	x
Nitzschia aff. *linearis* W. Smith	NITLINE	-	x	-	-
Nitzschia sp1	NITSP1	-	x	-	-
Nitzschia spp	NITSPP	x	x	x	x
Pinnularia brebissoni Kützing	PINBRE	x	-	x	x
Pinnularia sp 1	PINSP1	x	x	-	x
Pinnularia spp	PINSPP	-	x	-	-
Placoneis sp 1	PLASP1	x	-	-	-
Planothidium dubium (Grunow) Round & Bukhtiyarova	PLADUB	-	x	-	-
Planothidium frequentissimum (Lange-Bertalot) Lange-Bertalot	PLANFR	x	x	x	x
Planothidium lanceolatum (Brébisson) Round & Bukhtiyarova	PLANLA	x	x	x	x
Pseudostaurosira brevistriata Grunow in Van Heurck	PSEBRE	x	x	x	x
Pseudostaurosira pseudoconstruens (Marciniak) D. M. Williams & Round	PSEPSE	-	-	-	x
Rhopalodia gibba (Ehrenberg) O. Müller	RHOGIB	-	-	x	x
Rhopalodia gibberula (Ehrenberg) O. Müller	RHOGUL	x	-	x	x
Sellaphora pupula (Kützing) Meresckowsky	SELPUP	x	x	-	x
Stauroneis aff. *atacamae* Hustedt	STNATA	-	-	x	-
Staurosira construens Ehrenberg var. construens	STACON	x	-	x	-
Staurosira laucensis var. *vulpina* Lange-Bertalot et Rumrich	STALAV	x	-	x	-
Staurosira venter (Ehrenberg) Cleve & Möller	STAVEN	x	x	x	x
Staurosira sp 1	STASP1	x	x	x	x
Staurosira spp	STASPP	-	-	x	-
Staurosirella pinnata (Ehrenberg) Williams & Round	STLPIN	x	x	x	x
Tabularia aff. *laevis* Kützing	TABLAE	-	x	-	x
Ulnaria ulna (Kützing) Compère var. *ulna*	ULNULN	x	x	x	x
Ulnaria aff. *acus* (Kützing) Aboal	ULNACU	-	x	-	-
Unidentified Pennates	UNIDPE	x	x	x	x

Table 1: Presence and absence of the identified diatom taxa with their codes.

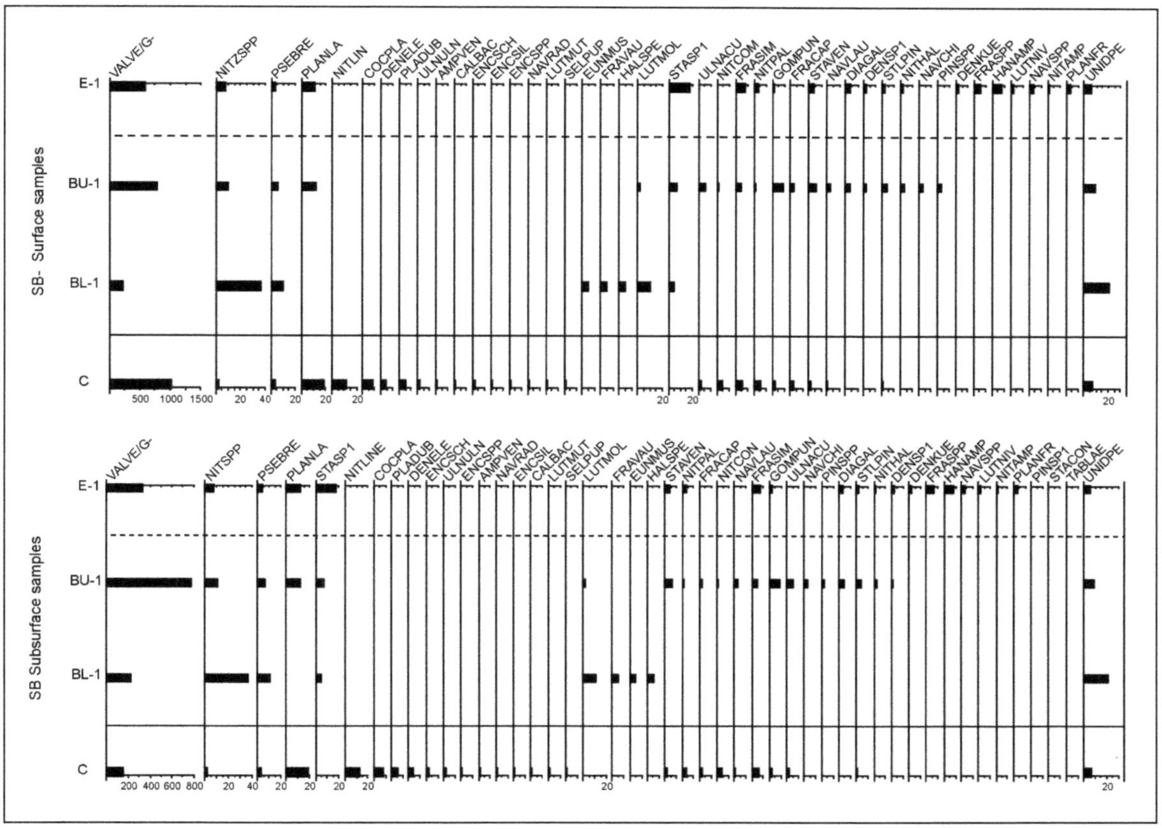

Figure 9: Relative abundance (%) of species in SB (UNIDPE: unidentified diatoms).

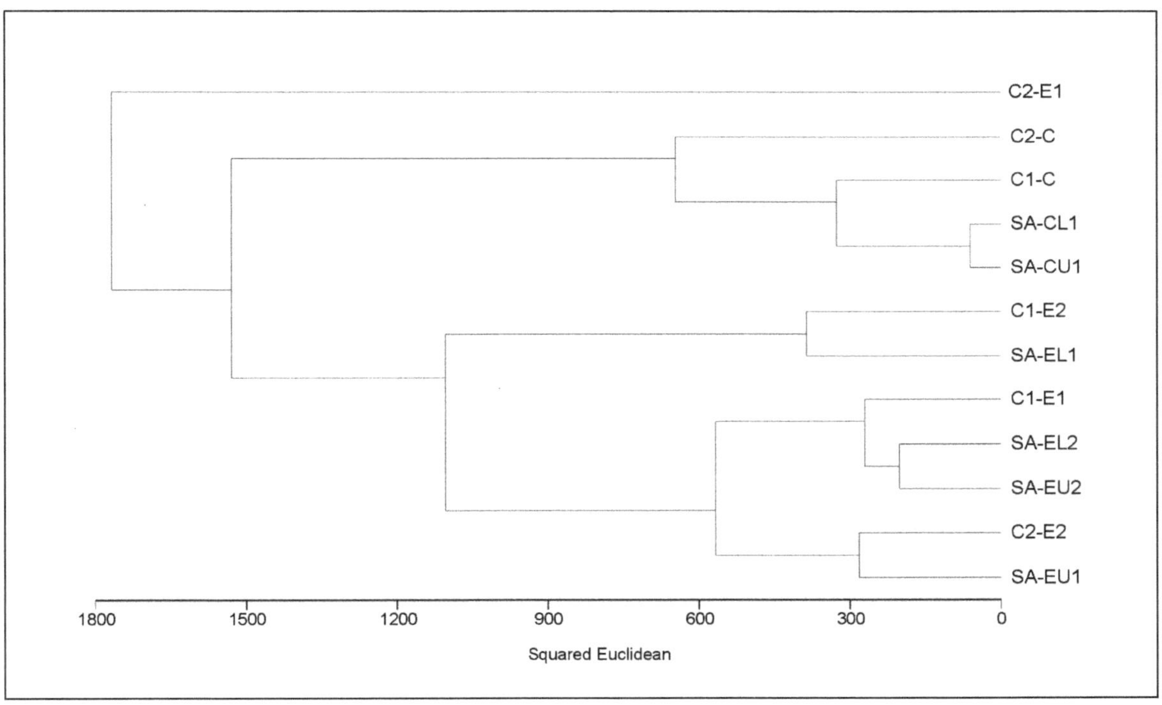

Figure 10: Dendrogram produced by cluster analysis of the diatom species composition in both the center and the external samples C1, C2, and SA.

It is likely that the erosion processes are responsible for the differences between the percentages seen in the two sectors of the furrow and those noticed in the modern canals. Therefore, we conclude that this line of analysis for investigating ancient canals is unreliable, at least until a significant number of replicated studies are available to statistically validate this supposition.

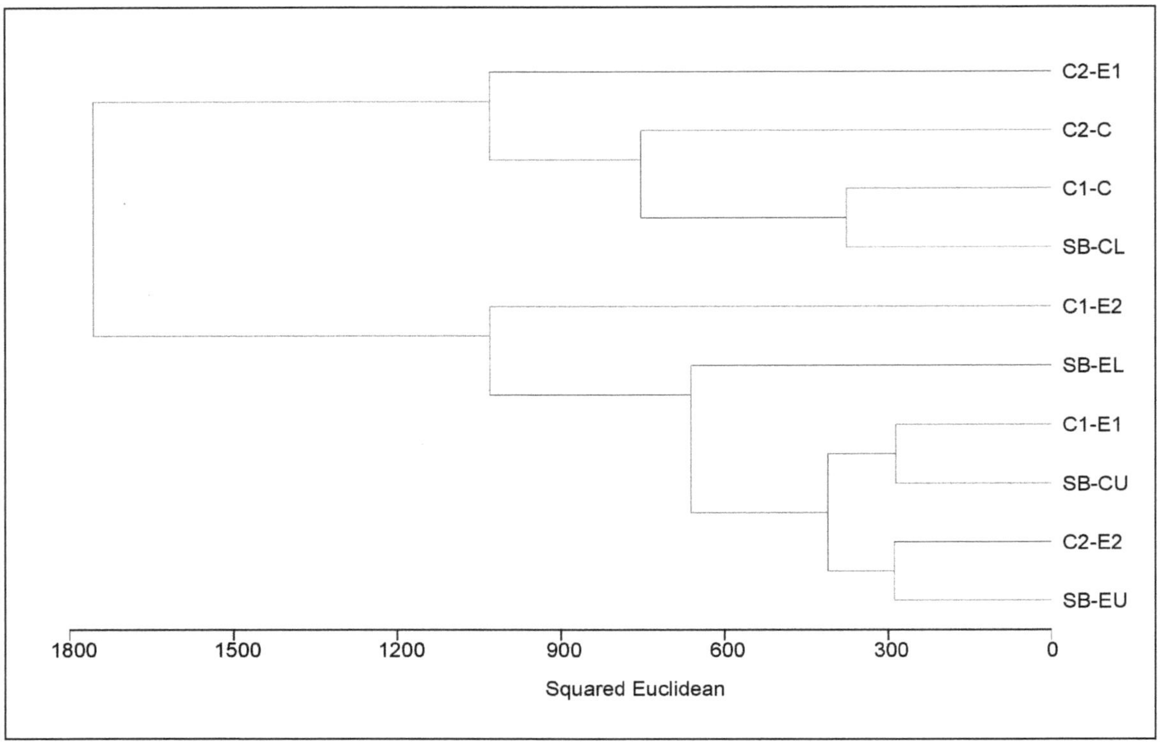

Figure 11: Dendrogram produced by cluster analysis of the diatom species composition in both the center and the external samples C1, C2, and SB.

	C1	C2	SA-U	SA-L	SB-U	SB-L
Df	6	6	6	6	3	3
X^2_t	12.59	12.59	12.59	12.59	7.81	7.81
X^2	46.18	61.16	7.58	31.3	16.48	3.05
H_0	R	R	NR	R	R	NR

Table 2: Results of the Pearson's X^2 Test. Df= degrees of freedom; H_0= null hypothesis; R= rejected, NR= not rejected.

CONCLUSIONS

The results of the statistical analysis of the unbroken and the broken valves obtained from different sampling points of the same feature are not significant, hence this analysis is not a suitable tool for researching past canal features. Our negative results are probably a consequence of continuous erosion on all sampled points, being this eolian action on exposed sectors, or hydric action of flux, which modified the signal given by the water activity. Nevertheless, the analysis of the composition and abundance of diatom assemblages and its comparison with modern analogues were found to be useful for identifying the existence of ancient canals and thus, of archaeological water management.

Both sectors of the furrow could be of fluvial origin. However, SA is the only sector where a clear signal of its channel function was found, partly due to its close similarity with modern canals. In addition, the highest diatom concentrations and the highest percentages of planktonic or tychoplanktonic species were obtained from the center of the furrow, which is the main sector for the permanence of stagnant water during the intervals when the canal became inactive.

This study has presented the first evidence for this type of archaeological hydraulic technology ever found in the locality and the sub-basin Las Pitas. The correct identification of past water management together with the archaeological context of Peñas Coloradas has enabled us to conclude that this locality was part of a productive agropastoral landscape.

We hope that these results will encourage further interdisciplinary analyses, which include other aquatic organisms, such as ostracods or chrysophytes. A phytolithic and palynological approach, in particular, would enable us to answer new questions, for example, whether the water was conducted to crop products for human consumption or livestock (pasture). A multiproxy analysis would also be a great aid for better understanding site-formation processes and the role of erosion in this type of hydric features.

ACKNOWLEDGEMENTS

We thank to Dr. R. Lombardo (FCEyN-UBA, CONICET) and Lic A. de Olmos (FFyL-UBA) for helping with the statistic analysis, to M.Sc. A. García Suárez (University of Reading) and Lic. M. Fernández (CADIC-CONICET) for helping with the English version and their commentaries, and Lic D. Echazú, Lic M.J. Ramon Mercau and Ms. S. Bustos for their continued support.

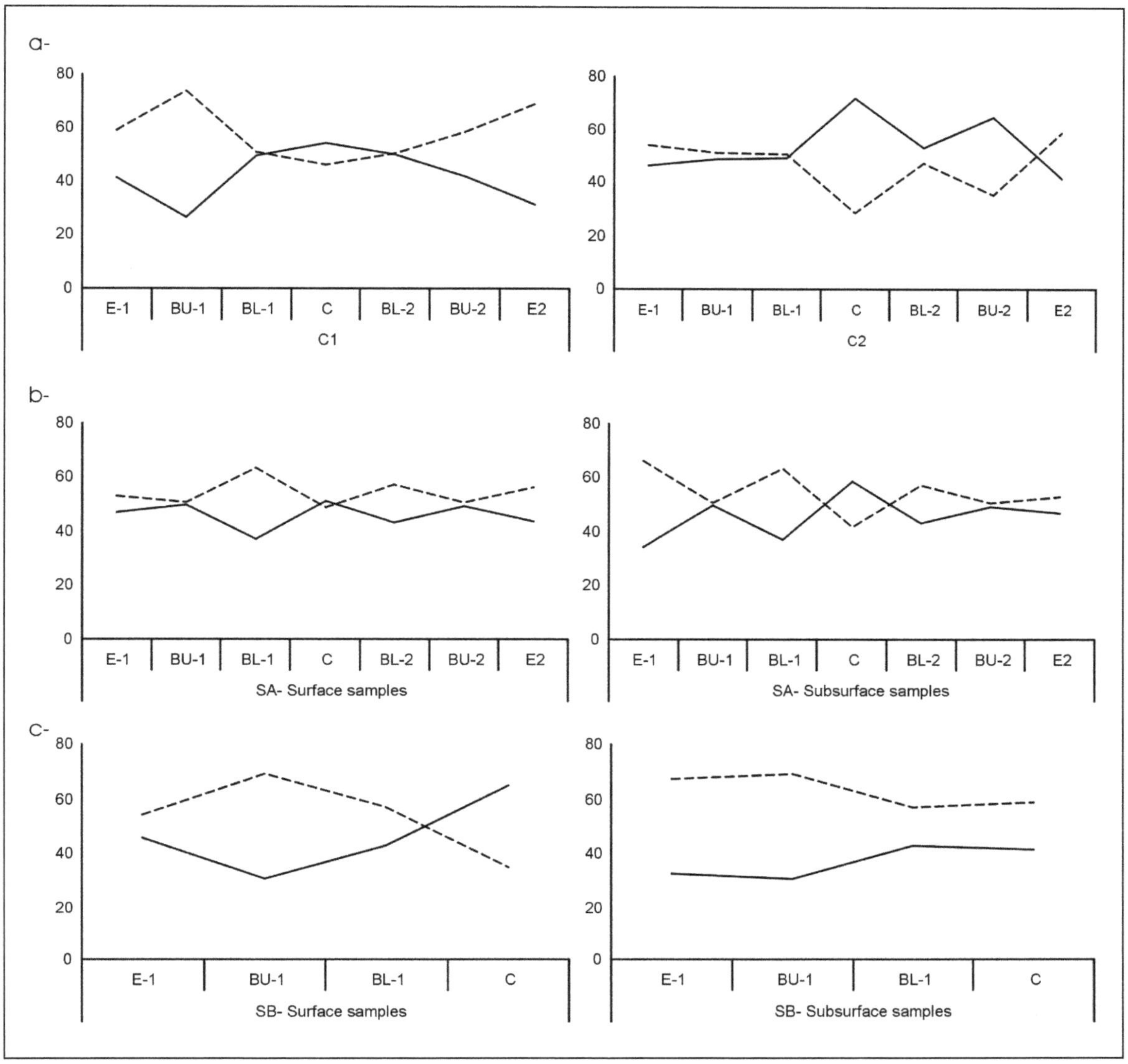

Figure 12: Distribution of broken (dotted lines) and unbroken (black lines) valves percentages through different sampled points: a- C1 and C2; b- SA; c- SB.

REFERENCES

ASCHERO, C.
1999. El arte rupestre del desierto Puneño y el Noroeste Argentino. In *Arte Rupestre en los Andes de Capriconio*, pp. 97-135. Museo Chileno de Arte Precolombino, Santiago de Chile.

2000. Figuras humanas, camélidos y espacios en la interacción circumpuneña. In *Arte en las Rocas. Arte Rupestre, Menhires y Piedras de Colores en Argentina*, edited by M.M. Podestá and M. De Hoyos, pp. 15-44. Sociedad Argentina de Antropología and Asociación Amigos del INAPL, Buenos Aires.

ASCHERO, C., A. MARTEL AND S. LÓPEZ CAMPENY
2006. Tramas en la piedra: rectángulos con diseños geométricos en Antofagasta de la Sierra (Puna meridional, Argentina). In *Tramas en la Piedra. Producción y Usos del Arte Rupestre*, edited by D. Fiore and M.M. Podestá, pp. 141-156. World Archaeological Congress, Sociedad Argentina de Antropología and Asociación Amigos del INAPL, Buenos Aires.

BABOT, M.
2009. La cocina, el taller y el ritual: explorando las trayectorias del procesamiento vegetal en el Noroeste Argentino. *Darwiniana* 47(1): 7-30.

BATHURST, R., D. ZORI AND J. BYOCK
2010. Diatoms as bioindicators of site use: locating turf structures from the Viking Age. *Journal of Archaeological Science* 37(11): 2920-2928.

BATTARBEE, R.
1986. Diatom analysis. In *Handbook of Holocene Palaeoecology and Palaeohydrology*, edited by B.E. Berglund, pp. 527-570. Chichester, John Willey & Sons.

1988. The use of diatom analysis in archaeology: a review. *Journal of Archaeological Science* 15(6): 621-644.

BAYMAN, J., M.R. PALACIOS-FEST, S.K. FISH AND L.W. HUCKEL
2004. The paleoecology and archaeology of long-term water storage in a Hohokam reservoir, southwestern Arizona, U.S.A. *Geoarchaeology* 19(2): 119-140.

BENEŠ, J., J. KAŠTOVSKÝ,, R. KOČÁROVÁ, P. KOČÁR, K. KUBEČKOVÁ, P. POKORNÝ AND P. STAREC
2002. Archaeobotany of the Old Prague Town defense system, Czech Republic: Archaeology, macro-remains, pollen, and diatoms. *Vegetation. History and Archaeobotony* 11: 107-119.

BRADBURY, J., S. COLMAN AND R. REYNOLDS
2004. The history of recent limnological changes and human impact on Upper Klamath Lake, Oregon. *Journal of Paleolimnology* 31: 151-165.

Cameron, N.
2007. Use in archaeology. *Encyclopedia of Quaternary Science*, pp. 523-529. Elias, S.A. Ed., Elsevier.

JUGGINS, S. AND N. CAMERON
1999. Diatoms and archaeology. In *The Diatoms: Application for the Environmental and Earth Sciences*, edited by E.F. Stoermer and J.P. Smol, pp. 389-401. Cambridge University Press, Cambridge.

COHEN, M.L.
2009a. Perspectivas espaciales desde el sitio Punta de la Peña 3 cumbre (PC3-c) Antofagasta de la Sierra. Catamarca. *Actas del 1° Congreso Latinoamericano y 2° Nacional de Arqueometría* 2: 477-487. Comisión Nacional de Energía Atómica (CNEA), Bs. As.

2009b. Fotografías y algo mas... imágenes de Antofagasta en 1923 como aporte para el planteo de hipótesis y la búsqueda de evidencias contrastadoras. *Libro de Resúmenes de las VIII Jornadas de Comunicaciones de la Facultad de Cs. Naturales e I.M.L. y 1° Interinstitucional*, pp. 109.

2010. *Prácticas sociales, estrategias de visibilidad y construcción de la cartografía social durante el lapso ca. 1000-1500 AD en Antofagasta de la Sierra, Catamarca. Perspectivas desde el Sitio Peñas Coloradas 3 Cumbre.* Unpublished PhD Dissertation, Facultad de Filosofía y Letras, Universidad de Buenos Aires.

DÍAZ, A.
2009. *Historias de tierra y agua: introducción a los espacios agrícolas prehispánicos de Laguna Blanca. El caso de estudio de la aldea arqueológica Piedra Negra (Dpto. Belén, Provincia de Catamarca).* Unpublished Graduate Thesis, Escuela de Arqueología, Universidad Nacional de Catamarca.

DÍAZ, C. AND N.I. MAIDANA
2006. A new monoraphid diatom genus: Haloroundia. *Nova Hedwigia* 130: 177-184.

Elías, A.
2010. *Informe de análisis de material lítico del sitio Peñas Coloradas 3 Bajo*. Ms.

FERNÁNDEZ, M. AND N.I. MAIDANA
2008. Diatomeas: su utilidad en el estudio de la Arqueología de Patagonia. (Poster). *VII Jornadas de Arqueología de la Patagonia*, pp. 28. Ushuaia.

FERNÁNDEZ, M. AND M. SALEMME
2010. Resultados preliminares sobre el análisis de diatomeas en sitios de la Meseta central de Santa Cruz (Patagonia Argentina). *Libro de Resúmenes del V Simposio del Hombre Temprano en América*, pp. 116. La Plata.

GRANA, L.
2007. *Diatomeas fósiles: un acercamiento a la relación entre la sociedad y el ambiente durante el Holoceno en Antofagasta de la Sierra, Puna Meridional.* Unpublished Graduate Thesis, Facultad de Filosofía y Letras, Universidad de Buenos Aires.

HADLEY K., M. DOUGLAS, R. MCGHEE, J. BLAIS AND J. SMOL
2010. Ecological inf uences of Thule Inuit whalers on high Arctic pond ecosystems: a comparative paleolimnological study from Bathurst Island (Nunavut, Canada). *Journal of Paleolimnoloy* 44: 85-93.

HORROCKS M., G. IRWIN, M. MCGLONE, S. NICHOL AND L.J. WILLIAMS
2003. Pollen, phytoliths and diatoms in prehistoric coprolites from Kohika, Bay of Plenty, New Zealand. *Journal of Archaeological Science* 30(1): 13-20.

JANSMA, M.J.
1990. Diatoms from Neolithic excavations on the former island of Schokland, Ijselmeerpolders. The Netherlands. *Diatom Research* 5(2): 301-309.

KLIGMANN, D.M.
2003. *Procesos de formación de sitios arqueológicos: tres casos de estudio en la Puna Meridional Catamarqueña Argentina.* Unpublished PhD Dissertation, Facultad de Filosofía y Letras, Universidad de Buenos Aires.

KLIGMANN, D.M. AND M. CALDERARI
2012. Diatoms and ceramic provenance: a cautionary tale. *Archaeometry* 54(1): 129-143.

KRAMMER, K. AND H. LANGE-BERTALOT
1986. *Bacillariophyceae 1. Teil: Naviculaceae. In Süsswasserflora von Mitteleuropa*, edited by H. Ettl, J. Gerloff, H. Heynig and D. Mollenhaue. G. Fischer Verlag, Jena.

1988. *Bacillariophyceae. 2. Teil: Bacillariaceae, Epithemiaceae, Surirellaceae. In Süsswasserflora von Mitteleuropa*, edited by H. Ettl, J. Gerloff, H. Heynig and D. Mollenhaue. G. Fischer Verlag, Jena.

1991a. *Bacillariophyceae. 3. Teil: Centrales, Fragilariaceae, Eunotiaceae. In Susswasserflora von Mitteleuropa*, edited by H. Ettl,, J. Gerloff, H. Heynig and D. Mollenhauer. G. Fischer Verlag, Jena.

1991b. *Bacillariophyceae. 4. Teil: Achnanthaceae, Kritische Ergänzungen zu Navicula (Lineolatae) und Gomphonema. In Süsswasserflora von Mitteleuropa*, edited by H. Ettl, J. Gerloff, H. Heynig and D. Mollenhauer. G. Fischer Verlag, Jena.

LANGE-BERTALOT, H.
2001. *Diatoms of Europe 2. Navicula sensu stricto. 10 Genera Separated from Navicula sensu lato. Frustulia*. A.R.G. Gantner Verlag.

LOWE, R.
1974. *Environmental requirements and pollution tolerance of freshwater diatoms*. EPA-670/4-74-005 Environmental Monitoring Series. CEPA, Cincinnati

MAIDANA, N.I. AND C. SEELIGMANN
2006. Diatomeas (Bacillariophyceae) de Ambientes Acuáticos de Altura de la Provincia de Catamarca , Argentina II. *Boletín de la Sociedad Argentina de Botánica* 41: 1-13.

MANNION, A.
1987. Fossil diatoms and their significance in archaeological research. *Journal of Archaeology* 6: 131-147.

MARTEL, A.
2009. Arte rupestre: construcción y significación del espacio en la Puna meridional argentina (Antofagasta de la Sierra, Catamarca). In *Crónicas sobre la Piedra. Arte Rupestre de las Américas*, edited by M. Sepúlveda, L. Briones and J. Chacama, pp. 271-280. Ediciones Universidad de Tarapacá, Arica.

2010. *Arte Rupestre de Pastores y Caravaneros. Estudio Contextual de las Representaciones Rupestres durante el Período Agroalfarero Tardío (900 d.C. - 1480 d.C.) en el Noroeste Argentino*. Unpublished PhD Dissertation, Facultad de Filosofía y Letras, Universidad de Buenos Aires.

MARTEL, A. AND C. ASCHERO
2007. Pastores en acción: imposición iconográfica vs. autonomía temática. In *Producción y Circulación Prehispánicas de Bienes en el Sur Andino. Colección Historia Social Precolombina*, compiled by A.E. Nielsen, M.C. Rivolta, V. Seldes, M.M. Vázquez and P.H. Mercolli, Tomo 2, pp. 329-349. Editorial Brujas, Córdoba.

MARTÍNEZ MACCHIAVELLO, J.C., G. DE LA FUENTE AND N. RATTO
1999. La utilización del análisis de diatomeas (Bacillariophyta) en la investigación arqueológica: una perspectiva interdisciplinaria. *Arqueología* 9: 49-69.

MORALES, M.
2010. *Arqueología Ambiental del Holoceno Temprano y Medio en la Puna Seca Argentina*. Unpublished PhD Dissertation, Facultad de Filosofía y Letras, Universidad de Buenos Aires.

NEELY, J.A., S.C. CARAN, B.M. WINSBOROUGH, F. R. SORENSEN AND S. VALASTRO
1995. An early Holocene hand-dug water well in the Tehuacan Valley of Puebla, Mexico. *Current Research in the Pleistocene* 12: 38-40.

NICHOLS, D.
1988. Infrared aerial photography and prehispanic irrigation at Teotihuacan: the Tlajinga canals. *Journal of Field Archaeology* 15(1): 17-27.

OGNJANOVA-RUMENOVA, N.
2008. Palaeoenvironment and archaeology: The use of diatom analysis in archaeology. In *Geoarchaeology and Archaeomineralogy*, edited by R.I. Kostov, B. Gaydarska, M. Gurova, pp. 291-294. Sofia.

OLIVERA, D. AND S. VIGLIANI
2000-2002. Proceso cultural, uso del espacio y producción agrícola en la Puna Meridional Argentina. *Cuadernos* 19: 459-481.

PALACIOS-FEST, M.R., J.B. MABRY, F. NIALS, J.P. HOLMLUND, E. MIKSA AND O.K. DAVIS
2001. Early irrigation systems in Southeastern Arizona: The ostracode perspective. *Journal of South American Earth Sciences* 14: 541-555.

PURDUE L., W. MILES, K. WOODSON, A. DARLING AND J.F. BERGER
2009. Micromophological study of irrigation canal sediments: Landscape evolution and hydraulic management in the middle Gila River Valley (Phoenix Basin, Arizona) during the Hohokam occupation. Quaternary *International* 216: 129-144.

QUESADA, M.
2006. El diseño de las redes de riego y las escalas sociales de la producción agrícola en el 1er milenio DC (Tebenquiche Chico, Puna de Atacama). *Estudios Atacameños* 31: 31-46.

ROSEN, A. AND S. WEINER
1994. Identifying ancient irrigation: a new method using opaline phytoliths from emmer wheat. *Journal of Archaeological Science* 21(1): 125-132.

ROUND, F.E., R.M CRAWFORD AND D.G. MANN
1990. *The Diatoms: Biology and Morphology of the Genera*. Cambridge University Press, Cambridge.

RUMRICH, U., H. LANGE-BERTALOT AND M. RUMRICH
2000. *Iconographia Diatolmologica 9. Diatomeen der Anden von Venezuela bis Patagonien/Tierra del Fuego*, edited by H. Lange-Bertalot. A.R.G. Gantner Verlag K.G., Germany.

SALMINCI, P.
2011. *Espacios Residenciales y Productivos. El Paisaje Arqueológico de Antofagasta de la Sierra entre los Siglos XI y XVI d.C.* Unpublished PhD Dissertation, Facultad de Filosofía y Letras, Universidad de Buenos Aires.

SEELIGMANN, C. AND N.I. MAIDANA
2003. Diatomeas de la provincia de Catamarca (Argentina). Boletín de la *Sociedad Argentina de Botánica* 48(1-2): 39-50.

SEELIGMANN, C., N.I. MAIDANA AND M. MORALES
2008. Diatomeas (Bacillariophyceae) de humedales de altura de la Provincia de Jujuy - Argentina. *Boletín de la Sociedad Argentina de Botánica* 43(1-2): 1-17.

STRAUB, F.
1990. Diatomées et reconstitution des environnements préhistoriques. *Archéologie Neuchathloise* 10, *Hauterive-Chapréveyers* 4: 17-30.

SOLÁ, P. AND M. MORALES
2007. Indicadores biológicos y mineralógicos en un tiesto estilo "Incaico" hallado en Susques, Puna argentina. *Intersecciones en Antropología* 8: 361-364.

TCHILINGUIRIAN, P. AND D. OLIVERA
2000. De aguas y tierras: aportes para la reactivación de campos. *Relaciones de la Sociedad Argentina de Antropología* XXV: 99-117.

2009. Geomorfología, paleoambientes Holocenos y geoarqueología: sitios Corral Grande y Arroyo Las Pitas. Antofagasta de la Sierra, Prov. de Catamarca, Puna Meridional. Arqueometría *1° Congreso Latinoamericano y 2° Nacional de Arqueometría*, Vol II, pp. 544-551. Bs. As.

TROMBOLD, C. AND I. ISRADE-ALCANTARA
2005. Paleoenvironment and plant cultivation on terraces at La Quemada, Zacatecas, Mexico: the pollen, phytolith and diatom evidence. *Journal of Archaeological Science* 32(3): 341-353

VAN DAM, H., A. MERTENES AND J. SINKELDAM
1994. A coded checklist and ecological indicator values of fresh water diatoms from the Netherlands. *Journal of Aquatic Ecology* 28: 117-133.

VIGLIANI, S.
2005. El sitio bajo del Coypar II: Las evidencias más tempranas (*ca.* 1000 AP) del proceso agropastoril en la puna meridional Argentina (Antofagasta de la sierra, Catamarca). *Andes* 16: 325-350.

ZAR, J.
1974. *Biostatical Analysis*. 3rd edition. Prentice-Hall. Englewood Cliffs, N.J.

[1] The main problem we face regarding the research of past irrigation features is the difficulty entailed in dating them. There are different suitable methodologies such as the direct measuring of organic materials associated with canals, luminescence dating (Purdue *et al.* 2009), and indirect measuring involving the ceramics associated with canal features, which we would later relate to ^{14}C samples from other sites to establish a chronology (Quesada 2006). So far, we have not been able to establish any of these associations, so we have therefore considered the archaeological context of the site to provide an estimated date for the features. Nevertheless, we hope that future excavations of the furrow will bring datable finds to light.

PALEOENVIRONMENTAL STUDIES OF THE QUEBRADA DE LAPAO, JUJUY PROVINCE, ARGENTINA (23° 22´ 01" S, 66° 21´ 52,8" W, 3650 M A.S.L.) FOR THE 9400-7300 YRS B.P. SPAN

Pablo Tchilinguirian[1], Marcelo R. Morales[2], Brenda Oxman[2] and Malena Pirola[3]

[1] CONICET - Instituto Nacional de Antropología y Pensamiento Latinoamericano - Facultad de Ciencias Exactas y Naturales, Universidad de Buenos Aires, [2] CONICET - Instituto de Arqueología, Facultad de Filosofía y Letras, Universidad de Buenos Aires, [3] Instituto de Arqueología, Facultad de Filosofía y Letras, Universidad de Buenos Aires

Abstract

Paleoenvironmental research carried out in the Andes highlands during the past decades has shown great variability in terms of the impact of broad scale climate changes in different environmental settings. In order to address this issue, we have undertaken systematic studies of several localities in the Puna of Argentina oriented to modeling past environmental scenarios for human adaptation in the area. In this paper we summarize the main results obtained by multi-proxy analysis -mainly diatoms, pollen, and geomorphology- carried out in the Quebrada de Lapao, located at 3650 m a.s.l. in the Dry Puna of the Jujuy Province. In general terms, a wetland environment that included a water body developed in the headwaters of Lapao between 9400 and 7700 yrs B.P. (i.e. between 10,500 and 8500 cal. yrs B.P.). This was followed by an alluvial event that formed a more unstable and brackish wetland between 7700 and 7550 yrs B.P. (i.e. 8500 and 8300 cal. yrs B.P.). These results also suggest that there was a complex array of local and regional factors involved in the fluvial responses to environmental change at Lapao, including: a) the catchment area of the system, b) the characteristics of climate changes, and c) the evolution of base level at the Pastos Chicos - De las Burras basin (i.e. the outflow system of Lapao). We consider that these results are central for archaeological studies in the area, in order to model past Puna landscapes in terms of differential stability and space-time availability of resources. From this perspective, the similarities and differences found among local environments present a promising opportunity in terms of spatial stratification for future archaeological modeling.

Key words: *Paleoenvironment - Dry Puna - Early Holocene - Mid-Holocene*

INTRODUCTION

During the last decades, a growing body of paleoenvironmental studies in the Andes highlands has allowed more detailed characterizations of the main climate tendencies and patterns during the last part of the Quaternary (*e.g.* Latorre *et al.* 2002, 2003, 2006; Quade *et al.* 2008; Rech *et al.* 2002, 2003; Servant and Servant-Vildary 2003).

These studies have also shown the existence of great variability in terms of the environmental impact of broad climatic changes (*i.e.* global) in different space-time scales. Historically, investigations carried out in the Puna of Argentina are scarce (*e.g.* Grana and Morales 2005; Morales 2004, 2011; Morales and Schittek 2008; Morales *et al.* 2008, 2009; Olivera *et al.* 2004; Oxman 2010; Tchilinguirian 2009; Yacobaccio and Morales 2005) compared to the broad corpus of data available for Northern Chile and Southern Bolivia (detailed summaries and systematizations of these studies and their results are available in Morales 2011, Tchilinguirian 2009 and Tchilinguirian and Morales 2012), but both regions share general regional trends and patterns, as well as local and sub-regional variability. To better understand the mechanisms and factors involved in the conformation of the observed paleoenvironmental variability in different spatial scales along the Andes, an effort towards the integration of the mentioned research should be done. In this sense, we stress that local elements such as topography, hydrogeology and geomorphology have conditioned the effects of broad-scale climate change on local environments by sometimes amplifying, and others minimizing, its impact.

The problem of the impact of climate change on the environment -in terms of duration, intensity and extension- is a key issue in archaeology, due to its role in modifying the abundance and space-time distribution of resources available for human groups. Since resource structure is germane for understanding human organizational patterns and trends, paleoenvironmental research oriented to modeling past resource structure is central in the studies of archaeological localities.

Quebrada de Lapao Grande (*i.e.* Quebrada de Lapao, LP) has recently become the focus of archaeological and paleoenvironmental research (Morales 2004, 2011; Oxman 2010; Tchilinguirian 2009; Yacobaccio and Morales 2005). The studied area is located 5 km to the north of the town of Susques, a small village in the Jujuy Province, Argentina (Figure 1). The first paleoenvironmental studies in the Quebrada de Lapao were diatom analysis of different sedimentary profiles located at the headwaters of the course (Yacobaccio and Morales 2005). One of them, LP 5 (Figure 2b) which was dated between ~9400 and ~7300 B.P., is the most thoroughly studied sequence (*i.e.* Morales 2011) in the Quebrada de Lapao. Archaeological studies in the area have included excavations in the Late Holocene residential site Puesto Demetria (Yacobaccio *et al.* 1997-1998) and ongoing distributional and morphological studies of lithic materials, tentatively assigned to the Early and Mid-Holocene, which appear as surface materials along one of the fluvial terraces (Hoguin pers. comm.).

This paper presents new data about the geomorphology and sedimentology of several sections along the Lapao drainage system and a pollen analysis of the LP 5 sequence. These new data are discussed in the light of the existing diatoms analysis (Morales 2011), whose results are also summarized in this paper. We consider that this work will contribute not only to the reconstruction of past environments of the region but also to the enrichment of future archaeological theoretical models and explanations.

THE PUNA OF ARGENTINA

The studied area is located in the Puna region, which comprises the arid highlands of Argentina located between 22° and 27° S and 3000 to 4500 m a.s.l (Figure 1). This area is characterized by high solar radiation, high daily temperature amplitude, marked seasonality in rainfall, and low atmospheric pressure. Precipitation in the area is largely governed -as in the rest of northwestern Argentina- by the South American MonsoonSystem (Garreaud and Falvey 2009; Garreaud *et al.* 2003). This system produces about 80% of the annual precipitation occurring in the Andes highlands (*ca.* 200 mm/yr in the studied area) between December and February (Bianchi and Yañez 1992; Marengo and Rogers 2001; Salati *et al.* 1979; Vuille and Keimig 2004).

The Puna desert biome presents a typical altitudinal variation in vegetation communities, ranging from "Tolar" (shrub steppe), composed mainly by Compositae including the Asteraceae, Solanaceae and other families (i.e. *Fabiana* spp. *Parastrephia* spp., *Baccharis* spp., Adesmia spp., etc.) located below 4000 m a.s.l., to "*Pajonal*" (highland grasslands) situated over 4000 m a.s.l. and composed mainly by the Poaceae family (*i.e. Festuca* spp., *Stipa* spp., etc.) (Cabrera 1976; Ruthsatz and Movia 1975). Between the Pajonal and the Tolar there is a narrow ecotonal belt -currently located at ~4000 m a.s.l.- with a mixture of both vegetation arrangements. Both vegetation ranges include a particular kind of community with a patched distribution, a wetland type environment called "*Vegas*" (usually mentioned as Andean peatlands or peatbog) (Morales, 2004, 2011; Squeo *et al.* 2006; Tchilinguirian 2009). They are constituted by soft grasses, dominated by the Juncaceae and accompanied by Cyperaceae and other families such as Plantaginaceae and Campanulaceae (*v.g. Scirpus atacamensis*, *Carex* spp., *Oxychloe andina*, *Juncus* spp., *Hypsella* spp., *Plantago* spp., etc.) (Braun Wilke *et al.* 1999). Primary productivity is concentrated in perennial and stable hydrological systems, such as primary basins and high valleys (Olivera 1997). Yacobaccio (1994) defined these patches as "nutrient concentration zones" (ZCN) because they account for the majority of the available regional biomass of this desert.

The most important wild animal resources for humans in the Puna include two camelids (vicuña -*Vicugna vicugna vicugna*- and guanaco -*Lama guanicoe cacsilensis*-), large rodents (*e.g.* viscachas and chinchillas, *Lagidium viscacia* and *Chichilla brevicaudata*, respectively), and a small cervid (the taruca, *Hippocamelus antisensis*).

QUEBRADA DE LAPAO

The region is an extensive Tertiary graven (Susques graven) located between the Taire and Los Cobres ranges in the Department of Susques, within the Dry Puna of the Jujuy province (Figure 1). The geology is composed of Ordovisic metamorphic rocks as well as Tertiary alluvial sediments covered by vulcanoclastic and Coranzulí ignimbrite rocks and Pliocene coarse fan deposits (Nullo 1988). The graven uplifted during the Pliocene by the Andean deformation, followed by a cycle of canyon cutting that produced the current features of the landscape.

The Lapao draining system comprises several quebradas (i.e. small canyon formations) and covers 110 km^2, stretching out over 15 km from the Taire mountain range (Cerro Lares: 4400 m a.s.l.) towards its confluence with the Pastos Chicos-De Las Burras river basin (3594 m a.s.l.). The Lapao system can be subdivided into three geomorphic areas: the upper (> 4000 m a.sl.) and middle course (4000-3650 m a.s.l.) formed by Quebrada Azul and Quebrada Cuevas respectively, and the lower course called Lapao Grande (*i.e.* Quebrada de Lapao), comprising a 5 km-long quebrada (starting at 3650 m a.s.l.) that ends at the village of Susques (3600 m a.s.l), where the Lapao stream flows into the Pastos Chicos river (Figures 1 and 2a).

The upper Lapao system covers the north of the studied area, in the eastern slope of the Taire mountain range, which has a strong gradient and numerous rock outcrops. The vegetation of this area is dominated by the *Pajonal* community, with vegetation coverage near 50% and shallow soils. The upper section of the drainage system concentrates most of the rainfall in the area, due to the orographic effect of the Taire mountain range on the easterly flow. Therefore, this region is dominated by run-off and erosion.

The middle Lapao area develops along two landforms: the eastern piedemont of Taire range and the ignimbrite Pliocene plateau. The piedemont comprises dissected alluvial fans composed by coarse conglomerate deposits of Pliocene age (Figure 1d). The plateau is a vast region that rises from 100 to 120 m above base level and is related with the Coranzuli caldera collapse. In this area, vegetation composition depends on the altitude with the *Tolar* community (shrub steppe) only below 3900 m a.s.l. and an ecotonal belt composed by a mixture of

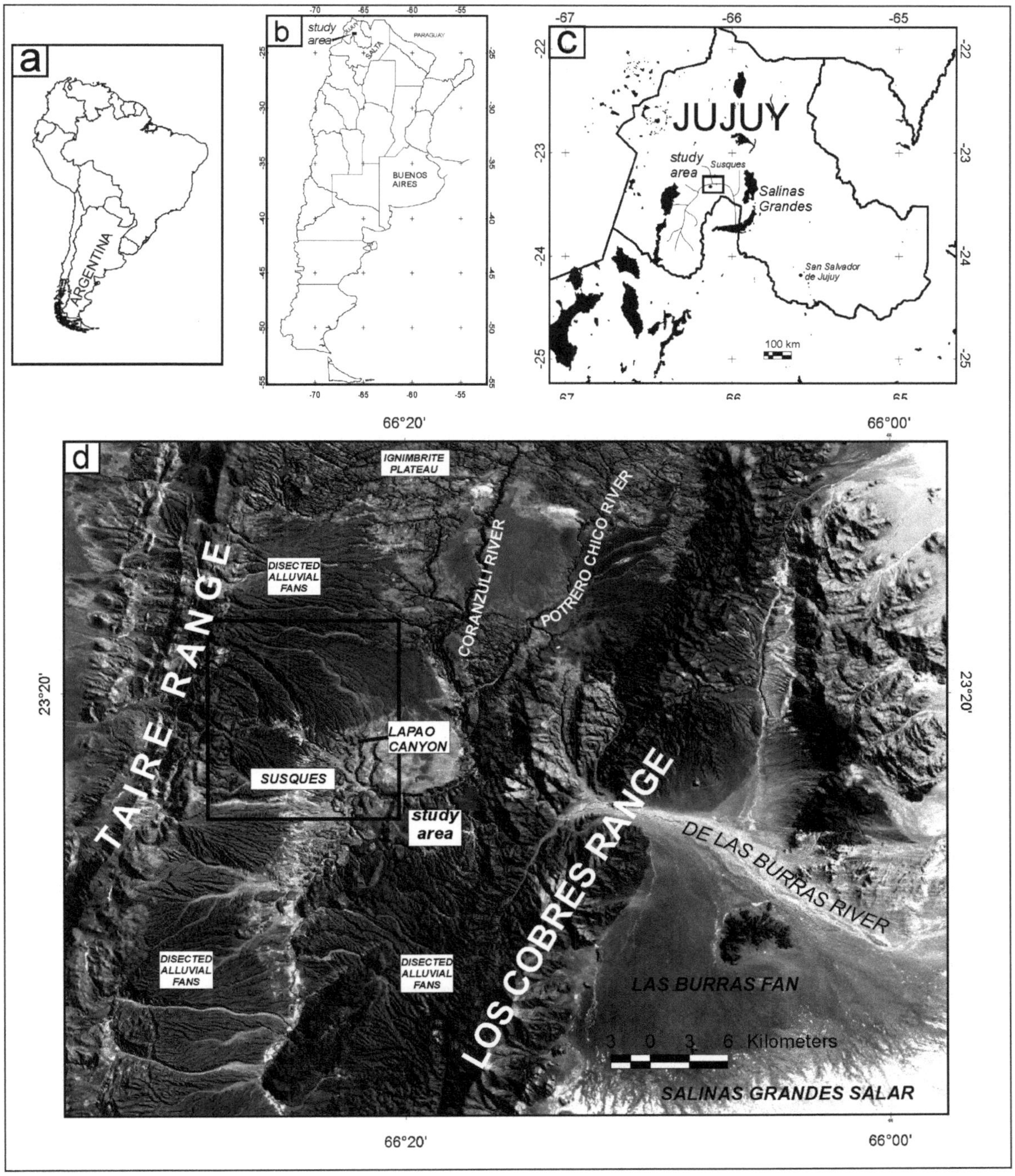

Figure 1. Study area. A) and B) illustrate the relation of the study area to South America and Argentina; C) Jujuy province, boxed area is shown in D), Lansat ETM image showing the study area and location of the most important range and geomorphic units.

Tolar-Pajonal communities (shrub and herbaceous steppe) between 3900 - 4100 m a.s.l.

The lower Lapao area, comprising the last 5 km of the drainage system, was the focus of the paleoenvironmental studies described in this paper. Its distinctive feature is that the creeks have excavated deep quebradas (50 and 90 m of depth) in the ignimbrite rocks (Figure 3). Soils are poorly developed Entisols (Torriorthents), while the vegetation is dominated by a shrub steppe (*Tolar*) with medium to low coverage (30-40%).

Organic wetlands, locally known as "Vegas", appear along the upper and lower Lapao system. These wetlands are composed by deep organic (Moore 1987) and sandy organic matter rich soil with hydromorphic and freeze features (the latter only in the upper section of the drainage basin), such as organic and ice mounds (Histosols, Criols and Crioacuents). The Vegas are associated with perennial streams fed by small groundwater reservoirs, which respond to high mountain precipitation and moisture coming from the east.

The streamflow and other hydrological data about the watercourses included in this basin are little known, other than the ephemeral and seasonal nature of the streams (i.e. existence of waterflow only in direct response to precipitation or spring activity) (Fairbridge 1968). Ephemeral watercourses are typical of arid and semi-arid areas of Argentina, where there are high-intensity convective storms. River swells are commonly associated with tractive currents and mass movement processes such as debris or mud flows, commonly referred to as "aluvión" or "volcán" that are also recorded in Lapao younger deposits.

METHODS AND TECHNIQUES

The geomorphology of the Lapao basin was studied through high-resolution satellite and Lansat imagery (Figure 1d). The study included the mapping of landforms along the drainage basin, with particular emphasis on the identification of levels and types of fluvial terraces. Using Lansat imagery and digital topography models (SRTM) active wetlands were detected in order to analyze their characteristics and spatial distribution.

Ten sedimentary profiles were studied along the upper, middle and lower area of the Quebrada de Lapao (Figure 2b). Sedimentary lithofacies were defined for all the deposits according to grain size, sedimentary structures, and body shapes following the procedures of Miall (1982, 1996) and Friend (1983), accompanied by biological components. Lithofacies, reflecting depositional process, were described and grouped into facies associations representing different sedimentary environments (Tables 2 and 3). Finally, we divided the five profiles in allostratigraphic units, separated by discontinuities.

In the lower area of the Lapao drainage basin, in the headwaters of Quebrada de Lapao, a sedimentary sequence -LP5 section (Figure 2b and 5)- was studied applying a multi-proxy approach. The numbers of samples analyzed in each case are specified in the results section for each proxy. The bulk organic matter of four samples coming from LP 5 were dated by regular ^{14}C (three) and AMS (one) techniques (Table 1). These dated layers are mainly composed of complete and broken plant epidermis fragments. Such terrestrial plant material will not suffer from 14C reservoir effect and is therefore highly efficient for dating. Another thirteen dates on bulk organic matter were taken in other terrace sites (LP13, LP12, LP11, LP7, LP6, LP4, LP2, LP1, El Tío, Figure 2b) and included in the geomorphological analysis of Quebrada de Lapao (Table 1). A summary of the information related to all of them is presented in the Table 1. In the case of LP5 a linear interpolation age-depth model (Bennett 1994) was applied to estimate an absolute date for each of the analyzed samples. Two biological proxies were also analyzed in the LP5 record: diatoms and pollen. The techniques involved in these studies are described below.

Diatom analysis followed the standard procedures suggested by (Battarbee 1986): oxidation of the organic matter with H_2O_2 30% and elimination of carbonates with HCl; elimination of the oxidant with three washes with distillated water. The treated material was mounted in slides with Naphrax®. Between 300 and 400 valves were counted and taxonomically assigned in each slide. The observation was carried out in a Reichert-Jung (Polyvar) binocular microscope (OM) under 1000x magnifications. To confirm the taxonomic assignment of some of the smallest diatoms a Phillips XL30 TMP Scanning Electronic Microscope was used, ranging between 5000 and 30000x.

Sample origen	14C date (yrs B.P.)	Lab code	Method	Calibrated Date 1σ (cal. yrs B.P.)	
				More probable range	Midpoint
Lapao 2 c.1	125 ± 40	UGA 7833	AMS	60 - 145	103
Lapao 12 A	250 ± 60	LP 987	Regular	270 - 330	300
Lapao 2 c.4	380 ± 40	UGA 7836	AMS	430 - 502	466
Lapao 2 c.3	400 ± 40	UGA 7835	AMS	438 - 508	473
Lapao 13 Sup	472 ± 33	AA79832	AMS	487 - 543	515
Lapao 13 Inf	653 ± 49	AA79831	AMS	548 - 675	612
El Tío	760 ± 40	LP 1078	Regular	671 - 712	692
Lapao 12 B	1120 ± 40	LP 1074	Regular	974 - 1060	1017
Lapao 2 c.2	1480 ± 50	UGA 7834	AMS	1316 - 1403	1360
Lapao 4	3680 ± 150	LP 1054	Regular	3829 - 4243	4036
Lapao 1	7550±90	LP 1025	Regular	8299 - 8426	8363
Lapao 6	7750±40	UGA 8727	AMS	8513 - 8586	8550
Lapao 5 D4	7770±80	LP 981	Regular	8446 - 8609	8528
Lapao 7	8230±40	UGA 8728	AMS	9129 - 9272	9201
Lapao 5 D3	8380±100	LP 1518	Regular	9281 - 9497	9389
Lapao 5 D2	8560±90	LP 1509	Regular	9469 - 9627	9548
Lapao 11	8980±100	LP 1763	Regular	9920 - 10071	9996
Lapao 5 D1	9380±110	n/a	Regular	10416 - 10755	10586

Table 1. Radiocarbon dates obtained in Quebrada de Lapao , their 1σ calibration range and midpoint using IntCal.09 curve (Reimer *et al.* 2009).

The taxonomical identifications were based on the atlases of Germain (1981), Krammer and Lange-Bertalot (1991-6), Lange-Bertalot (2001), Round *et al.* (1990), Rumrich *et al.* (2000) and several other related articles.

The ecological interpretation rested on the ratio between frequencies of species with different life form affinities (littoral / benthic and aerophilic) as a moisture index of

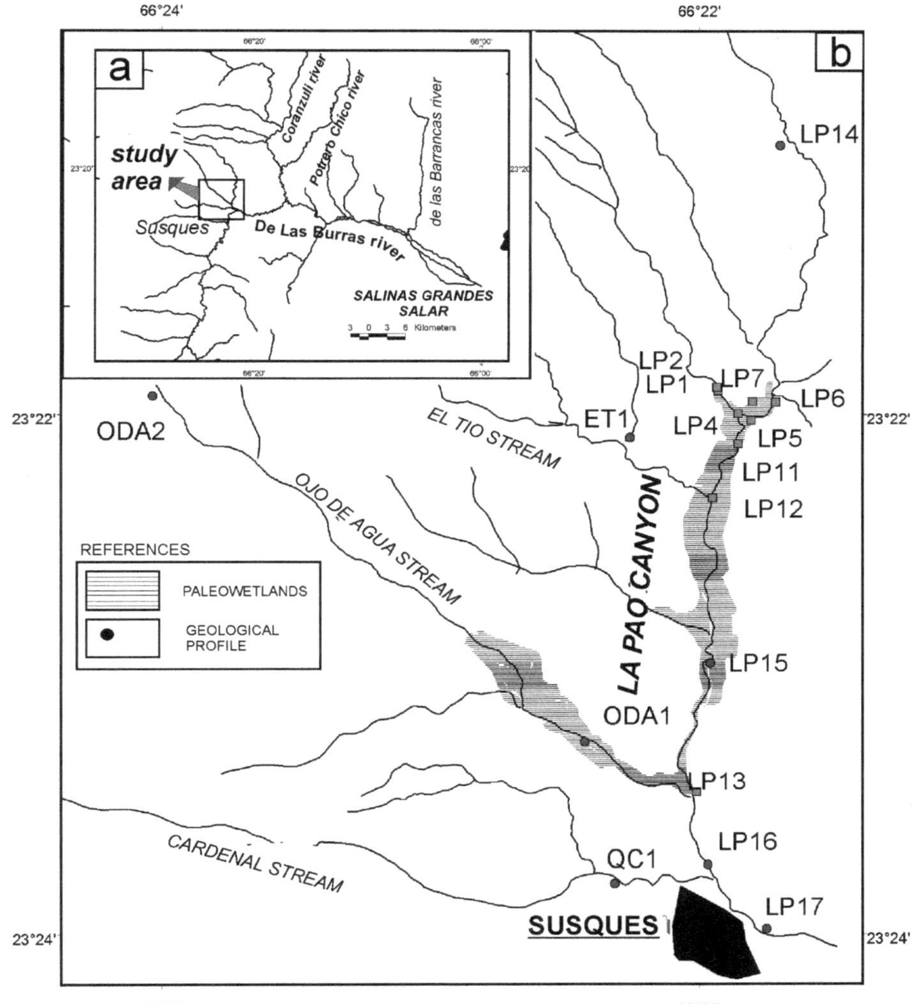

Figure 2. Location of study area: a, relation of the study sections to the catchment area; b, location of sedimentary profiles along the Lapao system.

Figure 3. Landforms and field aspect of the Lapao canyon, facing west at Puesto Demetria; 1, red claystone Tertiary sediments; 2, Ignimbrite cliff; 3, Landslide, blocks fall and colluvial sediments; 4, Terrace level I; 5, Terrace level II.

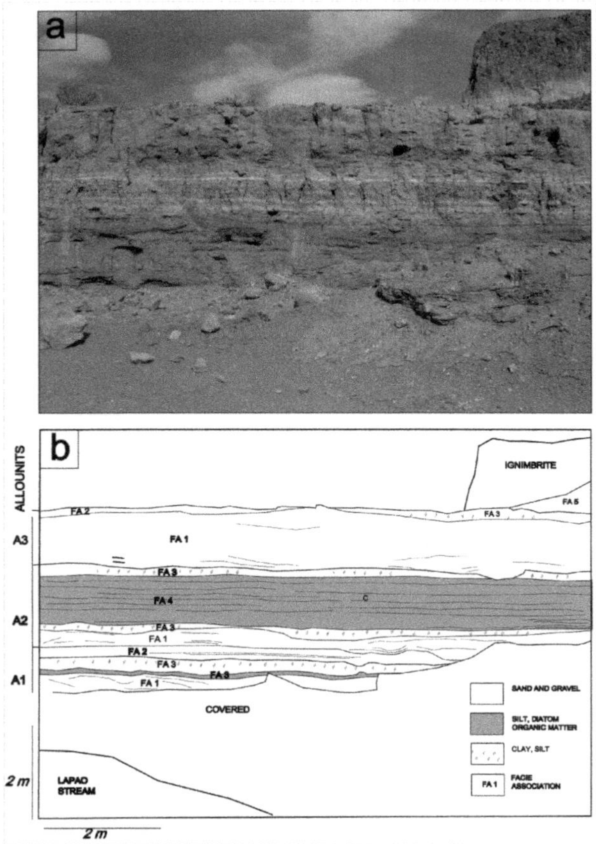

Figure 4. Facies tracing and field aspect of the terrace deposits at LP 6 section, facing east, showing the fluvial sediments and the paleowetland deposits of facies 4.

Figure 5: East view of LP5 deposits where the multi-proxy analysis was carried out.

the environment, and the ratio between frequencies of salt and fresh water diatoms according to their salinity affinities, obtained from the works of De Wolf (1982), Lowe (1974), Van Dam *et al.* (1994) and Vos and De Wolf (1993), among others.

Pollen samples were processed according to the standard protocol for Quaternary Pollen (Faegri and Iversen 1989). At least 200 pollen grains were counted and taxonomically assigned using a Zeiss-Axiolab microscope (OM) under 400x magnifications. The concentration of the material was established following the standard procedure of Stockmarr (1971). The quantification and statistical treatment of the species described was conducted with the TILIA and TGView software (Grimm 1987, 2004). The identification of pollen types was carried out using published Atlas (Heusser 1971; Lupo 1998; Markgraf and D' Antoni 1978) and the Pollen Reference Collection of the Laboratorio de Palinología de la Facultad de Ciencias Agrarias UNJu/Conicet.

The interpretation of pollen results was based on literature related to the regional vegetation communities present in the region today (Braun Wilke *et al.* 1999; Cabrera 1976; Ruthsatz and Movia 1975). The total abundance of pollen types and their frequency in each sample were the basic variables in the interpretation of the pollen record. Particularly, the ratio between the frequencies of herbaceous steppe and shrub pollen components was used as an environmental moisture index.

RESULTS

Geomorphology

The physiography of the Susques region has evolved chiefly by a process of general denudation. Fluvial erosion has cut below the ignimbrite rocks that make up most of the Susques Plateau uplands, into continental Tertiary claystone red beds. The distance and rate of retreat has varied and was more intensive in the lower Lapao basin, near the Pastos Chicos River, the local base level. Ongoing erosion in the lower basin has created a fantastic array of mesas, buttes, spires and pinnacles, some of which reach an elevation of 3700 m a.s.l. and stand 200 m above the general level of the rivers. Preservation of many of these features seems to have been due to the protection provided by resistant ignimbrites as cap rock to many of the erosion plateau remnants. Deep canyons excavated in ignimbrite rocks are common fluvial features in many tributaries like Susques, Portrero Chico and Coranzuli (Figure 2a). The ignimbrites escarpment erodes and retreats by a process of sapping, landslide and block fall off the cliff, with a resulting rapid removal of the unprotected lower Tertiary claystone. Fluvial erosion has reached the Paleozoic basement rocks in only a few places, particularly in the Los Angostos de las Burras in the uplift of Los Cobres range (Figure 1d). The outcomes of this process have been the production of deeply incised quebradas, where frequent damming events occur by lateral landslides such as those detected in the Coranzulí River basin. Other base level controls are associated with fan progradation at the end of de Las Burras River in Salinas Grandes or with Quaternary tectonics movements in the case of the Susques graven (Figure 1d). The excavation of the major valleys of Susques probably began during the late Pliocene, after the deposition of conglomerates of the Taire piedemont. Regional down cutting continued intermittently throughout the Quaternary cycle and is recorded by the terraces preserved along the large streams.

In the lower Lapao basin, the stream occupies a narrow, U-shaped valley excavated into soft Tertiary claystone and highly resistant ignimbrite rocks (Figure 3). Relief on the Quebrada de Lapao slopes is over 120 m along upper reaches, and is commonly 200 m or more at the confluence with the Pastos Chicos River. Ancient rotational slumps are located in the quebrada and occur when a slump block, composed of ignimbrite, slided along a concave-upward slip surface with rotation about an axis parallel to the slope. Fluvial terraces are located along the lower basin. These are elongated landforms that flank the sides of gravel floodplains and fluvial valleys. Four terracing episodes have been defined. Terrace I is aggradational (i.e., fill terraces; Bull 1991) and the maximum height of terrace treads along the Quebrada de Lapao is quite consistent at about 4 m above bankfull level. Several prominent, well-preserved fill terrace treads at this level were mapped between the Puesto Demetria (located in the headwaters of Quebrada de Lapao) and the confluence with Ojo de Agua, a lateral stream (Figure 2b). Downstream from this point, Terrace I was eroded. Terrace I comprises peats and diatomaceous silts interbedded with green clays and reddish colluvial silts dated between ~9400 and ~7500 ^{14}C yrs B.P. On the surface of Terrace I, many archaeological materials have been recorded along the headwaters of the Quebrada de Lapao that could be associated to Early and Mid-Holocene deposits in archaeological sites of the region (Yaccobacio and Morales 2005). Terrace II, III and IV are aggradational and unpaired. They are 3, 2 and 1 m above bankfull level respectively.

Sediments and paleosoils

Twelve litho-facies in the aggradational fluvial succession were identified and described (Table 2). These were grouped into six facies; channel sediments (Facies 1); crevasse and channel splay deposits (Facies 2); fine-grained overbank deposits (Facies 3); organic and diatom overbank sediments (Facies 4); colluvial wedge sediments (Facies 5), d) eolian sand sediments (Facies 6) and ephemeral braidplain deposits (Facies 7). The proportion of channel deposits to overbank deposits varies, but the latter are usually thicker due to their accumulation rates.

Facies 1 is formed by clast-supported, massive to crudely bedded fine gravel and coarse sands (litho-facies Gm, Gt, Sm) that appear in thin (10-50 cm thick) beds. Occasionally, very thin (1-3 cm) and discontinuous massive mud drapes (Fm) appear. It comprises narrow channel bodies with erosional surfaces. The sediments are interpreted as the product of bar accumulation by flood events.

Facies 2 is formed by massive coarse sand and fine gravel lenses (litho-facies Sm) that appear in thin layers (10 to 15 cm). These facies forms a tabular body that can be traced for tens of meters. The deposits spread laterally into FA 3 or FA 4 overbank fines (Figure 4). It is interpreted as crevasse splay and channel deposits.

Facies 3 is composed by a tabular body made up by thin (5 to 20 cm) lens shaped layers of light red, fine-grained, massive, bioturbated (burrows and roots) sediments (litho-facies Fr). It can also be horizontally laminated (litho-facies Fl). Thin lens shaped layers of massive and bioturbated diatom (litho-facies Dr) or horizontal laminated diatom (litho-facies D) also appear. Individual layers can be traced laterally for up to 5 m. Facies 3 accumulated by mud and diatom sediment suspension fallout in temporary small and shallow water bodies situated in the overbank. Soil features reveal a vegetated floodplain area.

Facies 4 is dominated by very fine lamination and massive white diatomite (litho-facie D, Dm, Figure 6a), bioturbated light red muds (litho-facies Fr), and organic paleosoils composed by well-preserved plant epidermis and vascular plant bundles (litho-facies C, Figure 6b). These deposits occur in a tabular body 30 cm thick that is traceable for hundreds of meters. Facies 4 is interpreted as organic soil developed in swamp sands along the overbank.

Facies 5 is composed by massive blocks and mud sediments in wedge bodies. This unit corresponds to colluvial wedges shed from adjacent Tertiary claystone and ignimbrite hillslopes.

Facies 6 is formed by very low angle, cross-laminated, well-sorted medium and fine sands (litho-facies Sl). In only occurs in late Holocene deposits associated with dunes and sand sheet landforms.

Facies 7 includes massive, horizontally crude, clast-supported stratified gravel with thin and discontinuous mud drapes. The dominance of massive beds suggests rapid sedimentation of flash flood events.

Allostratigraphy

The sediments of fluvial Terrace I have a three-stage process of aggradation that have produced staked allounits (units A1, A2, A3 and B). The process of vertical staking usually eliminates upper facies in older alluvial sediments. Units C and D correspond to alluvial fill of late Holocene Terraces II and III (Figure 7).

In particular, in the LP5 profile (Figure 8) the allounit A is subdivided in three subunits A1, A2 and A3 separated by reactivation surfaces. The lowermost units A1 and A2 are up to 1.5 m thick, which are presented as outcrops that extend over hundreds of meters. These outcrops are dominated by black and white organic materials and diatom muds (Facies 4), with minor development of overbank mud (Facies 3) and channel sediments (Facies 1 and 2). The organic material contains uncarbonized, well-preserved plant epidermis and vascular plant bundles of 1 to 2 mm in length. These organic layers represent backswamps with organic soils development. The samples obtained from these units were dated at 9380 ± 110 (unit A1 in LP5), at 8560 ± 90 and 8380 ± 100 yr B.P. (unit A2 in LP5) (Table 1). Unit A1 deposits were also dated in other Terrace I sedimentary profiles of Quebrada de Lapao, confirming their Early Holocene chronology at 8980 ± 100 yr B.P. in LP11 and at 8230 ± 40 yr B.P. in LP7 (Figure 8).

Figure 6. Field aspect of some layers of the Lapao fluvial deposits: a) thinly laminated diatom (litho-facies D); b) thinly laminated organic sediments (litho-facies C) with massive sand (Sm); c) sediments from floodplain tractive currents.

Unit A2 is composed by channel deposits overlain by the thick overbank diatom and organic paleosoils of Facies 4, and was dated at 7770 ± 80 in LP5. Other dates on peat deposits in LP6 (Figure 8) -7750 ± 40 yr B.P.- and LP1 -7550 ± 90 yr B.P.- confirmed the Mid-Holocene age of A2 unit.

Unit A3 is composed by channel, gravel-sand sediments (Facies 1). These sediments are mainly stratified and are composed by fine gravel and medium to coarse sands, in narrow channel bodies with little scouring at its base. These facies presents light colors and no organic matter.

Only unit A1 is recognizable at Puesto Demetria in the headwaters of the Quebrada de Lapao (LP1, LP7, LP6 section, Figures 2b and 7a). This implies that the paleo-wetland of A1 unit was at least 1 km long, much larger than the wetland at historical times, which is adjacent to the Puesto Demetria spring (LP2 site, Figure 2b). In a few places at Cardonal canyon (Figure 2b) the organic facies of unit A1 and A2 (QC2 profile, Figure 2b) were recognized. Unit A2 covers an area smaller than the A1 (from LP5 to LP 12 and in QC2, Figure 2b), possibly because it was eroded by an A3 event and late Holocene erosion. Unit A3 has a wide distribution and is located in LP6, LP15 and in Cardonal stream (QC1 profile, Figure 2b).

Unit B comprises extra fluvial sediment additions dominated by colluvial processes (Facies 6). A transverse concave profile is characteristic of this unit, as observed in LP6 and LP5. Colluvial materials from adjacent hillslopes and older terrace risers occlude the fluvial deposits, introducing changes not related to the hydrologic processes of fluvial sediments.

Unit C corresponds to fluvial Terrace II. This terrace shows down-valley continuity and is paired across the valley (from LP2 to LP17, Figures 2b and 7b). This unit is composed by light yellow, very coarse sands and fine gravels with massive to crude horizontal bedding with thin mud drapes (Facies 7). This facies may have been formed under high flow regimes or flash flood events. Fine-grained sediments occur very infrequently. However, they represent the final stage of vertical accreted fining-upward sediments, starting as Sh facies, and ending as white bioturbated or laminated diatom muds (litho-facies Dr, D) with little plant epidermis, thinly laminated sediments or as mineral, poorly developed paleosoils (Fr). A date obtained from LP4 (3680 ± 150 yrs B.P.) confirm a Late Holocene chronology for these deposits.

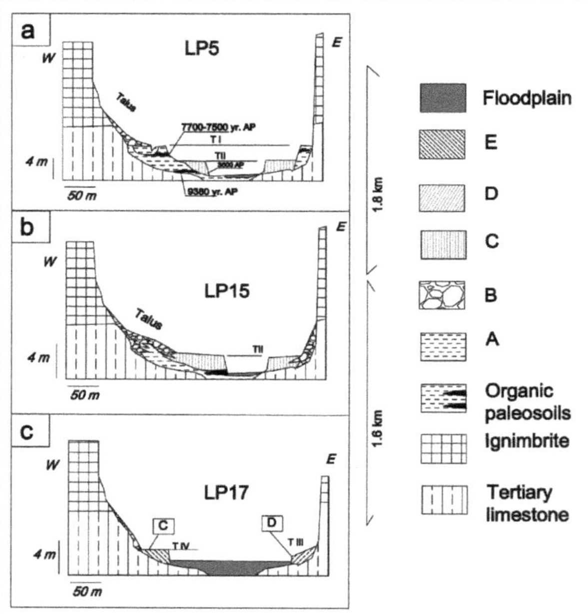

Figure 7: Cross-sections of Quebrada de Lapao; a) geomorphology in LP5 site; b) geomorphology in LP15 site; c) geomorphology in LP17 site.

Allounits D and E constitute fluvial Terraces III and IV. They are located in the proximity of the confluence of the Quebrada de Lapao with the Pastos Chicos- De Las Burras system (Figure 7c), and are composed by very coarse sands and gravels (litho-facies Sh, Gh) with no organic material nor traces of soil development. Thin lenses of organic material dated in Late Holocene and historic times (1120 to 250 yrs B.P.) have only been detected at LP2 (near Puesto Demetria spring), in the Cardonal-Quebrada de Lapao confluence and in El Tío paleospring (ET1 section, Figure 2b) (see radiocarbon dates in Table 1). The Late Holocene organic material is mainly composed by few carbonized, badly preserved plant epidermis remains of 1 mm in length in a sand-rich matrix.

Current alluvial sediments (allounit D) are composed by very coarse sands and gravels with massive and horizontal bedding (Facies 7, Figure 6c). These sediments are associated to a braided multichannel pattern and to ephemeral floods (litho-facies Gh,Sh). Near the confluence of Pastos Chicos and De Las Burras rivers (LP17 section) the modern alluvial sediments form a terminal alluvial fan with sand dunes (Facies 6) produced by western wind. Other aeolian deposits (Facies 3) and landforms cover fluvial Terraces I and II at the lower section of the Cardonal canyon (QC1 section) and the Ojo de Agua canyon (ODA2 section, Figure 2b).

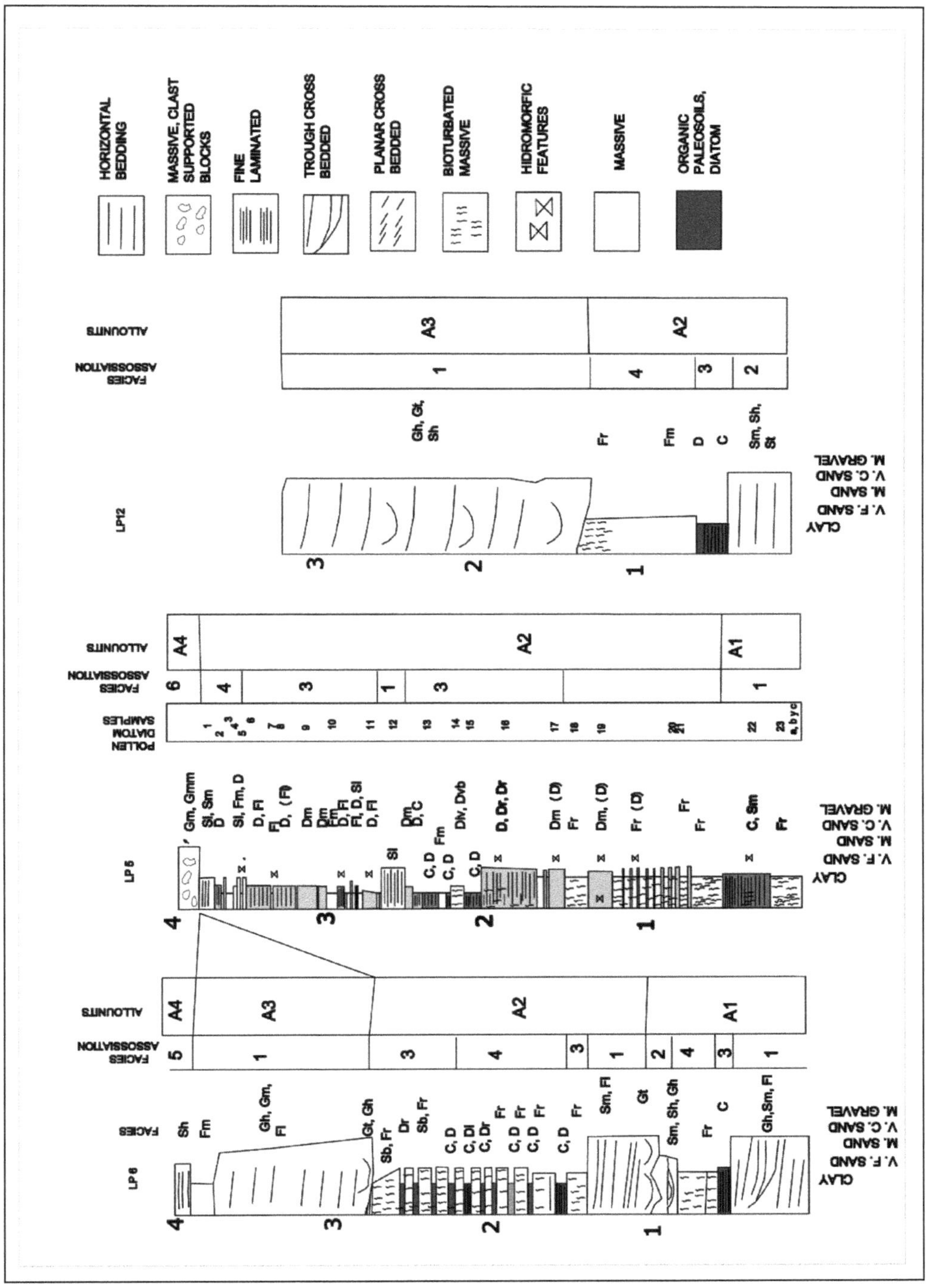

Figure 8. Sedimentary profiles LP6, LP5 and LP12 located along the outcrops of the terrace deposits exposed by channel incision along the riverbanks of the Lapao canyon. LP5 was the sequence studied by multi-proxy analysis.

Facies code (Miall 1996)	Description	Sedimentary structures	Interpretation
Gh	Gravel, clast supported	Horizontal bedding common clast imbrication	Channel fill, bar
Gmm	Gravel, matrix supported	Massive	Debris flows
Sh	Sand, very fine to very coarse	Horizontal bedding	Channel fill, bar
Sb	Sand, very fine to very coarse	Massive, bioturbated	Mineral palosoils in overbank
Sm	Sand, very fine to very coarse	Massive	Sand flows form lateral valleydunes
Sl	Sand, fine to medium	Laminar	Sand dunes
Sol	Sand, very fine to coarse with organic debris	Laminar	Overbank
Fol	Mud, clay, organic	Planar laminations	Overbank and backswamp
Fdl	Silt, diatomitic	Planar laminations	Overbank and backswamp
Fdb	Silt, diatomitic	Massive, bioturbated	Overbank and backswamp
Fdm	Silt, diatomitic	Massive	Overbank and backswamp
Fl	Mud, clay	Planar laminations	Ephemeral backswamp

Table 2. Summary of litho-facies described during the study of Lapao deposits.

Facies association	Dominant or characteristic lithofacies	Description	Two dimensional strata shape (Friend *et al.* 1979, Friend 1983) and paleocurrent direction	Depositional environment
1a	Sh, St, (Gh, Gt)	Horizontally stratified, planarcrossbeded, trough bedded coarse sands and stratified gravel lens	Channel, paleocurrent parallel to Pastos Chicos direction	Sandy braided channels
1b	Gh, Gt,	Horizontally stratified, planar crossbeded, trough bedded medium to fine gravels and coarse to medium sands	Channel, paleocurrent parallel to Lapao direction	Gravel braided channels
2a	Fl, Sl, (Fdl)	Laminated bedded fine to very fine sand and thin laminated mud. Reddish gray. Wedge layer of laminated diatomitic mud	Sheet and wedge	Playa lake system, ephemeral lagoons
2b	Fol, Ddl, Fdb (Sd)	Planar laminated organic black sediments interbbeded light gray laminated diatomitic muds and silts. Thin wedge layer of light red fine sands	Sheet	Backswamp with peat in low energy floodplain
3	Sm, Sl	Massive and Horizontally stratified sand	Wedge in valley transversal section	Colluvial distal deposits
4	Sl	Massive fine to medium sand, light gray	Sheet	Eolic sheet sands

Table 3. Description of dominant litho-facies and depositional environment of the FAs preserved in the exposed terraces of the Lapao system.

Pollen

Thirteen samples were processed for this study, but only seven that met pollen abundance requirements were included in the analysis. In general terms seven families (Poaceae, Asteraceae, Urticaceae, Cyperaceae, Chenopodiacea/Amarantaceae, Ephedraceae and Mimosaceae) and two species (*Myriophyllum quitense*, *Fabiana* sp.) were identified during the analysis. A few grains of Pteridophyta were also counted. Early Holocene samples were clearly dominated by typical taxa of an herbaceous steppe (*i.e.* Poaceae). Towards the end of this period (*ca.* 8600 yrs B.P.) an increase in a local moisture marker (*Myriophyllum quitense*) was detected. Mid-Holocene samples presented higher pollen counts and a greater taxonomic diversity, showing an increase of Asteraceae and Chenopodiacea / Amarantaceae towards the end of the record and including low quantities of Cyperaceae, a local moisture indicator. It is also worth noting that starting at 7700 yrs B.P. a strong reduction in pollen productivity was observed.

Based on the changes in pollen sample composition, two major environmental phases were identified in the LP 5 sequence (Figure 9). The lower section of the record - between ~9300 and ~8200 yrs B.P.- shows clear evidences of a moist environment that could be characterized as an herbaceous steppe, with an event of input of *Myriophyllum quitense* -an aquatic and/or wetland species- which is common today in stagnant waters or low energy systems.

These environmental characteristics suggest a downslope migration of high grassland vegetation assemblages to lower altitudes, almost 500 m below the present day boundary for this vegetation arrangement. The presence of *Myriophyllum quitense ca.* 8600 yrs B.P. as an indicator of local moisture is consistent with other diatom and sedimentology evidence that suggest the formation of a shallow pond.

The upper section of the record, between ~8600 and ~7500 yrs B.P., shows a compositional change, although Poaceae pollen still dominates. This change is observed by an increase in the frequency and diversity of the shrub steppe pollen types, mainly represented by the Asteraceae and Chenopodiacea / Amarantaceae families. This is accompanied by the presence of Cyperaceae that suggests shallow water conditions, at least during two events dated at ca. 7800 and 7600 yrs B.P. Integrating these elements, the pollen record of the upper section of LP5 could be reflecting a local wetland environment surrounded by drier conditions related to a regional drought process.

Diatoms

In the eleven samples analyzed from the LP 5 record 49 species of diatoms were identified. The most abundant taxa were the Fragilariaceae group (*i.e.* geni *Staurosira*, *Staurosirella*, *Pseudoestauroira*, *Fragilaria*, *Tabularia* and *Ulnaria*) which dominate the flora of eight samples and co-dominate four of them. In particular, the Amphora genus (i.e. *A. tucumana* and *A. veneta*) dominates three samples and co-dominates two, and other species like *Cocconeis placentula* and *Cymbella cistula* were also highly frequent, dominating in three samples in each case. Correlation (Pearson r) analysis constrained by sample depth was applied to define diatom flora zones (Morales 2011).

Zone A is dominated by littoral species like *Cocconeis placentula* and *Pseudostauroisira brevistriata*. This zone also present some benthic species like *Cymbela cistula* and other corresponding to Navicula and Nitzschia genus. Zone B is the largest segment of the record in terms of flora similarity. The samples included in it were clearly dominated by littoral forms due to the rise in the frequency of *Staurosira construens* var. *venter*, followed by *Staurosira construen* var. *subsalina*. Finally, samples from Zone C were largely dominated by benthic forms like *Amphora veneta* and *Amphora tucumana* (Figure 10).

In general terms, the study of diatoms from LP5 record suggests the presence of a changing wetland environment during the 9400-7300 yrs. B.P. span. This wetland alternates between the presence of: a) a profusely vegetated *Vega*-type landscape, and b) a shallow and irregularly available water body during the second half of Zone B (i.e. ~8600-8100 B.P.). This wetland also seems to have showed variable conditions in terms of salinity (Figure 11) and pH. However, in general terms it seems to have been slightly brackish and alkaline.

In particular, the period represented by Zone A (i.e. ~9000-8800 yrs. B.P.), the abundance of littoral and epiphytic diatoms suggests a *Vega*-type environment with a good vegetation cover. Diatoms of Zone B seem to show a gradual onset of moister conditions and suggest that the water body system reached its maximum humidity during the 8600-8100 yrs B.P. span, which was briefly interrupted by an event of drought near 8400 yrs B.P. Finally, Zone C shows a substantial change in this environment in less than 400 years. The previous conditions turn again into a more unstable environment, probably a marsh-like environment, copiously vegetated but progressively arid, as indicated by humidity and salinity indexes (Figure 11). These conditions continued until the abrupt interruption of the sedimentary sequence at ca. 7300 yrs B.P.

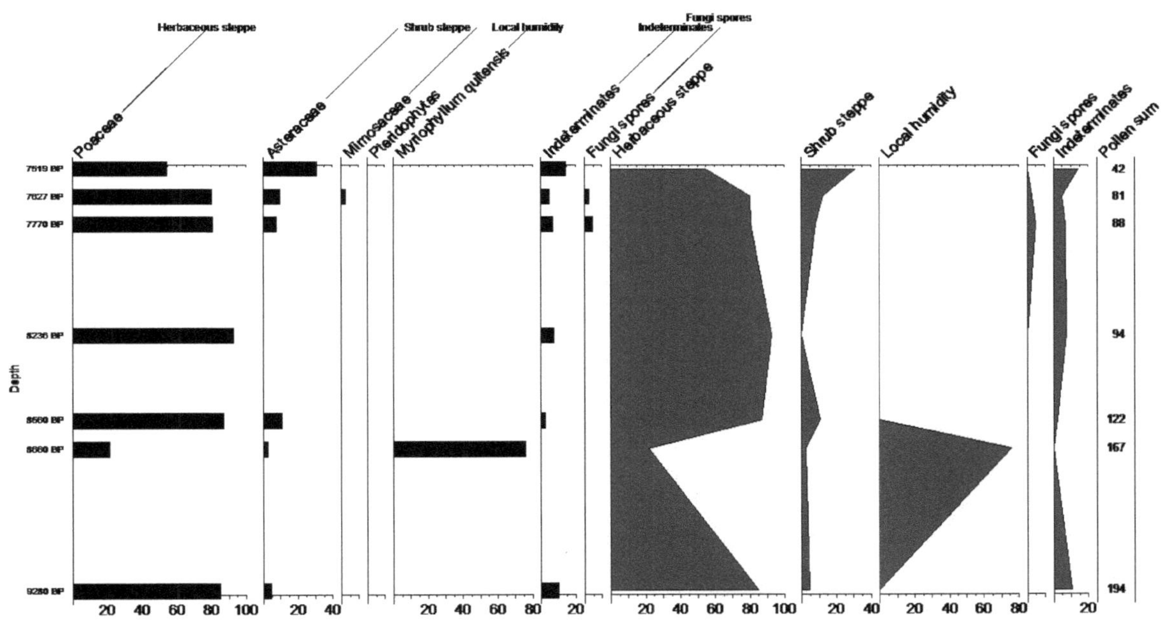

Figure 9. Pollen spectra, representing most frequent types (left) and vegetation assemblages (right).

DISCUSSION

The results obtained from the multi-proxy analysis carried out in this paper show that the paleoenvironmental evolution of the Quebrada de Lapao was complex, including changes in paleohydrology, vegetation assemblages and coverage. In general terms, a floodplain wetland environment developed in the headwaters of the quebrada between 9400 and 7500 yrs B.P. Firstly, the organic material from allounit A1, corresponding to the Early Holocene, evidence the presence of a large organic wetland that extended to the lower sector of the quebrada, near the confluence with Ojo de Agua stream (Figure 2b) and reached the 3635 m a.s.l. contour. Before ca. 7700 yrs B.P., the Early Holocene wetland was eroded and covered by a new alluvial event (allounit A2) that formed a new, more unstable and brackish, wetland environment between ca. 7700 and 7550 yrs B.P. Pollen and diatom analysis show that the most humid conditions of the sequence are bracketed between 8600 and 8000 yrs B.P., interpreted from the great abundance of littoral diatom species during this period and the occurrence of *Myriophyllum quitense* pollen around 8600 yrs B.P. The ecological requirements of the gastropod *Biomphalaria peregrine* found in these units is also indicative of the presence of

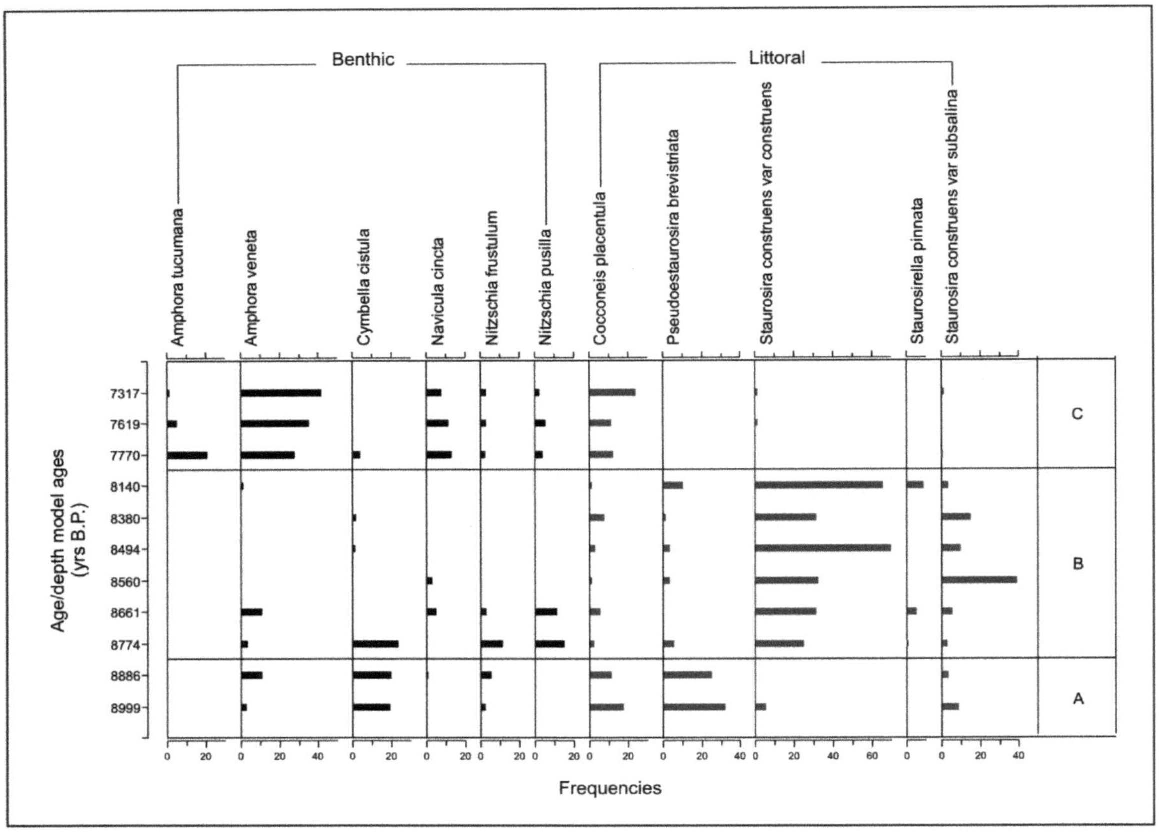

Figure 10. Most frequent diatom species registered in LP5 record, divided by life form (i.e. Benthic and Littoral) and ordered by the age/depth model. To the right, diatom flora zones divisions are indicated.

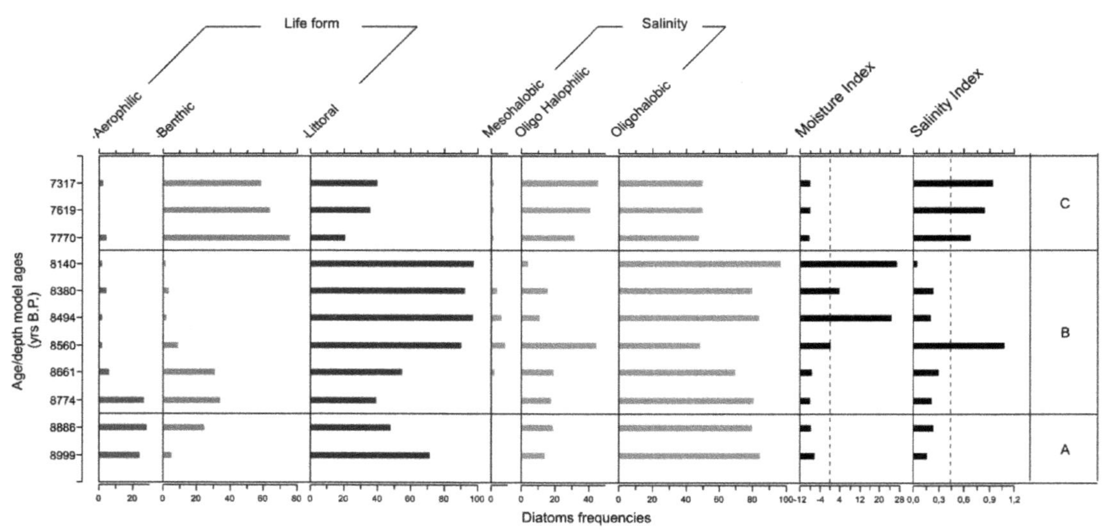

Figure 11. Summary of life form and salinity affinities of the diatom record of LP5 and the related Moisture and Salinity indexes calculated for the environment. The dotted lines represent the mean value of the series for each index. Diatom flora zones are indicated on the right margin.

a slightly brackish and alkaline water body with a broad littoral area (Yacobaccio and Morales 2005).

It should be noted that allounits A1 and A2 could be associated to auto-cyclical changes of the fluvial system due to its own migration and not necessarily due to significant external changes. However, the study of a sub-regional paleoenvironmental record near Quebrada de Lapao (*i.e.* Pastos Chicos) (Tchilinguirian *et al.* 2012) has yielded similar results, consistent with an almost 1000-year humid period during the beginnings of the Mid-Holocene, which suggests regional climatic conditions favoring the development of this kind of wetland environments.

In order to explain the presence of wetlands in the Quebrada de Lapao floodplain during the Early Holocene and the onset of the Mid-Holocene, several facts must be reviewed. On the one hand, the formation of organic matter deposits and organic paleosoils (OM and Facies 4) is closely related to the growth of littoral vegetation, which is clearly conditioned by moisture and temperature. For instance, low temperatures associated to near surface water tables generate anoxic conditions that improve the preservation of OM. The formation of OM deposits is also related to a positive imbalance between primary productivity and littoral vegetation biomass accumulation rate. At the same time, this imbalance must be greater than the sedimentary floodplain accumulation and fluvial tawleg incision rates. Both processes, the formation of OM deposits and organic paleosoils, are controlled by diverse factors such as the presence of: 1) constantly available water with minor flow variations that generates a geomorphic stability in floodplain environment, 2) near surface or emerged water-tables with 3) minor variations in water-table depth, and 4) low salinity and alkalinity values. The three last conditions associated with vadose water levels are important to develop the best organic soil forming and preserving conditions. Others studies suggest that the presence of paleowetlands in arid areas could be interpreted as: a) formed under moister climates (Nester *et al.* 2007; Rech *et al.* 2002, 2003) or b) during periods of precipitations with low intensity (*i.e.* non-stormy type precipitations), regularly distributed along the year (Servant and Servant-Vildary 2003).

Today, these elements related to wetland formation conditions are absent in the Quebrada de Lapao. Currently, erosive floodplain processes (incision) and ephemeral, deep and unstable water-table levels are the dominant features in the studied area. The presence of dunes in several places in Lapao system also prevents the development of organic soils. The characteristics observed in the modern and sub-modern vegas in the quebrada evidence poor OM content, clearly showing the differences between the formation conditions - and consequently, the environmental ones- of the Early and Mid-Holocene with those of modern times. Moreover, current OM sedimentation environments are placed in the southern slope of the Taire mountain range (24° 1' 24.01" S, 66° 34' 25.91" O) above 4300 m a.s.l. In the Andes highlands, above 4000 m a.s.l., wetlands are abundant because they are supplied by perennial springs with low fluctuations in their flow. These springs are charged by orographic rains (mostly restricted to austral summer) and local daily freeze-thaw melt-water. This hydrologic setting determines a low energy discharge and low seasonal variation in wetland streams, also reducing the erosion and inorganic deposition processes in the floodplain environment. These geomorphic conditions have been mentioned as a necessary feature and a primary condition to control organic soil development and/or paludization processes (Hall 1990).

Considering what has been mentioned above, a possible explanation for the conditions observed in the headwaters of the Quebrada de Lapao is related to the downslope expansion of the climatic and soil formation conditions observed today at the high-altitude Taire mountain range during the 9400 - 7300 yrs B.P. span, allowing a ponding and paludization process in the Lapao floodplain environment. The palynological record at Lapao also supports this idea, as evidenced by the rise in Poaceae pollen frequency suggesting a grassland steppe composition. Similar downslope migrations of vegetation assemblages of "pajonal" were also mentioned for other localities by Markgraf (1985) and Fernández *et al.* (1991) in the Dry Puna of Jujuy, and by Latorre *et al.* (2006) and Maldonado *et al.* (2005) in the Chilean slope.

The fluvial aggradation observed in A1 and A2 units seem to have occurred during a period of high water-table levels that allowed the formation of a stable wetland environment, including the development of a shallow water body (Morales 2011) framed in a grassland steppe. In turn, the sedimentation processes represented by units A3, D and E occurred with the onset of unstable water-table levels that generated an unstable wetland environment, framed in shrub steppe vegetation. These different paleoenviromental conditions during the fluvial aggradation processes could be explained not only by climate change, but also by base level changes. As previously mentioned, the base level of Lapao system is the Pastos Chicos - De las Burras River. This fluvial system crosses Los Cobres range Quaternary fault structures and has evidence of damming by its tributaries (Coranzuli River, Fig 2a). Other base level change is related with the evolution of De las Burras terminal fan progradation in the closed basin of Salinas Grandes. Environmental evolution in Lapao, which is one of these tributaries, is complex and partially depends on neotectonics and local geomorphologic processes. Therefore, we suggest that there is a complex array of factors which drove Quebrada de Lapao fluvial responses to environmental change, because climate change in Lapao catchment alone cannot account for all the modifications observed in the evolution of the system.

CONCLUDING REMARKS

In sum, the fluvial terraces in Quebrada de Lapao are an expression of contrasting erosional and depositional processes, played out at two watershed scales: Quebrada de Lapao (local) and Pastos Chicos - De Las Burras basin (sub-regional). The studied terraces were formed by the conjoining of several variables like hydrology of the spring located at the headwaters, changes in water-table levels and hillslope sediment flux regulated by changes in vegetation. The forces external to the Lapao system (*i.e.* base level of De Las Burras, neotectonics and regional climate) establishes unstable erosion and depositional conditions. Climate controls refer not only to mean elements like the precipitation intensity, seasonality, and inter-annual variability, but also to spring discharge and fluvial sediment fluxes. Consequently, these landforms not only yield high spatial resolution paleohydrologic data, but also provide information about the base level history of the whole basin.

The interruption of the wetland conditions around 7300 yrs B.P. in the Quebrada de Lapao is coherent with processes recorded at Pastos Chicos, located ~40 km to the south, although at Quebrada de Lapao they begin more abruptly and slightly earlier (Morales 2011; Tchilinguirian *et al.* 2012). We suggest that these discrepancies are due to the different catchment area of both hydrological systems, at least in part. The Pastos Chicos catchment, covering almost 1000 km^2, constitutes one of the largest and more stable systems in the western Dry Puna. For this reason, a buffer effect is likely to be reflected both in the chronology and the intensity of the signals. This issue has a major relevance for archaeological studies in the area. Future archaeological models must stratify the Puna landscape in terms of the catchment area, because of the differential stability and availability of primary productivity expected (and consequently prey availability) for each type of hydrological system during Mid-Holocene. From this perspective, not only the particular environmental conditions at different locations are important as scenarios for human adaptation. Their similarities and differences show an enormous potential in terms of archaeological modeling from a paleoecological perspective.

ACKNOWLEDGEMENTS

We want to thank Hugo D. Yacobaccio, Daniel E. Olivera and Liliana C. Lupo for their advice and comments on one early draft of this paper. We also thank two anonymous reviewers whose comments clearly have improved the final manuscript of this paper. We acknowledge the support of CONICET, ANPCyT and UBACyT who funded the research projects that framed the paleoenvironmental studies in Quebrada de Lapao.

REFERENCES

BATTARBEE, R.W.
1986. Diatom analysis. In *Handbook of Holocene Palaeoecology and Palaeohydrology*, edited by B.E. Berglund, pp. 527-570. J. Wiley and Sons, New York.

BENNETT, K.
1994. Confidence intervals for age estimates and deposition times in late-Quaternary sediment sequences. *The Holocene* 4(4): 337-348.

BIANCHI, A.R. AND C.E. YAÑEZ
1992. *Las precipitaciones en el noroeste argentino*. Instituto Nacional de Tecnología Agropecuaria (INTA).

BRAUN WILKE, R., L. PICCHETTI AND B. VILLAFAÑE
1999. *Pasturas Montanas de Jujuy*. Gofica Editora, Salta.

CABRERA, A.L.
1976. *Regiones fitogeográficas argentinas*. Series Regiones fitogeográficas argentinas, ACME, Buenos Aires.

DE WOLF, H.
1982. Method of coding of ecological data from diatoms for computer utilization. Meded / Netherlands. *Rijks Geol. Dinst*. 36(2): 95-98.

FAEGRI, K. AND J. IVERSEN
1989. *Textbook of Pollen Analysis*, edited by K. Faegri, P.E. Kaland and K. Krywinski. John Wiley & Sons, Chichester. 4th edition.

FAIRBRIDGE, R.W.
1968. *The Encyclopedia of Geomorphology*. Reinhold, New York.

FERNÁNDEZ, J., V. MARKGRAF, H.O. PANARELLO, M. ALBERO, F.E. ANGIOLINI, S. VALENCIO AND M. ARRIAGA
1991. Late Pleistocene/Early Holocene environments and climates, fauna, and human occupation in the Argentine Altiplano. *Geoarchaeology* 6(3): 251-272.

FRIEND, P.
1983. Towards the field classification of alluvial architecture or sequence. In *Modern and Ancient Fluvial Systems,* edited by J.D. Collinson and J. Lewin, pp. 345-354. Blackwell Publishing Ltd., Oxford.

FRIEND, P., M. SLATER AND R. WILLIAMS
1979. Vertical and lateral building of river sandstone bodies, Ebro Basin, Spain. *Journal of the Geological Society* 136: 39-46.

GARREAUD, R.D. AND M. FALVEY
2009. The coastal winds off western subtropical South America in future climate scenarios. *International Journal of Climatology* 29(4): 543-554.

GARREAUD, R., M. VUILLE AND A.C. CLEMENT
2003. The climate of the Altiplano: observed current conditions and mechanisms of past changes. *Palaeogeography, Palaeoclimatology, Palaeoecology* 194(1-3): 5-22.

GERMAIN, H.
1981. *Flore des Diatomées: Eaux Douces et Saumâtres*. Société Nouvelle des Éditions Boubée, Paris.

GRANA, L. AND M. MORALES
2005. Primeros resultados paleoambientales del análisis de diatomeas fósiles del Holoceno Medio y Tardío de la cuenca del Rio Miriguaca, Antofagasta de la Sierra, Puna Catamarqueña. In *Entre Pasados y Presentes. Trabajos de las VI Jornadas de Jóvenes Investigadores en Ciencias Antropológicas*, pp. 392-409. Instituto Nacional de Antropología y Pensamiento Latinoamericano, Buenos Aires.

GRIMM, E.
1987. CONISS: a FORTRAN 77 program for stratigraphically constrained cluster analysis by the method of incremental sum of squares. *Computers and Geosciences* 13: 13-35.

2004. *Tilia y TGView 2.0. 2*. Illinois State Museum, Springfield.

HALL, S.A.
1990. Channel trenching and climatic change in the southern US Great Plains. *Geology* 18(4): 342-345.

HEUSSER, C.J.
1971. *Pollen and Spores of Chile: Modern Types of the Pteridophyta, Gymnospermae and Angiospermae*. University of Arizona Press, Tucson.

KRAMMER, K. AND H. LANGE-BERTALOT
1991-6. *Bacillariophyceae*. Series Bacillariophyceae. Gustav Fischer Verlag, Stuttgart.

LANGE-BERTALOT, H.
2001. *Vol. 2: Navicula sensu stricto, 10 Genera Separated from Navicula sensu stricto, Frustulia*. Ruggell, Gantner.

LATORRE, C., J.L. BETANCOURT AND M.T.K. ARROYO
2006. Late Quaternary vegetation and climate history of a perennial river canyon in the Río Salado basin (22° S) of Northern Chile. *Quaternary Research* 65(3): 450-466.

LATORRE, C., J.L. BETANCOURT, K.A. RYLANDER AND J. QUADE
2002. Vegetation invasions into absolute desert: a 45,000 yr rodent midden record from the Calama–Salar de Atacama basins, northern Chile (lat 22° - 24° S). *Geological Society of America Bulletin* 114(3): 349-366.

LATORRE, C., J.L. BETANCOURT, K.A. RYLANDER, J. QUADE AND O. MATTHEI
2003. A vegetation history from the arid prepuna of northern Chile (22–23 S) over the last 13, 500 years. *Palaeogeography, Palaeoclimatology, Palaeo-ecology* 194(1-3): 223-246.

LOWE, R.L.
1974. *Environmental Requirements and Pollution Tolerance of Freshwater Diatoms*. United States Environmental Protection Agency.

LUPO, L.C.
1998. *Estudio sobre la lluvia polínica actual y la evolución del paisaje a traves de la vegetación durante el Holoceno en la cuenca del río Yavi, Borde Oriental de la Puna, Noroeste argentino*. Unpublished PhD Dissertation, Fakultat fur Geschichts-und Geowissenschaften Universität Bamberg.

MALDONADO, A., J.L. BETANCOURT, C. LATORRE AND C. VILLAGRAN
2005. Pollen analyses from a 50,000-yr rodent midden series in the southern Atacama Desert (25° 30' S). *Journal of Quaternary Science* 20(5): 493-507.

MARENGO, J.A. AND J.C. ROGERS
2001. Polar air outbreaks in the Americas: Assessments and impacts during modern and past climates. In *Interhemispheric Climate Linkages* edited by V. Markgraf, pp. 31-51. Academic Press, San Diego.

MARKGRAF, V.
1985. Paleoenvironmental history of the last 10,000 years in northwestern Argentina. *Zentralblatt für Geologie und Palaontologie* 11: 1739-1748.

MARKGRAF, V. AND H. D' ANTONI
1978. *Pollen Flora of Argentina. Modern Spore and Pollen Types of Pteridophyta, Gymnospermae, and Angiospermae*. University of Arizona Press, Tucson.

MIALL, A.D.
1982. *Analysis of Fluvial Depositional Systems*, Series Analysis of fluvial depositional systems. American Association of Petroleum Geologists, Tulsa.

1996. *The Geology of Fluvial Deposits*. Springer, Berlin.

MORALES, M.R.
2004. *Casi Invisibles. Diatomeas, ambientes locales y estrategias cazadoras-recolectoras durante la primera mitad del Holoceno en la Puna Desértica.* Unpublished Graduate Thesis, Facultad de Filosofía y Letras, Universidad de Buenos Aires.

2011. *Arqueología Ambiental del Holoceno Temprano y Medio en la Puna Seca Argentina. Modelos Paleoambientales Multi-escalas y sus Implicancias para la Arqueología de Cazadores-Recolectores*. British Archaeological Reports (BAR) S2295, South American Archaeology Series, BAR Publishing, Oxford.

MORALES, M.R., R. BARBERENA, J. BELARDI, L. BORRERO, V. CORTEGOSO, V. DURÁN, A. GUERCI, R. GOÑI, A. GIL AND G. NEME
2009. Reviewing human-environment interactions in arid regions of southern South America during the past 3000 years. *Palaeogeography, Palaeo-climatology, Palaeoecology* 281(3-4): 283-295.

MORALES, M.R., L.C. LUPO, B. OXMAN AND C.M. CÁCERES
2008. Nuevos registros paleoambientales de diferente escala espacial del Holoceno Temprano y medio del Dto. De Susques (Jujuy). In *Libro de resumenes de las 1° Jornadas del Área Puneña de los Andes Centro-Sur*, Tucumán, pp. 107-108.

MORALES, M.R. AND K. SCHITTEK
2008. Primeros resultados paleoambientales del Holoceno medio en Alto Tocomar (Puna Salteña):

interpretación local e implicancias regionales. In *Libro de resúmenes de las 1° Jornadas del Área Puneña de los Andes Centro-Sur*, Tucumán, pp. 109-110.

NESTER, P.L., E. GAYO, C. LATORRE, T.E. JORDAN AND N. BLANCO
2007. Perennial stream discharge in the hyperarid Atacama Desert of northern Chile during the latest Pleistocene. *Proceedings of the National Academy of Sciences* 104(50): 19724-19729.

NULLO, F.
1988. *Descripción geológica de la Hoja 4 ab Susques, Provincia de Jujuy*. SEGEMAR Argentina.

OLIVERA, D.E.
1997. La importancia del recurso Camelidae en la Puna de Atacama entre los 10.000 y 500 años AP. *Estudios Atacameños* 14: 29-41.

OLIVERA, D.E., P. TCHILINGUIRIAN AND L. GRANA
2004. Paleoambiente y arqueología en la Puna Meridional argentina: archivos ambientales, escalas de análisis y registro arqueológico. *Relaciones-Sociedad Argentina de Antropología* XXIX: 229-248.

OXMAN, B.I.
2010. *Una perspectiva paleoecológica de las primeras ocupaciones de la Puna Seca Argentina: análisis polínico de perfiles naturales holocénicos ubicados en el Dto. de Susques, Provincia de Jujuy, Argentina*. Unpublished Graduate Thesis, Facultad de Filosofía y Letras, Universidad de Buenos Aires.

QUADE, J., J.A. RECH, J.L. BETANCOURT, C. LATORRE, B. QUADE, K.A. RYLANDER AND T. FISHER
2008. Paleowetlands and regional climate change in the central Atacama Desert, northern Chile. *Quaternary Research* 69(3): 343-360.

RECH, J.A., J.S. PIGATI, J. QUADE AND J.L. BETANCOURT
2003. Re-evaluation of mid-Holocene deposits at Quebrada Puripica, northern Chile. *Palaeogeography, Palaeoclimatology, Palaeoecology* 194(1-3): 207-222.

RECH, J.A., J. QUADE AND J.L. BETANCOURT
2002. Late Quaternary paleohydrology of the central Atacama Desert (lat 22°-24°S), Chile. *Bulletin of the Geological Society of America* 114: 334-338.

REIMER, P.J., M.G.L. BAILLIE, E. BARD, A. BAYLISS, J.W. BECK, P.G. BLACKWELL, C.B. RAMSEY, C.E. BUCK, G.S. BURR AND R.L. EDWARDS
2009. IntCal09 and Marine09 radiocarbon age calibration curves, 0–50,000 years cal BP. *Radiocarbon* 51: 1111-1150.

ROUND, F.E., R.M. CRAWFORD AND D.G. MANN
1990. *The Diatoms: Biology and Morphology of the Genera*. Cambridge University Press, Cambridge.

RUMRICH, U., H. LANGE-BERTALOT AND M. RUMRICH
2000. Diatoms of the Andes from Venezuela to Patagonia / Tierra del Fuego. *Iconographia Diatomologica* 9: 1-649.

RUTHSATZ, B. AND C.P. MOVIA
1975. *Relevamiento de las Estepas Andinas del Este de la Provincia de Jujuy*. FECIC.

SALATI, E., A. DALL'OLIO, E. MATSUI AND J.R. GAT
1979. Recycling of water in the Amazon basin: an isotopic study. *Water Resources Research* 15(5): 1250-1258.

SERVANT, M. AND S. SERVANT-VILDARY
2003. Holocene precipitation and atmospheric changes inferred from river paleowetlands in the Bolivian Andes. *Palaeogeography, Palaeoclimatology, Palaeoecology* 194(1-3): 187-206.

SQUEO, F.A., B.G. WARNER, R. ARAVENA AND D. ESPINOZA
2006. Bofedales: high altitude peatlands of the central Andes. *Revista Chilena de Historia Natural* 79: 245-255.

STOCKMARR, J.
1971. Tablets with spores used in absolute pollen analysis. *Pollen et Spores* 13: 615-621.

TCHILINGUIRIAN, P.
2009. *Paleoambientes holocenos en la Puna Austral, Provincia de Catamarca (27°S): implicancias geoarqueológicas*. Unpublished PhD Dissertation, Universidad de Buenos Aires.

TCHILINGUIRIAN, P., M.R. MORALES, B. OXMAN, L. LUPO, D. OLIVERA AND H. YACOBACCIO
2012. Early to Middle Holocene transition in the Pastos Chicos record, dry Puna of Argentina. *Quaternary International*. In press. Available online: 13 March 2012.

VAN DAM, H., A. MERTENS AND J. SINKELDAM
1994. A coded checklist and ecological indicator values of freshwater diatoms from the Netherlands. *Aquatic Ecology* 28(1): 117-133.

VOS, P.C. AND H. WOLF
1993. Diatoms as a tool for reconstructing sedimentary environments in coastal wetlands; methodological aspects. *Hydrobiologia* 269-270(1): 285-296.

VUILLE, M. AND F. KEIMIG
2004. Interannual variability of summertime convective cloudiness and precipitation in the central Andes derived from ISCCP-B3 data. *Journal of Climate* 17(17): 3334–3348.

YACOBACCIO, H.
1994. Biomasa animal y consumo en el Pleistoceno-Holoceno Surandino. *Arqueología* 4: 43-71.

YACOBACCIO, H.D., C.M. MADERO, M.P. MALMIERCA AND M.C. REIGADAS
1997-1998. Caza, domesticación y pastoreo de camélidos en la Puna Argentina. *Relaciones de la Sociedad Argentina de Antropología* XXII-XXIII: 389-418.

YACOBACCIO, H.D. AND M.R. MORALES
2005. Mid-Holocene environment and human occupation of the Puna (Susques, Argentina). *Quaternary International* 132(1): 5-14.

POLLEN ANALYSIS OF PASTOS CHICOS: PALEOENVIRONMENTAL AND ARCHEOLOGICAL IMPLICATIONS DURING THE HOLOCENE IN THE DRY PUNA OF ARGENTINA

Brenda I. Oxman[1] and Hugo D. Yacobaccio[1]

[1] CONICET - Instituto de Arqueología, Facultad de Filosofía y Letras, Universidad de Buenos Aires

Abstract

Historically, the models accounting the human colonization of the Puna were based on assumptions and broad generalizations about the environmental context in which human groups had developed. In this sense, the main objective of this paper is to characterize and precise the environmental scenario of the earliest human occupations in the Dry Puna of Argentina. For this purpose, we provide the results of the pollen analyses from different Early and Middle-Holocene profiles located in Pastos Chicos - Las Burras river basin (23° 40' S; 66° 25' W; 3890 m a.s.l.). The comparison with another sequence from the sedimentary profile located in Lapao ravine, Lapao 5 profile (23° 36' S, 66° 36' W; 3650 m a.s.l.), assists to understand the variability in the performance of the climate system in different temporal and spatial scales. These results show two paleoenvironmental moments: a) ~ 9300 - 7300 yr BP, a moist and stable period represented by a vegetation composed for grasses, and b) ~ 7300 - 6300 yr BP, a drier environment interrupted by a short-time events of moisture. The comparison with Lapao 5 sequence shows similar trends, although with some chronological discrepancies. More moisture events are recorded towards ~ 7700 yr BP; after this date and until the beginning of the Late Holocene, drier conditions were installed. Based on this information, the models proposed were evaluated and new hypothesis are generated about the subsistence patterns of hunter-gatherers who had inhabited the area during that period and then are evaluated with the presently known archaeological record.

Key words: *Pollen - Paleoenvironment - Hunter-gatherer groups - Holocene - Puna*

Many laws regulate variation, some few of which can be dimly seen...
Charles Darwin, The Origin of Species (1859)

INTRODUCTION

At present, the evidence of early human occupation in the Dry Puna comes mainly from caves and rockshelters, located in gorges and protected valleys, dating around 11,000 yr BP (Table 1).

Site	Radiocarbon Age BP	Calibrated Range one sigma BP(*)	Reference
Cueva Yavi	9.760 ± 160	11.061 - 11.348	Krapovickas 1987-88
Inca Cueva 4	10.620 ± 140	12.564 - 12.817	Aguerre *et al.* 1975
Huachichocana	10.200 ± 420	11.267 - 12.400	Fernández Distel 1986
Pintoscayoc 1	10.720 ± 150	12.409 - 12.451	Hernández Llosas 2000
Alero Cuevas	9.650 ± 100	11.005 - 11.025	López 2008
León Huasi 1	10.550 ± 300	12.044 - 12.843	Fernández Distel 1989
Hornillos 2	9.710 ± 270	10.650 - 11.407	Yacobaccio *et al.* 2009

Table 1. Earliest radiocarbon dates from archaeological sites for Early Holocene in the The Dry Puna Argentina. (*) Calib 5.01 INTCal 4.0.

Scholars from different theoretical frameworks (Human Ecology and Processual perspective, among others) tried to explain the diversity in use and function of the early archeological sites during the colonization of the region. In general terms, there are two tendencies:

- First, there are some authors who use the Optimal Foraging Theory. Within this group, two different hypotheses can be discriminated. One is the idea that during the Early Holocene hunter-gatherer groups developed a long-term strategy, in order to include as many potential ecological patches as possible, located at different altitudes (Yacobaccio 1990). From this perspective, the maximum diversity of archaeofaunas is expected (camelids and rodents; among others). The other idea, derived from Darwinian archeology, considers that the high consumption of small mammals could be understood as a "sub-optimal or bad adaptive responses", in terms of the investment and energetic ratio return (Muscio 1999).

- Second, another group of authors endorse the ecological model of land use proposed by Borrero (1994-1995) for the Patagonian region. This model consists of three stages employing biogeographical categories: Exploration - Colonization - Effective Occupation of the Space. The exploration is characterized as a stage in which small bands make casual headings to know the availability of the surrounding environment. The expected archeological record for this moment is sparse, spread out in the regional space. Colonization implies that several bands or groups settled in certain areas, which would generate an archeological concentration of evidences related to patches of available resources. The stage of effective occupation of the space assumes planning management of resources from residential sites to specific locations, according to the structure of available resources. This model has been used by Hernández Llosas (2000) to explain the archeological record of Pintoscayoc I site. She understands the temporal sub-segment dated in *ca.* 10,000 yr BP, where she explained some remains of rodents (mostly Chinchillidae and Caviidae), artiodactyls (mostly camelids), associated with lithic artifacts showing an expidiency technique of manufacture, as an exploration stage (Hernández Llosas 2000). On the other hand, Aschero (2000) from Inca Cueva 4 and Huachichocana III evidence argues that the use and exploitation of lithic raw materials, and functional complementation and re-use of sites, suggests some stability in the use of resources in certain areas for the Early Holocene. This evidence would be consistent with the expectations developed in the model proposed by Borrero (1994-1995) for the advanced stage of initial colonization (Aschero 2000). This does not mean that the area had not been explored earlier, but that the archaeological evidence we have today suggests that the early occupations of the Puna do not fit in an exploration phase, but in a colonization which implied the redundant use of more productive areas (Aschero 2011).

In this context, the incorporation of paleoenvironmental studies helps to evaluate the proposed models. The scarcity of systematic paleoenvironmental studies in the area has turned it difficult to understand the environmental scenario in which human groups developed the colonization of the Puna. In this way, this study contributes to the demystification of the Puna as an unattractive area for human occupation, and achieves a better understanding of the strategies developed by human groups within certain specific contexts. For this reason, this research fulfills the need to develop a broader database on the particular conditions of this environment.

In regard to the paleopalynological background in the Argentine Dry Puna, from evidence of El Aguilar (Markgraf 1985), Yavi (Lupo 1998) and Barro Negro (Fernández *et al.* 1991), three phases can be suggested, although with some chronological discrepancies: a) a cold and wet phase between 10,000 and 7500 yr BP, with a dominance of the Poaceae represented in a Herbaceaous steppe and b) a dry phase between 7500 and 4000 yr BP, with elements of the Shrub steppe (Asteraceae, Chenopodaceae, *Ephedra* sp., among others) and c) *circa* 4000 yr BP the present conditions were established.

Fortunately, in the last few years the number of paleoenvironmental studies, carried out from different lines of evidence (such as pollen, diatoms, sediment analysis, among others), has increased our knowledge about the variability of the impact of global climate changes in different spatial scales (Morales 2004, 2011; Morales and Schitteck 2008; Oxman 2010; Tchilinguirián 2009; Yacobaccio and Morales 2005). The goal of this paper is to continue with this way of research and advance in the study of the paleoenvironmental conditions that may have influenced the use of space during the Holocene by hunter-gatherer groups.

Within this framework, the pollen analysis can provide information about changes in the composition of the regional vegetation and even in some cases it is possible to define the conditions of humidity and temperature in which different species had developed. At the same time, the composition of the vegetation is a principal component in the conformation of environmental scenario, thus conditioning the behavior of a variety of living organisms. These studies contribute to the task of building an ecological framework in which different organisms interact with each other.

THE STUDY AREA

The study area is located in the Puna of Jujuy, which comprises the arid highlands of NW Argentina, between 22° and 24° S, and from 3000 to 4500 m a.s.l. (Figure 1). This area is defined as a highland desert biome, where several NE-SW oriented mountain ranges are observed. This biome includes an altitudinal variation of the "tolar" (shrub steppe) ubicated between 3600 - 3900 m a.s.l., "pajonal" communities (highland grasslands) ubicated between 4100 - 4700 m a.s.l. (Cabrera, 1976), mixed steppe (ecotone) between 3900 - 4100 m a.s.l., with patchy wetlands ("vegas") occurring in both of them (Borgnia *et al.* 2006). The Puna is characterized by high solar radiation due to its high altitude, wide daily thermal amplitude, marked seasonality of rainfall (but never more than 400 mm/year), and low atmospheric pressure. Primary productivity is mainly focused on stable hydrological systems such as primary basins, high valleys (Olivera 1997), and wetlands. Several permanent freshwater basins, salt lakes, pans and playas constitute the general hydrological net. A few rivers and several springs, irregularly distributed all over the landscape, are the main freshwater sources, which are a critical resource for human and ungulate populations. Summer precipitation in Northern Argentina is largely governed by the so-called South American Monsoon-like System (Garreaud *et al.* 2009). This system produces about 80% of the annual precipitation falling in the Andean highlands between December and February (Vuille and Keimig 2004). In turn, these conditions determine a heterogeneous distribution of plant and animal resources. Some patches defined as "nutrient concentration zones" (NCZ) (Yacobaccio 1994) contain the majority of the available regional biomass. The most important animal food sources for humans in the Puna included several medium-sized mammals, the vicuña (*Vicugna vicugna*) and the guanaco (*Lama guanicoe*); and some rodents, vizcachas (*Lagidium viscacia*) and chinchillas (*Chinchilla laniger*), and a cervid, the Taruca (*Hippocamelus antisensis*).

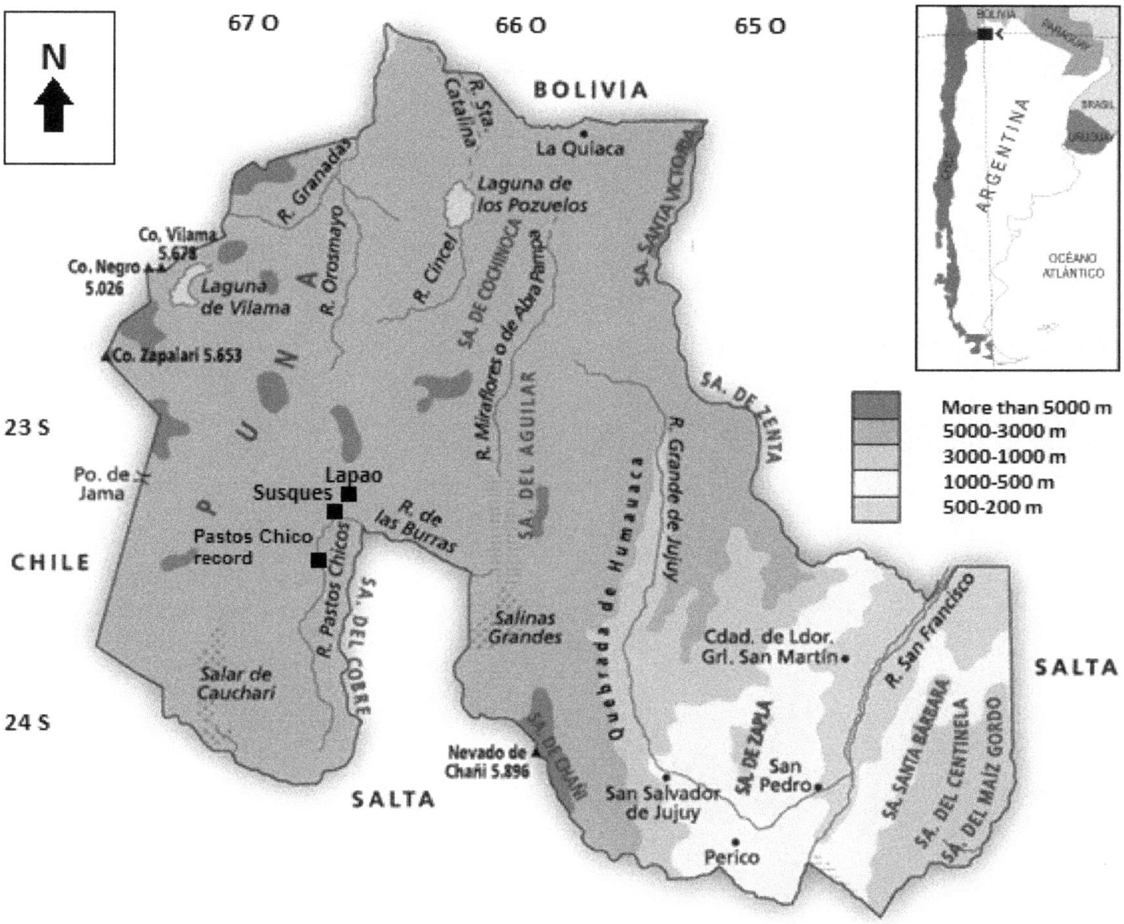

Figure 1: Location map of the Pastos Chicos River in the Jujuy Province, Argentina.

DESCRIPTION OF THE PROFILES

Fieldwork for this study has been conducted in the area of Pastos Chicos River, the main watercourse of a larger hydric system, called Pastos Chicos - Las Burras (Figure 2). The geomorphology report of the Pastos Chicos River (Tchilinguirian 2008) described the area as steady all over the year, although with significant variations in water volume due to rain falling in the upper basin. Pastos Chicos River catchment covers 988 km^2 and occupies a north-south oriented tertiary tectonic depression bounded on the west by the Taire mountain range (5120 to 4200 m a.s.l.) and the Los Cobres Mountains (4200 to 4500 m a.s.l.) in the East.

Detailed field studies of the Holocene deposits were performed on outcrops exposed over an area of more than 20 km through longitudinal Pastos Chicos river profile in order to indentify lithofacies and lithofacies associations. The Pastos Chicos valley has two different geomorphic sections along its North-South axes. The northern section is occupied by a Playa-lake system composed by very fine sediment, reddish sands and mud with parallel lamination, whereas in the middle section, the Pastos Chicos valley is excavated in Tertiary sedimentary and pyroclastic rocks. The profile selected for this investigation is located in this section. Into the Pastos Chicos River, three levels of fluvial terraces were developed on both sides of the river (TI, TII and TIII). In this opportunity, we only describe the Holocene deposits of the TI Alloformation II (Figures 2 and 3).

The geomorphologic description of the profile separates different deposition events based on the presence of erosive discontinuities. Terrace I is + 6 m above the current level of the river watercourse; sedimentary deposits are quaternary aged and they are called: "Alloformation Pastos Chicos II". This Alloformation is separated from others by means of higher level erosive discontinuities. Four sub-sedimentary units have been distinguished within the Alloformation Pastos Chicos II, separated by erosive discontinuities of minor importance called Unit A, B, C, D and E. Further, within each Unit there are different types of sediments and sedimentary structures that can be grouped into sedimentary lithofacies. Several distinctive lithofacies were defined in terms of relative scales of strata thickness and internal structures, generally allowing gathering information regarding depositional form and process in relatively small scales (Bridge 1993; Tchilinguirian 2008).

Twenty-eight samples from two different profiles of Pastos Chicos record were taken for geological, pollen and diatoms analyses and only half of them were processed (distributed along the whole profile) in this first stage of the investigation for the pollen analysis.

Figure 2. Geomorphology of Pastos Chicos River Valley. 1: Terrace level with outcrops of Quaternary age 2: Terrace lower level, 3: Permanent River channel (Tchilinguirian 2008).

For chronological purposes three samples were selected for dating. The bulk organic matter of two of these samples -PCH2-M2 and PCH1-M3- were dated to 7900 ± 100 and 8900 ± 130 yr BP using conventional 14C, respectively. Another age of 6935 ± 69 yr BP was obtained from a bird bone (PCH2-M15). Based on These dates and the strong stratigraphic correlation, between profiles, a linear interpolation age-depth model (Bennett 1994) was created to estimate the relative age of each analyzed sample (Table 2) (Oxman 2010). This model have been applied -assuming a constant sedimentation rate between dates- due to the similarities observed in sedimentary accumulation and in pollen counts in the chronological packages (11.49 yr/cm between both peat dates, and 6.34 yr/cm between the 7900 ± 100 peat sample and the 6935 ± 69 date of the bone sample) of this sequence (unit A). The chronology of the samples placed in units B and C have been considered as post 4200 yr BP due to the stratigraphic correlation with other dated sequences in the locality (Tchilinguirian *et al.* 2012).

MATERIALS AND METHODS

As mentioned above, this paper uses pollen analysis as proxy data to infer particular types of vegetation and, indirectly, the climatic conditions (temperature and relative humidity) under which plant communities developed. This kind of paleoenvironmental information is complementary with other lines of evidence, such as the studies of sediments, fluvial geomorphology and analysis of diatoms, which also allow inferring moisture conditions, though on a more restricted scale (Dincauze 2000).

The fossil samples were processed according to the Standard Protocol for Quaternary Pollen (Faegri and Iversen 1989). The laboratory stage consisted in the observation of the samples under Zeiss-Axiolab biological microscope and due to low pollen concentration in general of all samples, at least 200 grains per sample were counted. The quantification and statistical treatment of the species described were performed using TILIA software (Grimm 1987, 2004). The identification of pollen types was carried out using the existing Atlas (Heusser 1971; Markgraf and D' Antoni 1978) and the pollen reference catalog of the Palynology laboratory of the Facultad de Ciencias Agrarias UNJu/Conicet (Lupo 1998).

RESULTS

Of the 14 extracted samples for pollen analysis, four of them were sterile and 10 were fertile and could be subjected to pollen analysis. The identification of pollen grains usually was performed at the family level, except in cases in which it was possible to reach a greater degree of precision allowing identification to a genera level. In this case has been identified: one genus (Pteridophyta), six families (Poaceae, Asteraceae, Fabaceae, Chenopodiaceae, Solanaceae and Mimosaceae) and two species (*Carex* sp. and *Ephedra* sp.).

In general terms, Poaceae family dominates most of the pollen samples, followed by Asteraceae. The other taxa are significantly less frequent. The only exception is a sample around 6300 yr BP (M16) where Pteridophyta spores are highly abundant.

Based on the changes in the relative frequency of pollen types, two major environmental phases were identified. The first part of the record, between ~ 9200 and 7300 yr BP shows a clear evidence of a moist environment - possibly a paleowetland- with clear dominance of herbaceous species (Poaceae). In ~ 7300 yr BP (M10) an important change in the pollen spectra composition was detected. This change was represented by an increase in the diversity of pollen types. However, the Poaceae family still dominates over other taxa. Towards ~ 7300 yr BP (M10), the sequence showed a gradual and steady increase in the frequency and diversity of the regional vegetation represented by shrub steppe species represented by Asteraceae, Chenopodiaceae, Fabeaceae, Solanaceae and Verbenaceae over the local vegetation which is represented by moist condition indicators such as fern spores and grasses (Poaceae). Around 6300 yr BP (M16), the pollen spectra change in the composition species with a marked presence of the herbaceous steppe mainly composed by Asteraceae, *Ephedra* sp., Mymosaceae plus Poaceae and Pteridophyta spores. This composition that had not been previously recorded, is interpreted as an increase in local moisture conditions within a shrub steppe context. Post 4200 is the only case in which the Asteraceae dominate the composition of the sample, which represents a more arid environment than previously presented (Figure 4).

Sample	Depth	Chronology	Sample analyzed
PCH2 M20	402		X
PCH2 M19	362		
PCH2 M18	331	Post 4200	X
PCH2 M17 Bis	311		
PCH2 M17	308		
PCH2 M16	280	6319	X
PCH2 M15	270	**6935 ± 69**	
PCH2 M14	260	6998	X
PCH2 M13	250	7062	
PCH2 M12	240	7125	X
PCH2 M11	230	7189	
PCH2 M10	220	7252	X
PCH2 M9	210	7316	
PCH2 M8	200	7379	X
PCH2 M7	190	7443	
PCH2 M6	180	7506	X
PCH2 M5	170	7570	
PCH2 M4	168	7583	X
PCH2 M3	142	7748	
PCH2 M2	118	**7900 ± 100**	X
PCH1 M8	118	7900	
PCH1 M7	108	8015	X
PCH1 M6	91	8210	
PCH1 M5	63	8532	X
PCH1 M4	33	8877	
PCH1 M3	31	**8900 ± 130**	X
PCH1 M2	20	9130	
PCH1 M1	0	9256	X

Table 2. Ages and radiocarbon dates (those with 1σ) of the model and samples analyzed.

In sum, the interpretation of the pollen analysis indicated two different paleoenvironmental moments: a) the first one, between ~ 9200 yr BP and 7300 yr BP (PCH1-M1 to PCH2-M8), shows dominance of a herbaceous steppe, this could indicate colder and moister conditions, which might have produced by a down-slope displacement of the herbaceous steppe vegetation towards a shrub steppe landscape that currently dominates; b) the second phase, after ~ 7300 yr BP and post 4200 yr BP (M10 to M18), shows a subtle increasing diversity in the composition around 7300 yr BP (M10) and a subsequent increase in the abundance of the elements related to the shrub steppe; suggesting a gradual settlement of a dry landscape.

EVIDENCE FROM OTHER PROXY DATA

Complementation with other lines of evidence on the same samples assists in specifying our paleoenvironmental interpretations (Dincauze 2000). Following these criteria, the diatom, geological and sediment analysis have been included in this analysis (Morales 2004, 2011; Tchilinguirián 2008, 2009).

In general terms, the geological analysis indicate that at the base of the Holocene deposits, the presence of an extended thick alluvial organic-fine sediment (units A, Figure 4) outcrops on the border of the valley. This is a prominent unit feature of the stratigraphic scheme, representing an important Early to Middle Holocene low energy fluvial system with organic paleowetlands, oxbow lakes and backswamp muds (Tchilinguirian *et al.* 2009).

In Pastos Chicos I profile, the base of unit A is overlaid by tertiary limestones, whereas the upper contact with unit B is a channel scoured. The lowermost unit A is up to 1 to 3 m thick over hundred meters long an outcrop and is dominated by greyish green colors. The lower section of the sequence (subunit A1) consists of a horizontally bedding, pale green medium and massive sand of the river channel. Unit A2 consists of fine laminated, hard, black, organic layers interbedded with thin light gray laminated diatomaceous silt. The organic material contains uncharred, well-preserved plant epidermis and vascular plant bundles of 1 to 2 mm length. These organic layers represent backswamps with organic soils development. Unit A3 is composed of massive, grey light, very fine sand and diatomaceous silt interbedded with laminated dark grey medium organic sand. The subunit A4 (2 m) is characterized by fine laminated, green mud and clays interbedded with thin (2 cm) layers of massive and white diatomite. No organic matter found, and thus remains undated.

Unit B is composed of channel facies in the lower section (subunit B1), overbank facies in the middle section (subunit B2) and carbonate pond deposits of facies, located in the uppermost section of the sequence (subunit B3).

The diatom's assemblages in Pastos Chicos record (Morales 2004, 2011) suggest two paleoenvironmental moments. The first one, between ~ 9200 and 7300 yr BP could be interpreted as a moist and relatively stable environment. This moment would resemble a wetland with broad vegetated littoral areas presenting extensive littoral forms of the Fragilariaceae group -usually a species with broad tolerance to salinity fluctuations- interrupted by one drier pulse around 8000 yr BP, as evidenced by the negative anomalies in the moisture index. The conditions inferred from diatom analysis are coincident with those suggested by the geomorphologic analysis, which suggests a floodplain environment that is moist during most part of the year and with development of paleo-soils. The second moment, between ~ 7300 and 6000 yr BP, suggests a wetland environment which is much drier and more salty due to the arid conditions, but episodically interrupted by some moister pulses. From ~ 7300 to 6000 yr BP drier conditions have prevailed, characterizing the second environmental moment. At least two events with moister conditions were detected under these dry conditions, the first one at ~ 7000 yr BP and the second at ~ 6300 yr BP. It is also detectable the presence in this particular alofacie dry conditions - probably the drier moment in the studied sequence- between ~ 7500 and 7000 yr BP, as suggested by the rise in aerophilic diatoms and the values of the salinity index obtained from diatom's ecological affinities (Morales 2004, 2011).

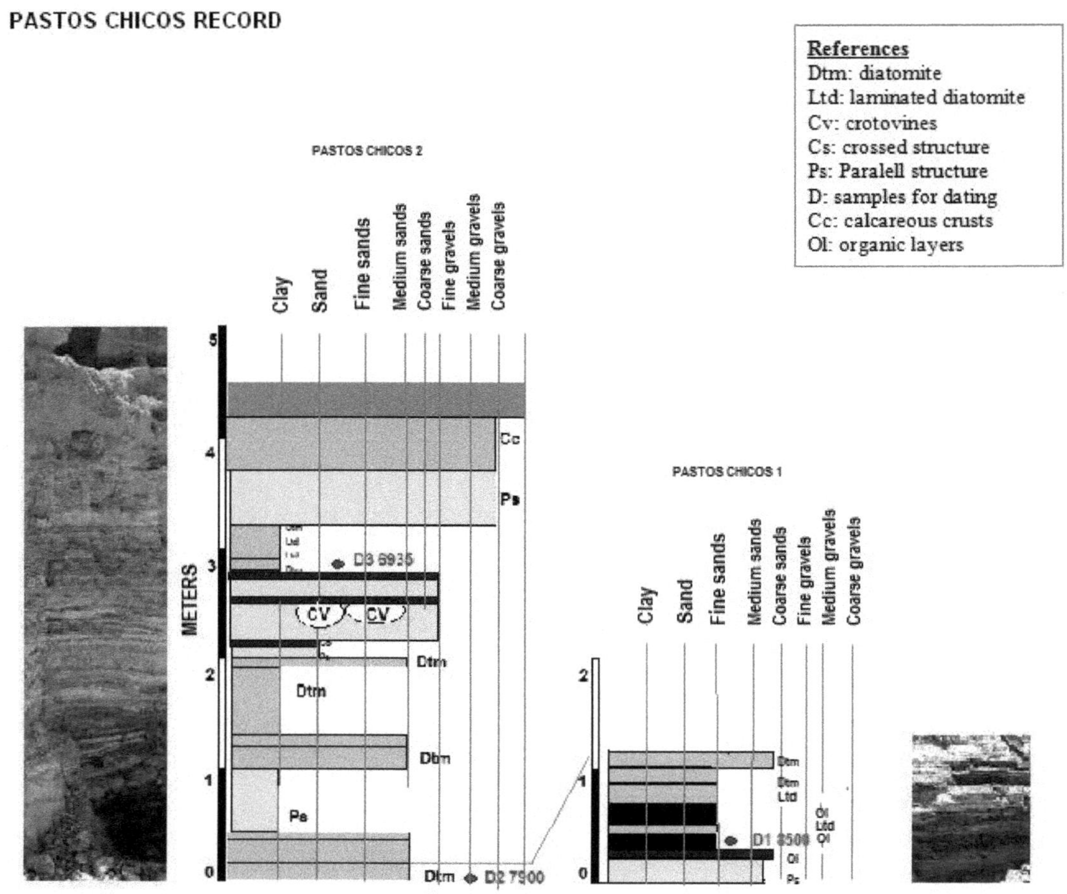

Figure 3. Photographs and drawings of PCh1 and PCh2 profiles (modified from Morales 2011).

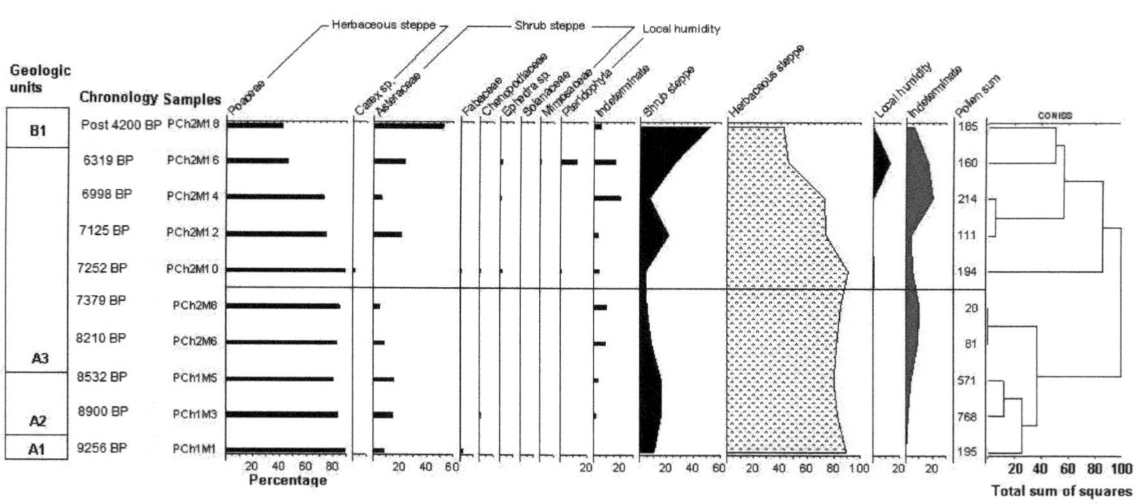

Figure 4. Pollen diagram from Pastos Chicos record.

The comparison of Pastos Chicos record with one of his tributaries, Lapao 5 (L5) (Tchilinguirian *et al.* in this volume), allows the generation of some hypotheses about the response of different catchment environments to the global environmental changes. In the case of L5 the information generated by pollen analysis, indicates an extent of the highland grasslands (Poaceae) between ~ 9000 and 7770 yr BP, which would evidence wetter conditions than today. This was followed by the formation of a mixed steppe (ecotone) and drier conditions than the ones that were present in the previous period, similar to the present. In regard to the diatoms analysis, three paleoenvironmental zones were identified: 1) Between ~ 9300 and 8900 yr BP, characterized by the presence of wetland with broad vegetated littoral areas represented by the high frequencies of Littoral species with a minor percentages of the benthic species; 2) Between ~ 8900 and 7770 yr BP, installed wetter conditions than the previous period, although some variations. Between ~ 8500 and 8100 yr BP the peak of moisture throughout the profile is registered, and is characterized by the development of a palustre system, interrupted by a considerably drier moment around 8400 yr BP; and 3) between ~ 7700 and 7400 yr BP again are installed drier conditions than the previous period, represented by the dominance of species of benthic and epiphyte that indicate the conditions of the wetland, that remained until 7400 yr BP (Oxman 2010).

The result of the study of these cases (PCH and L5) show that the paleoenvironmental conditions between ~ 9300 and 7700/7300 yr BP would have been more stable and moister than the current conditions. Meanwhile, during the period ranging from ~ 8000/7300 to 6300 and post 4200 yr BP, the conditions were drier and more unstable, though with punctual moisture episodes, which can be described as a wetland considerably drier than the one of the previous period.

On the other hand, when we compare the sites, two different situations emerge. First, the Pastos Chicos record shows that the changes seem to initiate before and disappear afterwards in the extended area systems, while presenting higher stability in the long term. We think that this could be due to the geomorphological control of the broad moisture catchments area (~ 1000 km^2) that generates an average signal of the tributaries, reflecting mostly a signal of regional scale. Second, in the case of L5, the system responds to short-term changes with particular intensity, as the small area basins (~ 17 km^2) tend to reflect the changes locally produced with higher intensity (Tchilinguirian 2009; Tchilinguirian *et al.* 2009).

This information agree with the model proposed by Morales (2011), which points out that during the period ranging from 10,000 to 8000 yr BP, resources would have become abundant in most sectors of the dry Puna, with the exception of those areas above 4000 m a.s.l. to the West of the Pastos Chicos river basin, to the West of Olaroz - Cauchari basin, Rosario de Susques river high basin and around Laguna Vilama. On the contrary, the area located north of Guayatayoc lagoon, would gather most of the Dry Puna resources. Though more abundant during the period from 10,000 to 9000 yr BP, wherein the NCZ (Yacobaccio 1994) would be much more frequent and concentrated in basins without snowmelt water input, with different catchments than during the next period. From 9000 to 8000 yr BP it was possible for the minor catchments basins to have presented water deficit events and for NCZ to become more frequently available in basins with snowmelt water input of elevations higher than 4000 m a.s.l. Those areas located above 5000 m a.s.l. would have been covered by snow during most part of the year, so they would not have involved an important supply of resources (Morales 2011).

CONCLUDING REMARKS

Regarding the regional archeological record from the Early and Middle Holocene, some patterns may be delineated in order to evaluate the different models proposed for explaining the first occupations of the Dry Puna Argentina (Aschero 2000; Grosjean 1994; Olivera 1997; Yacobaccio 1991, among others).

During the Early Holocene has been recorded a more humid environment, less fragmented and more stable than nowadays, whereas the Middle Holocene had regional drier conditions (in these cases after 8000/7000 yr BP), more fragmentation of the environment with restricted loci of productive patches which were the focus of main amount of resources (Fernández *et al.* 1991; Lupo 1998; Markgraf 1985; Yacobaccio and Morales 2005, among others).

Regarding to the Early Holocene, the archaeological record shows:

- Strategic site location, near water sources, firewood and animal resources.
- Most part of the sites are located in caves or rockshelters, in gorges covered from the wind, near water resources and generally in ecotone sectors.
- Resources are used with relation to their local abundance (according to the reference site); zooarcheological remains differ from site to site following the specific availability of local resources.
- Some variability is also observed in the use of wild plants, such as tubers, tuberous roots, grasses, etc either for feeding purposes or for technological use.
- Lithic raw materials are generally local, the projectile points are triangular, and the artifacts show a low to medium energy investment.

In contrast, the Middle Holocene archaeological record shows:

- An increase in the relative abundance of camelids in most of the sites, and a decrease in the other animal resources.
- The lithic technology shows an increase of the morphological types of projectile points, which more

diversity than in the earlier period (new forms appear, like lanceolate or stemmed). There is also a greater diversity of other artifact types such as milling stones. There is also an increase in the use of obsidians from the Zapaleri/Laguna Blanca source.

Based on the paleoenvironmental data generated, the archeological evidences compiled and the available information referring to mobility and subsistence strategies by hunter-gatherer groups (Bettinger 1987; Binford 1980, 2001; Bousman 1993; Kelly 1995), now we can return to the discussion of the models proposed in order to explain the first occupations of the Dry Puna.

We argue that the extension of the high grasses (Herbaceous steppe) to lower elevation, controlled by humidity, was the result of the increase in precipitation (Betancourt *et al.* 2000). Consequently, the increase of the environmental primary yield would have allowed a higher amount of animal biomass. Related to this, the distance between patches would have diminished, which in turn would have provided hunter-gatherer groups with a higher energetic supply of resources in shorter distances. Also, it has been observed climatic changes impacts on the environment in different spatial and temporal scales, leading to variability, sometimes shortages, in resources supply. In this sense, it is also expected some variability in the management of animal and plant resources, whose availability depends on environmental and climatic factors (rainfall patterns, humidity and temperature; among others). A high mobility is the strategy for obtaining resources from different ecological patches. For this reason, we interpreted that the consumption of small mammals, for example, should not be seen as bad adaptive responses (as argued from Optimal Foraging Theory), but as short-term responses in a stage of recognition of the dynamics of the environment, in which better environmental conditions allowed the opportunistic strategy of obtaining local resources, thus explaining the high variability observed in the abundance of animal resources in different sites. These short-term responses would be the beginning of the designing of the long-term responses, the planned circuit and the typical Andean complementation of the resources from different altitude levels (proposed by Yacobaccio 2010).

In relation to the model proposed by Borrero (1994/1995), and used by Hernández Llosas (2000) and Aschero (2000), we argue that the characteristics of the archaeological record of the Argentine Puna make difficult, not only to recognize, but also to accept the idea of an "exploration stage" (Aschero 2011), which is contradicted on environmental and archeological grounds. In short, the evidence indicates the presence of hunter-gatherer groups with high residential mobility. The great productivity of many vegetation patches allowed an opportunistic exploitation of animal resources (Binford 1980). For that reason, we propose to use the concepts of "Dispersal" in order to denote the spreading out of individuals or groups which filled up the available vacant habitat, and the term "Colonization" as to the major extension of a population habitat or range that includes an established occupation of areas previously unoccupied or occupied (Dillehay 2000; Yacobaccio 2010). Thus, although we can hardly find evidence of this initial dispersion, we may assume that some knowledge in the management of resources is the result of previous experiences developed in the area. The record indicates some degree of knowledge and transmission of information on the management of environmental resources, a process that exceeds the capacity of a single generation (Meltzer 2002). For this reason, we interpret that the earliest known archaeological record in the Argentine Puna as befits a time of colonization.

Based on the conclusions derived from paleoenvironmental data and the knowledge of how different environments respond to broader climate change, we may derive some new hypotheses. On a broader scale, we can consider altitudinal divisions, which have a range of differential production areas. Regarding the altitude, three geo-environments could be ranked: valleys and ravines, pre-Puna and highlands or Altiplano. With these parameters established, we propose an initial period of dispersion of people, which could be moving either from the South Bolivian lowlands or the valleys and ravines, into a vertical direction to the other environments. These new areas would result from an extension of the circuitry of mobility, which included also the Puna of Chile. Each region collects various ecological attributes that might influence hunter-gatherer settlement patterns and decisions concerning hunting (Muscio 1999; Oxman 2010). The higher areas (above 4000 masl) were available for human occupation later in time than the lower ones (between 3800 and 3200 m a.s.l) (Morales and Schitteck 2008). However, the evaluation of paleoenvironmental and archaeological evidence reveal that when the highest areas were available they were rapidly occupied. This also reveals to us that those areas were unfit for long-term human´s occupation, but were known and visited by hunter-gatherer groups, as seen from the obsidian distribution, because this raw material ´s source is located in the high Altiplano area.

ACKNOWLEDGEMENTS

We are grateful to Liliana Lupo for her collaboration and support to carry out this research. We also want to express our sincere thanks to Marcelo Morales, Débora Kligmann, Lorena Grana, Jimena Alberti, Gabriel López, Laura Lombardi, Rodolphe Hoguin and Lucas Sgrecia for their helpful comments, and to Paul Tchilinguirián for the illustrations and photographs of the profile, and their geomorphologcal advice.

REFERENCES

AGUERRE, A.M., A. FERNÁNDEZ DISTEL AND C.A. ASCHERO
1975. Comentario sobre nuevas fechas en la arqueología precerámica de la Provincia de Jujuy.

Relaciones de la Sociedad Argentina de Antropología. Nueva Serie, Tomo IX: 211-214.

ASCHERO, C.A.
2000. El poblamiento del territorio 1. In *Nueva Historia Argentina: Los Pueblos Originarios y la Conquista*, edited by Myriam Tarragó, pp. 17-60. Editorial Sudamericana, Buenos Aires.

2011. Holoceno medio en la Puna argentina: dos puntos de observación para la cultura material y una perspectiva regional. In *Poblaciones Humanas y Ambientes en el Noroeste Argentino durante el Holoceno Medio*, edited by M. Mondini, J.G. Martínez, H.J. Muscio and M.B. Marconetto, pp. 33-44. Gráfica Corintios, Córdoba.

BENNETT, K.D.
1994. Confidence intervals for age estimates and deposition times in late-Quaternary sediment sequences. *The Holocene* 4(4): 337-348.

BETANCOAURT, J.L., C.L. QUADE AND K.A. RYLANDER
2000. 22,000-year record of monsoonal precipitation from northern Chile's Atacama desert. *Science* 289(5484): 1542-1546.

BETTINGER, R.L.
1987. Archaeological approaches to hunter-gatherers. *Annual Review of Anthropology* 16: 121-142.

BINFORD, L.R.
1980. Willow smoke and dogs' tails: Hunter-gatherer settlement systems and society for American Archaeology. *American Antiquity* 45(1): 4-20.

2001. Where do research problems come from? *American Antiquity* 66(4): 669-678.

BORGNIA M, A. MAGGI, M. ARRIAGA, B. AUED, B. VILÁ AND M. CASSINI
2006. Caracterización de la vegetación en la Reserva de Biósfera Laguna Blanca (Catamarca, Argentina). *Ecología Austral* 16: 29-45.

BORRERO, L.A.
1994/1995. Arqueología de la Patagonia. *Palimpsesto. Revista de Arqueología* 4: 9-69.

BRIDGE, J.S.
1993. Description and interpretation of fluvial deposits: a critical perspective. *Sedimentology* 40: 801-810.

BOUSMAN, B.
1993. Hunter gatherer adaptations, economic risk and tool design. Lithic technology 18(1-2): 59-86.

CABRERA, A.
1976. Regiones fitogeográficas argentinas. In *Enciclopedia Argentina de Agricultura y Jardinería*, Tomo 2-1. Second edition. ACME, Buenos Aires.

DILLEHAY, T.
2000. *The Settlement of the Americas: A New Prehistory*. Basic Books, New York.

DINCAUZE, D.F.
2000. *Environmental Archaeology: Principles and Practice*. University of Cambridge Press, Cambridge.

FAEGRI, K. AND J. IVERSEN
1989. *Textbook of Pollen Analysis*, edited by K. Faegri, P.E. Kaland and K. Krywinski. John Wiley & Sons, Chichester. 4th edition.

FERNÁNDEZ, J., V. MARKGRAF, H.O. PANARELLO, M.A. ANGIOLINI, F.E. VALENCIO AND M.S. ARRIAGA
1991. Late Pleistocene/Early Holocene environments and climates, fauna, and human occupation in the argentine Altiplano. *Geoarchaeology: An International Journal* 6(3): 251-272.

FERNÁNDEZ DISTEL, A.
1986. Las cuevas de Huachichocana, su posición dentro el precerámico con agricultura incipiente del Noroeste argentino. *Beitrage Zur Allegemeinen und vergleichenden Archaologie*, Band 8, Verlag Phillip von Zabern Mainz Am Reim: 353-430.

1989. Una nueva cueva con maíz acerámico en el N. O. Agentino: León Huasi 1, excavación. *Comunicaciones Científicas* 1: 4-17.

GARREAUD, R.D., M. VUILLE, R. COMPAGNUCCI AND J. MARENGO
2009. Present-day South American Climate. *Palaeogeography, Palaeoclimatology, Palaeoecology* 281(3-4): 180-195.

GRIMM, E.C.
1987. CONISS: a FORTRAN 77 program for stratigraphically constrained cluster analysis by the method of incremental sum of squares. *Computers & Geosciences* 13(1): 13-35.

2004. *TGView 2.0.2*. Illinois State Museum. Research and Collection Center. Springfield, Illinois.

GROSJEAN, M.
1994. Paleohydrology of the Laguna Lejía (North Chilean Altiplano) and climatic implication for late-glacial times. *Paleogeography, Paleoclimatology, Paleoecology* 109(1): 89-100.

HERNÁNDEZ LLOSAS, M.I.
2000. Quebradas altas de Humahuaca a través del tiempo: el caso de Pintoscayoc. *Estudios Sociales del N.O.A* 4(2): 167-224.

HEUSSER, C.J.
1971. *Pollen and spores of Chile. Modern type of the Pteridophytas, Gymnospermae and Angiospermae*. The University of Arizona Press, Tucson.

KELLY, R.L.
1995. *The Foraging Spectrum: Diversity in Hunter-Gatherers Lifeways.* Smithsonian Institution Press, Washington DC.

KRAPOVICKAS, P.
1987-88. Noticia. Nuevos fechados radiocarbónicos para el sector oriental de la puna y la Quebrada de Humahuaca. *Runa* XVII-XVIII: 207-219.

LÓPEZ, G.E.J.
2008. *Arqueología de Cazadores y Pastores en Tierras Altas. Ocupaciones Humanas a lo largo del Holoceno en Pastos Grandes, Puna de Salta, Argentina.* BAR International Series 1854, Oxford.

LUPO, L.C.
1998. *Estudio sobre la lluvia polínica actual y la evolución del paisaje a través de la vegetación durante el Holoceno en la cuenca del Río Yavi. Borde Oriental de la Puna, Noroeste argentino.* Unpublished PhD Dissertation, Bamberg University, Germany.

MARKGRAF, V.
1985. Paleoenvironmental history of the last 10,000 years in northwestern Argentina. *Zentralblatt für Geologie und Paläontologie* 11/12: 1739-1749.

MARKGRAF, V. AND H.L. D' ANTONI
1978. *Pollen Flora of Argentina.* The University of Arizona Press, Tucson.

MELTZER, D.J.
2002. What do we do when no one´s been there before? Thoughts on the exploration and colonization of new lands. In *The First Americans. The Pleistocene Colonization of the New World*, edited by N. Jablonski, pp. 27-35. University of California University Press.

MORALES, M.R.
2004. *Casi invisible: Diatomeas, ambientes locales y grupos humanos durante la primera mitad del Holocéno en la Puna Desértica.* Unpublished Graduate Thesis, Facultad de Filosofía y Letras, Universidad de Buenos Aires.

2011. *Arqueología Ambiental del Holoceno Temprano y Medio en la Puna Seca Argentina. Modelos Paleoambientales Multi-Escalas y sus Implicancias para la Arqueología de Cazadores-Recolectores.* British Archaeological Research South American Archaeology Series No. 15, Oxford.

MORALES, M.R. AND K. SCHITTECK
2008. Primeros resultados paleoambientales del Holoceno Medio en Alto Tocomar (Puna Salteña): interpretación local e implicancias regionales. *Libro de resúmenes de las Jornadas de Arqueología del área puneña de los Andes Centro-Sur*, pp. 109. ISES - UNT.

MUSCIO, H.
1999. *Colonización humana del NOA y variación en el consumo de los recursos: la ecología de los cazadores recolectores de la Puna durante la transición Pleistoceno-Holoceno.* Unpublished PhD Dissertation, Facultad de Filosofía y Letras, Universidad de Buenos Aires.

OLIVERA, D.E.
1997. La importancia del recurso Camelidae en la Puna de Atacama entre los 10.000 y los 500 AP. *Estudios Atacameños* 14: 29-41.

OXMAN, B.I.
2010. *Una perspectiva paleoecológica de las primeras ocupaciones de la Puna Seca Argentina: análisis polínico de perfiles naturales holocénicos ubicados en el Dto. de Susques, Pcia de Jujuy.* Unpublished Graduate Thesis, Facultad de Filosofía y Letras, Universidad Buenos Aires.

TCHILINGUIRIAN, P.
2008. *Informe geomorfológico del sector medio de la cuenca Pastos Chicos, Susques, Jujuy.* Unpublished report prepared for Dr. Hugo Yacobaccio.

2009. *Paleoambientes Holocenos en la Puna Austral (27°S): implicaciones geoarqueológicas.* Unpublished PhD Dissertation, Facultad de Ciencias Exactas y Naturales, Universidad de Buenos Aires.

TCHILINGUIRIAN, P., M.R. MORALES, L. LUPO, D.E. OLIVERA, B. OXMAN AND H. YACOBACCIO
2009. Síntesis paleoambiental del lapso 9000 - 7300 AP. En la Quebrada de Lapao, Jujuy (23.1°, 66,4°W, 3650 m). Paper presented at the *IV Congreso Argentino de Cuaternario y Geomorfología - XII Congresso da Associação Brasileira de Estudos do Quaternário*, La Plata, Buenos Aires.

TCHILINGUIRIAN, P., M.R. MORALES, B. OXMAN, L.C. LUPO, D.E. OLIVERA AND H.D. YACOBACCIO
2012. Early to Middle Holocene transition in the Pastos Chicos record, dry Puna of Argentina. *Quaternary International*. In press.

VUILLE, M. AND F. KEIMIG
2004. Interannual variability of summertime convective cloudiness and precipitation in the central Andes derived from ISCCP-B3 data. *Journal of Climate* 17(17): 3334-3348.

YACOBACCIO, H.D.
1990. *Sistemas de asentamiento de los cazadores-recolectores tempranos de los Andes centro-sur.* Unpublished PhD Dissertation, Facultad de Filosofía y Letras, Universidad de Buenos Aires.

1994. Biomasa animal y consumo en el Pleistoceno-Holoceno Surandino. *Revista de Arqueología* 4: 43-71.

2010. The Paleoindian and Archaic of Central and South America. In *The Cambridge World Prehistory*, edited by C. Renfrew and P. Bahn. In press.

YACOBACCIO, H.D. AND M.R. MORALES
2005. Mid-Holocene environment and human occupation at Susques (Puna de Atacama, Argentina). *Quaternary International* 132(1): 5-14.

YACOBACCIO H.D., M.P. CATÁ, P. SOLÁ AND M.S. ALONSO
2008. Estudio arqueológico y fisicoquímico de pinturas rupestres en Hornillos 2 (Puna de Jujuy). *Estudios Atacameños. Arqueología y Antropología Surandinas* 36: 5-28.

IN PURSUIT OF THE FIRE. CONTRIBUTIONS OF MICROCHARCOAL ANALYSIS TO THE ARCHAEOLOGY OF THE AMBATO VALLEY (CATAMARCA)

Henrik B. Lindskoug[1]

[1] CONICET - Museo de Antropología, Facultad de Filosofía y Humanidades, Universidad Nacional de Córdoba

Abstract

In this study microcharcoal is used as an indicator of the presence of fires in the past in order to evaluate the last phase of the Aguada occupation of the Ambato Valley, Catamarca Province. Microcharcoal is used as a paleoenvironmental indicator, and tools and methods are applied that were developed in pedoanthracology to analyze sediment samples obtained from the Ambato Valley. The preliminary studies carried out are presented in order to generate information about past wildfires and construct paleoenvironmental data for the area. Seventeen sites throughout the Ambato Valley were sampled in order to obtain paleoenvironmental information about past wildfire regimes.

Key words: *Microcharcoal - Pedoanthracology - Ambato*

INTRODUCTION

Several years of study in the Ambato Valley have provided rich infôrmation about the Aguada occupation. The contexts we associate with the final moments of the Aguada occupation in Ambato show, for the most part, situations that suggest an abrupt abandonment. Such contexts are characterized by the presence of burned and collapsed roofs, which generated broken ceramics pots, some still with content. Further, objects were distributed in activity areas as if they were in use, such as sets of pots, mortars and pestles, uncleaned fireplaces, and so on. Collapsed walls are also associated with these contexts. This scenario suggests an unplanned abandonment of the site (Gastaldi 2010; Laguens 2006). With respect to the absolute dating of these events, all yielded dates around 1000 ± 100 AD (Gordillo 2005; Marconetto 2007) indicating the contemporaneity of the events.

Recently, we have started to orient studies towards understanding the paleoenvironmental context of the last phase of the Aguada occupation of the valley. To obtain more information about why these events took place we have turned to pedoanthracology. Pedoanthracology is a relatively recent field of study and its application to cases in Argentina is a new development in the archaeology of the region. The field is based on the identification and dating of microscopic charcoal recovered in soil, and offers a complementary method for the reconstruction of vegetation based on the method of pollen records, providing a high degree of local resolution (Di Pasquale *et al.* 2008; Thinon 1978).

However, in the fields of pedoanthracology and microcharcoal studies there is no single universal technique applied, neither to the extracting, preparation nor counting of the samples. This prompted us to conduct an experimental study, testing new methods and indicating new techniques that can be applied, especially since the use of many different methods complicates the comparison of results obtained by different researchers.

The analysis of sediments sampled outside the archaeological sites allows microcharcoal frequencies to be monitored. These may indicate fire events in the past and supplement information already available on drought periods in the region (Marconetto 2009). Such analyses also provide information about the nature of the fires that affected the archaeological sites (Lindskoug and Mors 2010). We hypothesized that forest fires could have affected the region at the end of the first millennium, with the resulting burning of the settlements. We are aware, however, that conditions in the natural environment were not the only reason for the abandonment. Social and cultural factors in the society must have had a significant role in this process. Thus, we plan to test the hypothesis of an unfavorable environmental context for the population occupying the region with its likely effect on the natural resources.

ENVIRONMENTAL SETTING: THE AMBATO VALLEY

The Ambato Valley is located in Catamarca province, northwestern Argentina. The valley runs north to south; to the west is the Ambato or Manchao mountain range (4050 m a.s.l.), and to the east the Sierras Graciana-Balcozna (1850 m a.s.l.). The southern limit is the Catamarca Valley, and the northern boundary the Altos de Singuil. The Los Puestos River runs across the fluvial plain on the valley bottom. The valley forms part of the geological province called the Northwestern Pampeanas Ranges ("Sierras Pampeanas") and the area is characterized by narrow valleys and bolsons, alternating with high mountain ranges and forests. The geological components of the Ambato Valley are basement rock, mainly banded gneises, migmatites and schists outcrops, which are often intruded by pegmatites and tonalite-granodiorite bodies. Quaternary sediments along with small relicts of Tertiary sedimentary rocks fill the intermountain valley (Blasco *et al.* 1994). The region forms part of the "Monte" and "Chaqueña" phytogeographical provinces (Cabrera 1976). The valley is characterized by a warm continental climate, with precipitation ranging from 500-800 mm annually.

PEDOANTHRACOLOGY AND MICRO-CHARCOALS

The field of pedoanthracology is based on the identification and dating of microscopic charcoal recovered in soils and offers a complementary method for the paleoenvironmental reconstruction of vegetation based on that of pollen records (Di Pasquale *et al.* 2008; Thinon 1978). The first steps of microcharcoal research can be found in pollen studies. Many researchers count microcharcoal particles when counting pollen on microscope slides and thereby obtain information about past fire regimes. The pioneering work of Iversen (1941) in Denmark is an example. Many of these studies were carried out on lake sediments. The first person to define pedanthracology was Thinon (1978), who set up a framework and methodology for this relatively recent discipline. Later, Carcaillet and Thinon (1996) developed techniques for studying "dry" sediments in mountainous areas. However, there is no universal method for microcharcoal analysis, nor for the counting of particles. Various authors have developed their own techniques (see, for example, discussions in Patterson *et al.* 1987). Clark (1988a) and Whitlock and Millspaugh (1996) have developed techniques and theoretical models of charcoal dispersion after fire events and the accumulation during and after such events. Whitlock has also studied past fire events and climatic change from microcharcoals extracted from wet sediments in the USA. Other researchers working on the same problematic have made important contributions (Turner *et al.* 2008b, 2010; see contributions in Scott *et al.* 2000; Palaeo 164 [2000]). In the case of Argentina most studies have been carried out in Patagonia. Huber and Markgraf (2003) have studied the European impact on natural fire regimes and vegetation dynamics in southern Patagonia. Other studies in Patagonia that have focused on the relationship between human impact and climatic change include: Habertzel *et al.* 2006; Heusser 1987, 1994, 1999, 2003; Huber *et al.* 2004; Kitzberger *et al.* 1997; Kitzberger and Veblen 1999, 2003; Markgraf and Anderson 1994; Sottile 2008; Veblen *et al.* 1999; Whitlock *et al.* 2007.

There exists a difference between archaeoanthracology (the samples obtained are the result of the selection of species for human use) and pedoanthracology. The latter is based on the identification and dating of charcoal recovered from paleosoils. The charcoal studied results from natural fires or fires deliberately started to increase crop yields. In some cases fire might have been used as a hunting strategy, as indicated by ethnographic examples, or from other kinds of land clearing activities (Mistry *et al.* 2005). The advantage of pedoanthracology is that the samples are not distorted by the selection of species by people. In cases where identification is possible due to the size of the sample, identification is done otherwise only counting is done (Abdoun *et al.* 2000; Talon *et al.* 1998).

SOME REMARKS ABOUT CHARCOAL RECOVERED IN SOILS

According to Patterson *et al.* (1987) this type of analysis should address several issues. There are several processes that affect the material, including combustion and charcoal production, taphonomic processes that affect the material, dispersion, deposition and preservation. In addition, all these processes are affected by soil sampling. Other factors that may interfere with the analysis are the mode of the analysis/sampling and the preparation and counting problems that will be discussed later in this paper.

The selection of sampling areas outside of archaeological sites should be guided by the recent geological history of the region in order to achieve the best areas that operate as natural sediment traps. Another issue in the process of sampling is the possible mixing of layers due to natural processes such as bioturbation, which can be frequent, making it an important variable to consider when sampling the material in the field.

The size of microcharcoals may indicate distance to the source of the fires. "Large fragments" generally indicate a source nearby and "small fragments" are normally scattered by the wind larger distances. Thus, according to the sizes of the fragment it is possible to determine if the fire is local or regional, as discussed by Patterson *et al.* (1987) and Clark (1998a). Clark (1988a) has developed models and formulas for charcoal transport, although it should be noted that they are models and there are several factors, which are hard to consider in models that affect the processes of deposition and fragmentation of the charcoal particles. Moreover, studies by Nichols *et al.* (2000) have shown that water transportation transports larger particles a greater distance than smaller particles, which makes the above statement problematic.

A major problem is the identification of plant microcharcoals by microscope; some researchers count all black elements in the samples, which in many cases may be another material such as black mineral content. Charcoal can be determined if one can observe plant anatomy, but this often depends on the sediments. Some microcharcoals might be "dirty", that is their anatomy is concealed by heavy clay sediments, for example, making them difficult to identify. A matrix analysis is also necessary so one can estimate the presence of black or "dark" minerals that interfere with the count. According to experimental studies by Umbanhowar and McGrath (1998) it is possible to differentiate three types of microcharcoals associated with different types of vegetation. Microcharcoals from grasses are significantly longer (562 μm) and have a greater length:width ratio (3.62) than charcoal from leaves (380 μm; 1.91) or wood (348 μm; 2.13) (Umbanhowar and McGrath 1998). However, the anatomy of the cells, if observable, is the best indicator of "real" microcharcoals originating from past wildfires.

MATERIALS AND METHODS

In the Field

The sediments used in this study were collected during one fieldwork season in our study area. In total 17 stations were sampled from north to south, located in different areas of the Ambato Valley (see Figure 1). The location of the stations was carefully selected so that stations worked as sediment traps due to the problem of the transport of microcharcoal depending on natural factors such as wind and water (rain) conditions. The sampling took place in locations that operated as sediment traps in colluvial soils; the criteria for selection were based on location and inclination of the terrain. Some stations were sampled on hillsides on prehistoric agricultural terraces that functioned as sediment traps. The locations of the stations ranged from the valley bottom to river terraces, and included present and past river channels (see Table 1). Some samples were taken close to prehistoric settlements. The altitude of sampling locations ranged between 1061 m a.s.l. and 1259 m a.s.l.

Locations of the Stations

In total 17 stations were sampled in the Ambato Valley, the most northerly station is number 10, and Station 14 is the most southerly. There is a concentration of the sampled stations in the center of the valley, close to several large archaeological settlements (Figure 1). The stations sampled in the Ambato valley are taken from the valley bottom (the fluvial plain), the majority from the first and second terrace of the Los Puestos River, some were sampled in the piedmont zone. Stations 1, 2, 8, 9, 12, 13 and 14 were sampled in archaeological terraces or their immediately vicinity (see Table 1). Stations 8 and 9 were sampled in the archaeological terrace system Los Varela. Station 14 was sampled in a large archaeological terrace system located in the south of the Valley. Stations 1 and 2 are also located in the southern part of the valley and were also sampled in an archaeological terrace system. Stations 3 and 4 were sampled on the second terrace of the Los Puestos River. Station 3 is located 100 m south of Piedras Blancas site and 250 m north of Iglesia de los Indios site, while Station 4 is located in the middle of the patio of Iglesia de los Indios. Station 5 was sampled in the valley bottom and was not directly associated with an archaeological site. Station 6 was sampled on a river terrace close to the archaeological site of Cerco de Palos, 20 m from the paleo riverbed. Station 7 was located in the paleo riverbed. Station 10 is the most northern station and was sampled close to the source of the Los Puestos River on the second terrace of the Los Puestos River on the border to the piedmont zone. Station 11 is located between the first and second terrace of the Los Puestos River close to the edge of the terrace. Stations 12 and 13 were sampled close to the large archaeological settlement system, which includes a prehistoric terrace system on the Molina property. Station 15 is located close to Saaverda. Station 16 is located on the second terrace of the Los Puestos River, 7 m from the edge of the terrace. Station 17 was also sampled on the same terrace, 15 m from the edge of the terrace and very close to the archaeological site Martínez 2. Several stations can be grouped together based on geomorphic and vegetation criteria/divisions, altitude, and their location inside or outside archaeological sites. Stations 1 and 2 were both sampled in the piedmont area on the edge of the third river terrace. Both belong to the vegetation zone P3P. Stations 3, 4, 5, 6 and 7 can also be grouped together; all were sampled on the fluvial floodplain and belong either to the LLBg or LLB vegetation zone. Station 5, however, was not directly associated with an archaeological site. Stations 8 and 9 also share characteristics; both were sampled in the piedmont zone and belong to the P3MPA vegetation zone, and are consequently grouped together. Stations 12, 13 and 14 can also be grouped together as they were all sampled in the piedmont zone and belong to the P2B vegetation zone. The last stations to be grouped together are 16 and 17, both sampled on the fluvial floodplain on the second terrace of the Los Puestos River. Both belong to the LLBg vegetation zone. Stations 10, 11 and 15 are not grouped together, however, since they cannot be fitted in with the other stations.

Recollection of Reference Soil Samples

To help out in the identification process in the laboratory we collected reference soil samples after wildland fires, both in our area of investigation and in the Calamuchita Valley, Córdoba province. This was done to help the identification process of the microcharcoals. The samples were collected in several different areas: 1) areas directly affected by the fire; 2) areas located close to the area affected by the fire, around 300 m from the fire source; and 3) areas located further away from zones affected by the fire. Samples from Córdoba were collected a few days after the fire had passed. The samples collected in the Ambato Valley were collected 8 months after the fire. Since almost no rain had fallen after this fire, signs of the fire were clearly visible and a lot of microcharcoal was visible on the surface, clearly indicating which areas had been affected by the fire. In the case of the Ambato Valley, samples were collected on the hillsides, river bottom and river terraces, both inside and outside archaeological sites. Interviewing the local population, we obtained information about the extent and insensitivity of the fire that took place in December 2009. The stations directly affected by the fire were Stations 1 to 7 and 14. All reference samples were collected only on the surface and put in ziplock bags with information about geographical positions. This small reference collection proved to be highly helpful in the identification process of microcharcoals in the laboratory.

Sediment Extraction in the Field

The sediments were extracted with a sediment drill, a mechanical model with reinforced extensions, a cylindrical point (20 mm x 400 mm), and a maximum depth of 3 m. Normally the sediment cone was divided into sections of 10 cm; sometimes subdivisions were made after a preliminary ocular analysis demonstrated interesting changes or features in the soil sample. The minimum depth of the drilling was 40 cm and the maximum depth 235 cm. Stations were subdivided into between 3 and 26 samples, the average ranging from 8-17. In total 226 samples were collected from the 17 stations. The samples were collected in zip lock plastic bags and labeled, indicating station and sample.

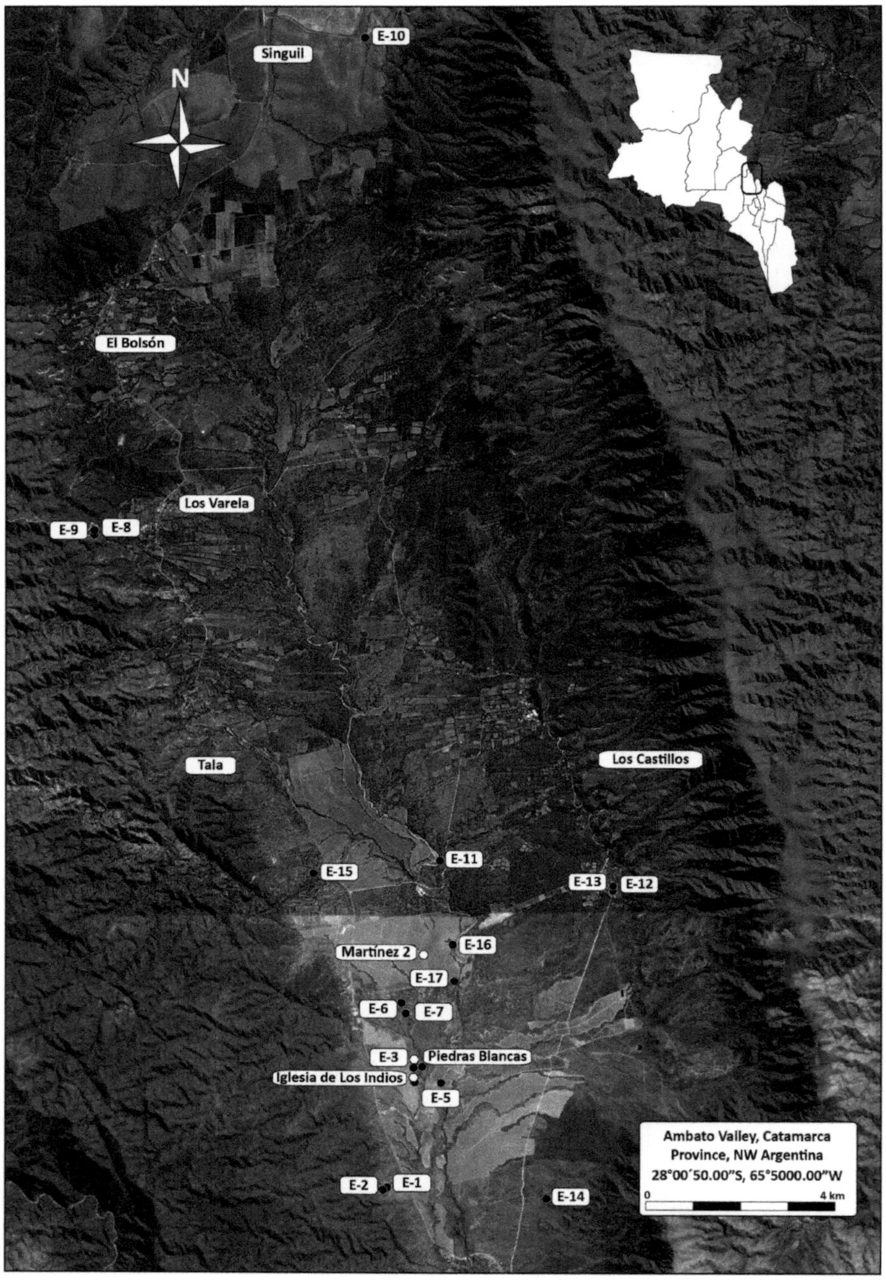

Figure 1: Map showing the stations sampled in the Ambato Valley and key archaeological sites mentioned in the text.

A preliminary classification of the sediments was done in the field according to color, form, particle size and sorting using a Petroleum Grain size chart (INTEQ). At every station, information about the local vegetation, terrain and factors that could affect the sample was recorded on a sample sheet (see Table 1).

In the Laboratory

Several problems were encountered during analysis of the soil samples. The first problem encountered was methodological -as there are no universal techniques in pedoanthracology and microcharcoal studies for the extraction, preparation and counting of samples- we had to develop our own methods. A major drawback of developing new methods is that comparison with other studies is complicated. The second problem encountered was related to the materials -our soil samples- extracted in the field. All samples contained a great many black elements, most of which were not microcharcoals but rather mineral content, which severally complicated the identification of fire signals in the samples obtained.

In relation to the methodological problem, the most common method for reconstructing fire histories has been the counting of charcoal particles by palynologist when counting pollen in the preparation of pollen slides (Whitlock and Millspaugh 1996). In this way it is possible to reconstruct paleoclimates based on information from both pollen and microcharcoal sources from the same sample, which is time saving. Turner (2007) undertook an interesting evaluation of several different methods and found several problems with the above method. Elements larger than 180 μm are eliminated in the cleaning process in pollen preparation, and as the process is rigorous and implies several steps it is likely that it results in fragmented microcharcoals and consequently an overestimation of the number of charcoals in the sample (Clark 1984 in Turner 2007: 36).

The methods normally used are: 1) Petrographic Thin Section (Clark 1988b); 2) Oregon Sieving Technique (Millspaugh and Whitlock 1995); 3) Bleaching and Filtering (Rhodes 1996; 1998); 4) Density Separation (Clark 1984; according to Turner (2007: 44) this technique hasn't been sufficiently refined yet); 5) Density Separation and Bleaching (developed by Turner *et al.* 2008a on the basis of Clark's (1984) technique); 6) and a technique employed by Carcalliet and Thinon (1996) that implies the digging of large trenches to obtain samples for flotation which are later dated by AMS. The material recovered is normally of larger size than in the other methods mentioned.

Station	Site and brief description	Altitude (m a.s.l.)	Geomorphology	Vegetation after De la Orden and Quiroga (1997)
1	South of Luchi terrace system	1087	Piedmont. Limit of the third terrace of the Los Puestos River	Grassland with Shrubbery (P3P)
2	South of Luchi terrace system, archaeological cultivation terrace (1 m high)	1084	Piedmont. Limit of the third terrace of the Los Puestos River	Grassland with Shrubbery (P3P)
3	Close to Piedras Blancas archaeological site	1078	Fluvial floodplain. Second terrace of the Los Puestos River	Gallery-Forest Flatland (LLBg)
4	Patio of Iglesia de los Indios	1076	Fluvial floodplain. Second terrace of the Los Puestos River	Gallery-Forest Flatland (LLBg)
5	Inside the Rinconada field, valley bottom close to present cultivated land	1061	Fluvial floodplain. Second terrace of the Los Puestos River	Forest Flatland (LLB)
6	Close to Cerco de Palos archaeological site and paleoriverbed	1087	Fluvial floodplain. Second terrace of the Los Puestos River	Gallery-Forest Flatland (LLBg)
7	Cerco de Palos, inside of the paleoriverbed	1083	Fluvial floodplain. Second terrace of the Los Puestos River	Gallery-Forest Flatland (LLBg)
8	Las Varelas archaeological terrace system, inside of a cultivation terrace	1257	Piedmont	Forest with Shubbery (P3MPA)
9	Las Varelas archaeological terrace system, inside of a cultivation terrace	1259	Piedmont	Forest with Shubbery (P3MPA)
10	Altos de Singuil, close to where Los Puestos River emerges	1226	Almost piedmont. Second terrace of the Los Puestos River	Shrubbery (P4A)
11	Between the first and second terrace of Los Puestos River, on the road to Los Castillos, close to the edge of the terrace	1106	Fluvial floodplain. Limit between the fist and second terrace of the Los Puestos River	Gallery-Forest Flatland (LLBg)
12	Close to the Molina house, inside of the archaeological settlement, on an archaeological cultivation terrace	1224	Piedmont	Open Forest (P2B)
13	Close to the Molina house, inside of the archaeological settlement	1225	Piedmont	Open Forest (P2B)
14	Close to large archaeological settlement south in the Ambato Valley, in archaeological terrace system	1112	Piedmont	Open Forest (P2B)
15	Close to Saavedra	1126	Almost piedmont. Third terrace of the Los Puestos River	Grassland with Shrubbery (P3P)
16	Close to Martínez 2 archaeological site, Ex. Martínez field, now Nuezes de Catamarca, close to terrace edge	1089	Fluvial floodplain. Second terrace of the Los Puestos River	Gallery-Forest Flatland (LLBg)
17	Close to Martínez 2 archaeological site, Ex. Martínez field, now Nuezes de Catamarca, close to terrace edge	1085	Fluvial floodplain. Second terrace of the Los Puestos River	Gallery-Forest Flatland (LLBg)

Table 1: Brief description of the stations sampled in the Ambato Valley.

The other problem, as mentioned, is the quantification or counting of particles. Three different standard methods are normally used: 1) Absolute Abundance Measure; 2) Point Count; and 3) Area Measurement. The Absolute Abundance Measure involves the counting of all particles, regardless of their size (Patterson *et al.* 1987). The Point Count method was developed by Clark (1982) and is based on the number of hits using an eyepiece graticule, normally 11 points. The particles that are not in the field of view are ignored. This method was developed to allow a rapid measurement of charcoal content. The Area Measurement method was developed by Waddington (1969) and is widely used. The method is based on a gridded eyepiece graticule and quantifies the percentage area occupied by microcharcoal. Each charcoal particle is measured and assigned to a size class to provide additional information about the source of the fire, transport of the particles, etc. This method is time-consuming but since it involves the measurement of the particles it can also provide important transport information. However, this method also has many disadvantages, as discussed by Turner (2007) and others. Several researchers disagree about the relevance of the size classes for microscopic charcoal, especially because of the fragility of microcharcoal (see discussion in Tuner 2007). Another method for estimating the microcharcoal particles is Image Software Analysis, first done by Horn *et al.* (1992) and later others. However, this method is not sufficiently developed yet and there are several problems with counting since the software cannot discriminate between charcoal remains and other black particles found in the microscope images. Consequently, we developed our own method of extraction and preparation of the samples.

The sediments were left to dry in the laboratory and were then analyzed with a microscope (Motic BA200) in transmitted light with 40x, 100x, 200x, 400x magnifications to obtain evidence of microcharcoals. Photos were also taken with a Motic camera (Moticam 1000, 1.3M Pixel) of the different features found.

The features identified in the samples were phytoliths, black vegetal structures, black minerals, microcharcoals, fibres, elongated silica structures, and indeterminate black elements (see Figures 2 and 3). In the analysis we used the category black vegetal structure when the cellular structure of the microcharcoal was clearly visible. This was done to differentiate this category from the category of microcharcoals where the cellular structure is not visible but the sample can be identified as a microcharcoal because of its form, colour and shape, especially notable on the edges of the microcharcoal (see Figure 4). We used the same identification criteria as Turner *et al.* (2008a). Tuner *el al.* (2008a) argue that accurate identification of mircocharcoal can be problematic for several reasons. For example, there are many black elements in sediments that can resemble charcoal, such as pyrite, dark plant fragments and insect cuticles. They also state that the actual identification is subjective and strongly depends on the experience of the analyst. We used the same identification criteria as Turner el al (2008a) in order to improve the identification of microcharcoals. These diagnostic criteria are: 1) jet black; 2) angular, straight edges; 3) straight but fuzzy edges; 4) blue hue; 5) the presence of cellular structure. If a combination of these criteria were applicable the elements were classified as a microcharcoal or black vegetal structure depending on the cellular structure. In this first step of our analysis we only evaluated presence and absence in the samples of the features mentioned above. The next step in our research will be to count the elements identified in the samples.

The procedure for the preparation of the samples was the following: a small standard amount of soil/sediment was prepared on a slide for microscope analysis; five drips of sewing machine oil were added; the sample was left for approximately 1 minute to disperse the oil, at which point a glass cover (20 x 20 mm, 0.13-0.17 mm thick) was placed on top. The samples were then sealed with clear nail varnish to produce a semi-permanent sample that could be used for several analyses. After the sample had been left to dry it was ready to be analyzed (see also Lindskoug 2010).

In relation to difficulties with the material extracted in the field, we noticed that in the first stage of the analysis we had problems identifying the microcharcoals as there were concentrations of both black biotite and black mica in many of the samples. We experimented with flotation with distilled water and centrifugation of the samples using filer paper to abstract the microcharcoal. The latter proved to be very time consuming so it was decided to use the former preparation technique. One reason we elected not to use chemical treatment of the samples was our concern that we would clean them and thereby loose the fire signal. A more rigorous chemical treatment implies several more steps in the cleaning process, which leads to greater mechanical damage to the microcharcoals. Since the microcharcoals are easily fragmented this would increase the microcharcoal abundance in each sample and lead to an overestimation of microcharcoals, affecting the consequent counting of the microcharcoal particles in each sample. The process of chemical cleaning is both costly and very time-consuming. It was decided to process more samples and not use chemical cleaning. Moreover, we lack the proper laboratory equipment to use chemical solutions. This concern also meant we elected to use elements with a black vegetal structure (carbonized elements with a cellular structure) as a securer indicator of fire than elements classified as microcharcoals. If elements with a black vegetal structure were registered this was also indicative of microcharcoals in our classification of the samples in the laboratory.

RESULTS

In total 217 of the 226 samples from the 17 stations recovered in the field were analyzed (see Table 2 and Figure 5). Some of the superficial reference samples were not analyzed since they were recovered to help the identification process and we knew they contained evidence of the December 2009 fire. These samples were extremely useful in identifying elements from fire events, although they were little impacted by taphonomic processes, such as dispersion, deposition and preservation. All the 17 stations analyzed contained fire signals. One hundred and sixty-two samples contained microcharcoals; 109 samples contained fire signals with identifiable black vegetal structures. All 17 stations contained either microcharcoals or black vegetal structures in the surface samples, and the only station that did not produce identifiable elements with black vegetal structure in the surface sample was Station 17. We lost the surface samples (the first 10 cm) at Stations 6, 9, 11, 13 14, 15 and 16 during the field extraction due to loose soil, which fell out of the sediment drill. Nonetheless, in the samples that followed in all stations black vegetal structure were present. We also lost parts of Stations 6 and 12 due to loose sediments. We chose to use the category of the black vegetal structure, which showed traces of plant anatomy, as the indicator of forest fire as it seemed to be a better and more secure indicator than the category of microcharcoals. If the category of microcharcoals were used it would have been hard to distinguish between the different levels or depths, since almost all of the samples analyzed contained microcharcoals. As stated above of the 217 samples analyzed, 162 contained microcharcoals and 109 elements with a black vegetal structure. This is why the fire pulse or rate seems to be better indicated by the category black vegetal structure than by microcharcoals. In addition, the black vegetal structure is a reliable indicator of plant combustion.

The results of the analysis of the particles with black vegetal structure are presented in Figure 5 and Table 2, indicating presence/absence in each level/sample of each station and the number of samples per station. The results in percentages of the total number of samples with black vegetal structure in each station are presented in Table 2. If we analyze the stations grouped together starting with Stations 1 and 2, located close to each other in the piedmont zone, they show various similarities -in the first 10 cm, with the black vegetal structure present but absent in the following section. The total quantity of samples with black vegetal structure present is between 35-44%. The next group, corresponding to Stations 3, 4, 5, 6 and 7, located on the fluvial plain, indicates a very similar pattern in the upper part of the samples: the first

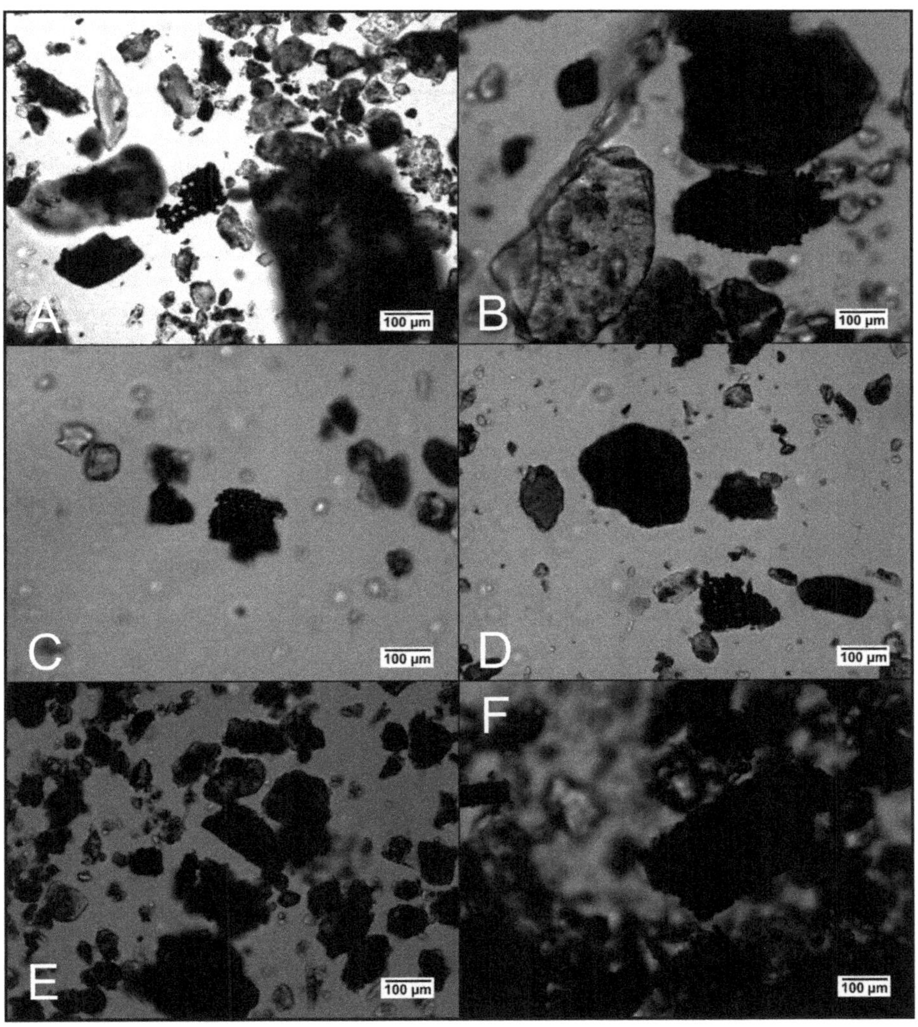

Figure 2: Photos of Microcharcoals with black vegetal (cellular) structure (200x). A: Station 16, sample 1. B: Station 16, sample 5. C: Station 16, sample 5. D: Station 16, sample 5. E: Station 17, sample 7. F: Station 17, sample 9.

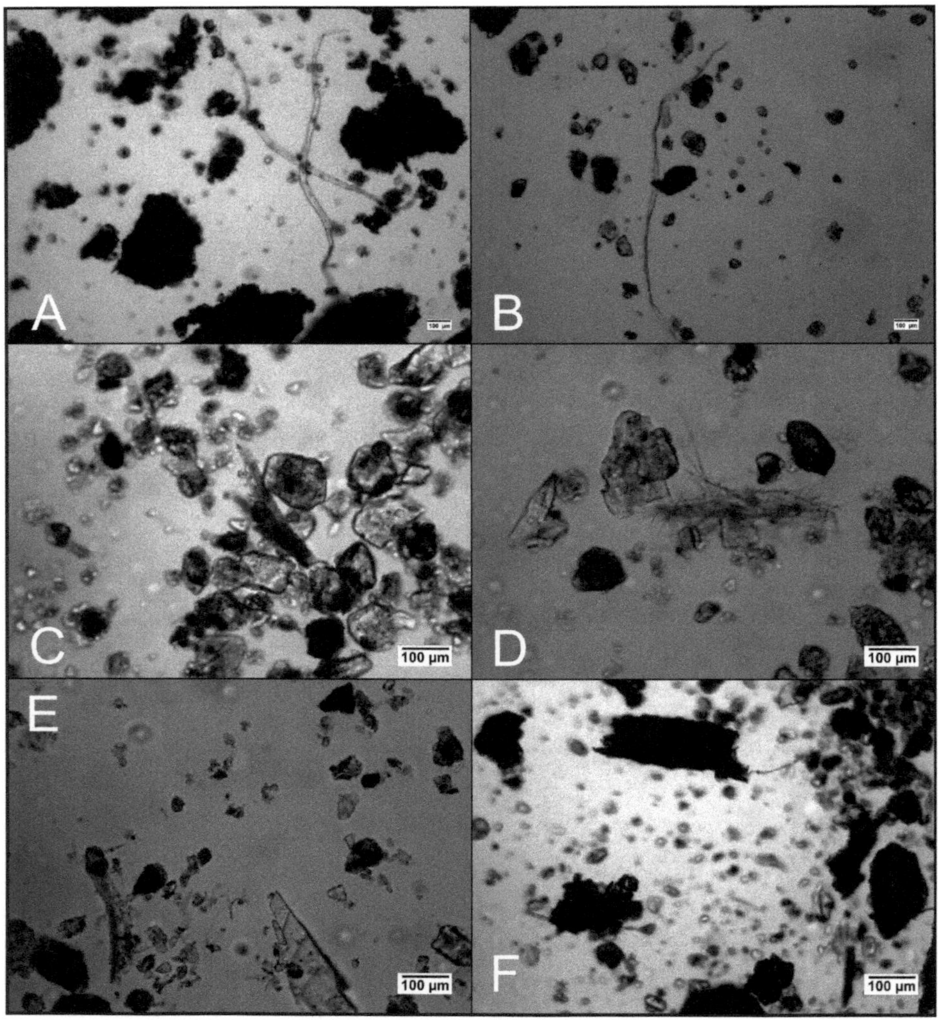

Figure 3: Photos of features identified in samples A: Fibers from Station 17, sample 8 (100x). B: Fibers from Station 17, sample 11 (100x). C: Articulated phytoliths in plant tissue, Station 2, sample 6 (200x). D: Raphides, Station 2, sample 6 (200x). E: Articulated phytoliths in plant tissue and microcharcoal, Station 2, sample 6 (200x). F: Microcharcoal, Station 10, sample 1 (100x)

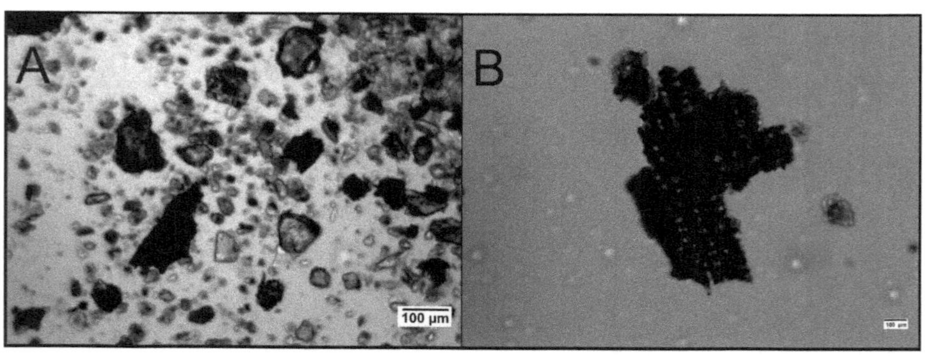

Figure 4: Photos of microcharcoal (A) and black vegetal (cellular) structure (B). A: Station 10, sample 1 (100x). B: Station 5, sample 8 (200x).

20 cm have black vegetal structure present, an abrupt absence between 20 and 40 cm in the majority of stations, followed by the presence of black vegetal structure until around 70 cm, again, in the majority of stations. The total quantity of samples with presence of black vegetal structure ranges from 35-78% with an average of 55%. Stations 3, 4, 5 are the most similar, showing several points in common throughout the sampled section. The next group, corresponding to Stations 8 and 9, both sampled in the Varela archaeological terrace system in the piedmont zone, shows great variation. Station 8 has a total of 90%

Station	Number of samples	Number of samples with black vegetal structure	% of samples with black vegetal structure of total samples
Station 1	23	8	35%
Station 2	9	4	44%
Station 3	23	12	52%
Station 4	20	7	35%
Station 5	16	6	38%
Station 6	9	7	78%
Station 7	8	6	75%
Station 8	10	9	90%
Station 9	4	2	50%
Station 10	22	9	41%
Station 11	15	8	53%
Station 12	10	3	30%
Station 13	3	1	33%
Station 14	11	3	27%
Station 15	11	11	100%
Station 16	11	9	82%
Station 17	12	4	33%
	Total number of samples	**Total samples with black vegetal structure**	**Total % of all samples with black vegetal structure**
	217	109	50%

Table 2: Number of samples collected and analyzed with black vegetal structure.

samples with black vegetal structure while Station 9 has 50%, meaning the two stations have extremely different compositions. Stations 12, 13 and 14 were grouped together since they were all located in the piedmont zone in the southern part of the valley and associated to archaeological settlements and terraces. The upper parts of the stations all include the presence of the black vegetal structure, followed by a varying period of its absence according to the station. Few elements with black vegetal structure were found -between 27 and 33% of the total quantity of the samples. The last stations grouped together were 16 and 17, both located on the fluvial floodplain close to the Martínez sites, and were very different. The total quantity of samples with black vegetal structure ranges from 33% (Station 17) to 82% (Station 16), with little correlation between stations.

Analyzing the stations affected by the wildfire we see that Stations 1 to 5 and 7 were affected. In addition, Stations 6 and 14 showed signs on the surface that they had been affected. Station 12 was not affected by the wildfire in 2009 but the surface sample indicates a fire signal. Stations 6, 11, 13, 14, 15 and 16 were not affected either but we lack surface samples (0-10 cm) from these stations. However, Station 15 showed evidence of fire in the samples extracted during fieldwork. Station 17 was located very close to the zone affected by the fire, within approximately 50 m, but no elements with identified black vegetal structures were found in the surface samples, although microcharcoals were found.

Fire signals were found in surface samples to 235 cm, which is the deepest sample in the study. From 0 cm to 185 cm fire signals were identified in the different stations sampled; there is then a gap from 185 cm to 225 cm, but this is probably because few of the stations are sampled to this depth.

A fire interval can be detected at 40-50 cm in Stations 1 to 11, except Station 6 where we lost the sample in the extraction process; we also lost Station 12 at the same height. We also found indications of fire at 40-50 cm in Stations 15 and 16. The only two stations that did not demonstrate a fire interval at this depth were 14 and 17. Station 13 only reached a depth of 40 cm maximum so was not taken into account. The stations sampled on the fluvial plain close to archaeological sites we registered a presence of fire signals of 7 out of 9 stations at the interval 40-50 cm the stations also indicates a "silence" or absence of fire signals in the previous samples. A similar phenomenon can be detected at 80 cm in 6 of 9 stations sampled and in 7 of 8 stations with analyzed sediments demonstrates a fire signal at this depth.

DISCUSSION

It is important to note that several stations (1, 6, 7, 8, 15 and 16) show long intervals of fire episodes, having a homogeneous profile. However, this might be a result of migration of microcharcoals in the stratas or it might depend on other local geological taphonomic processes. Several smaller fire signals or intervals are found in the sampled stations, including Stations 1, 4, 5, 9, 10, 11, 12, 14 and 17. The strongest fire signals are found in Stations 8 and 15; Station 8 produced only one sample with no signs of fire signals and in Station 15 all the samples recovered show fire signals. In the deepest section of Station 3, which is the deepest sample (225-235 cm), we found fire signals. In Station 12 we lost 10 cm (between 40-50 cm) during sampling. Even in Station 13 the shallowest of all the stations, sampled to a depth of only 40 cm, we found fire signals.

Our study has shown that fires occurred in the environmental setting in the Ambato Valley and that it is possible to identify them. Whether the fires were natural or anthropic is still to be determined. What is certain is that human activity in an area tends to increase fire activity. Further studies of the samples are needed to obtain more paleoenvironmental information about the ecosystem and natural environment in the valley. With more information on the paleoenvironmental context we will have more evidence for the reasons behind the abandonment of the Ambato Valley by the Aguada. Even so, we are certain that the environmental variables were not the only factors in the abandonment process. We argue that social factors might have had an important role in the process as suggested by other authors (González 1998; González and Pérez Gollán 1972; Marconetto and Laguens 2013). Natural and social/cultural factors can cause instability or stress within a society and work as a catalyser of pre-existing problems leading to increased conflict in both the social and natural orders, such as conflicts over land, internal conflicts between different groups in a society, and even external conflict with other groups occupying the same area or the vicinity. These factors (natural and cultural/social) affect one another in a complex system that creates tensions internal to the society and within the ecosystem (the natural environment) occupied by both human populations and their natural recourses. Thus, we further argue that we should continue to test the hypothesis of unfavourable environmental context but not exclude the social and cultural context in attempting to understand the disappearance of the Aguada from the Ambato Valley.

Interesting variations were observed in the stations analyzed. The stations sampled in the Ambato Valley all show different characteristics as far as formation and composition, but also share several similarities. However, one must keep in mind the different taphonomic processes mentioned above that influenced the microcharcoals. All the stations sampled in the valley show strong fire signals, indicating that fire is an integrated part of the local ecosystem. For all stations except 17 and the one for which we lost the top part we have identified microcharcoals with a clear visual anatomy demonstrating a black vegetal structure, which strongly indicates fires. This is shown by the most recent fire episode in the Ambato Valley, in December 2009; a very high percentage of elements with a black vegetal structure or anatomy was observed in all surface samples. Since little precipitation had fallen, rain, had not yet washed away and transported the microcharcoals from where the fire had occurred. It was observed that the fire went through both Piedras Blancas and Iglesia de los Indios archaeological sites and passed within approximately 250 m of Martínez 2 site, indicating that a forest fire in the area can affect all three settlements at the same time. This shows how wildfires can affect large areas and several settlements; however to further study the intensity of the fire and how it affected prehistoric settlements, it is vital to perform a profound study of the archaeological settlements in particular.

There are differences among the stations -some show short fire intervals while others indicate a continuum. This might have to do with migrations of the elements in the soil. Stations 10 and 11 show many short fire intervals while Stations 1, 2, 4, 5 and 12 have fewer intervals. Stations 7, 8, 15 and 16 show a continuum while Station 3 and 6 also indicate a continuum, although not as marked as the other stations. Stations 9, 13, 14 and 17 indicate a more mixed composition of fire intervals.

The Stations 1 and 2 group indicates how fire might have been used in the agricultural terrace system, although whether of natural or anthropic origins has still to be determined. Nevertheless, the other stations (12, 13 and 14) sampled in similar conditions but further south in the valley show very little evidence of fire. We found fire signals in all the stations on archaeological terraces in the piedmont zone. Stations 12, 13 and 14, which worked well as sediment traps for elements or sediments washed down from uphill, as indicated by the fire signal in the first samples. However, deeper in the samples there are very few fire signals. This might be the result of leaching, wind and water erosion, and might also indicate that ash was used to fertilize the soil -which remains to be investigated. Thus, it can be argued that the terrace systems can function as sediment traps.

Stations 3, 4, 5, 6, and 7 sampled close to archaeological settlements in the valley bottom show very interesting fire intervals with abrupt cuts in the sections, and many of them share a number of similarities. The fire signal intervals (0-20 cm and 40-70 cm) followed by intervals absent of fire signals might indicate regularities in the fire events in the valley. If this can be correlated to the phase of abandonment we will be one step further to understanding the processes that lead to the final abandonment of the valley by the Aguada people.

Regarding the location of the terraces on hillside (piedmont zone), terraces and on flat surfaces in the valley bottom there are no associations with the fire signals identified and there are no notable differences between stations sampled in this respect. A more careful analysis is needed, which will be the next step in our research. We are also waiting for the first AMS dates for the stations to be able to better correlate the fire signals to chronological time and geological position (depth) of the samples. Furthermore, we will also quantify the black vegetal structures in the samples to find out where the strongest fire signals are to be found in order to obtain more information about the past wild fire regimes in the area.

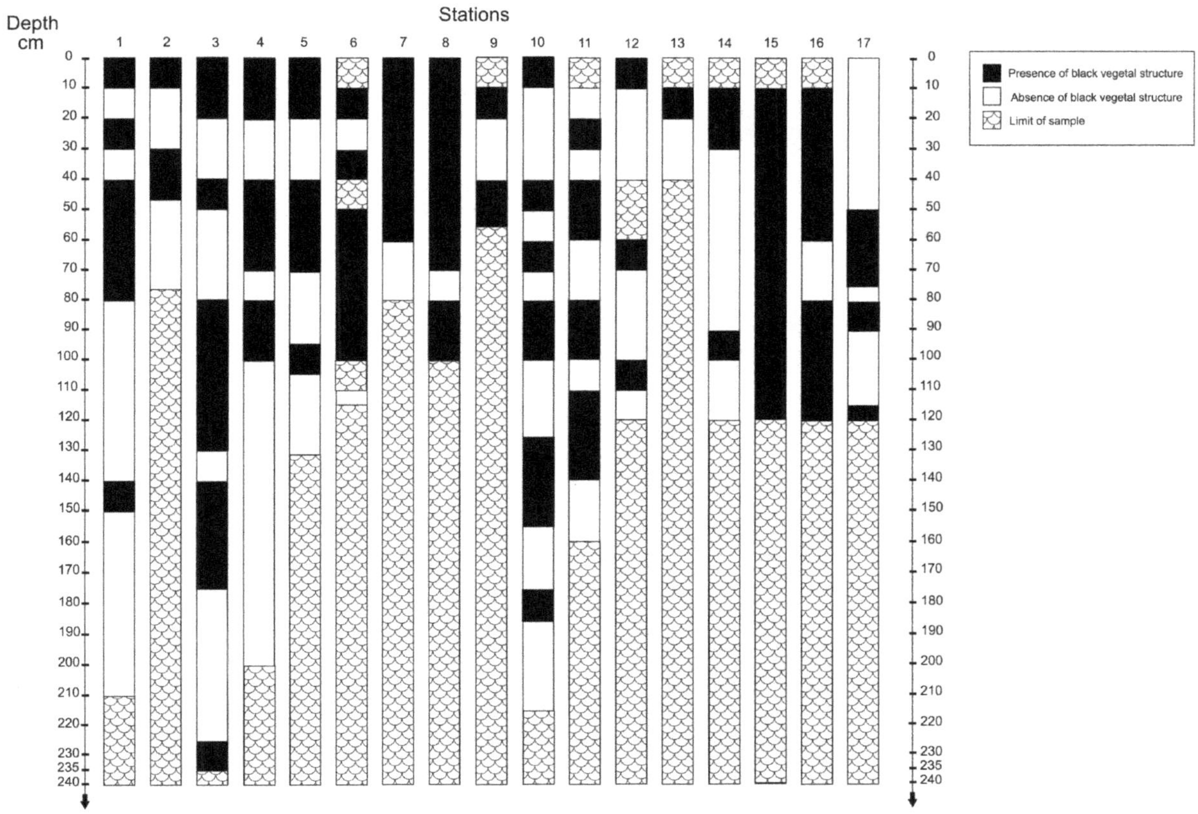

Figure 5: Presence/absence of elements with black vegetal structure, i.e. carbonized vegetal matter in the stations sampled in the Ambato Valley.

CONCLUSIONS

Our study has shown that natural fires have been a quite a common event in the area and that fire forms a natural part of the ecosystem of the Ambato Valley, as microcharcoal has been found in all of the sites sampled, although in different quantities at different levels. The quantity of microcharcoals at the different levels indicates that there have been periods with a higher frequency of fires. Whether this is related to human activity, natural processes, or perhaps a mixture of both, remains to be determined. Moreover, there is the possibility of identifying different fire "signs" in terms of Marquer (2010). Our study also contributes methodologically to the study of microcharcoals and presents problems encountered during the study and their possible solutions. The techniques used show that signs of fire are possible to identify in the sediment history using the tools suggested. The observed variables in this study have given us new insights into the paleoenvironmental conditions and context in the valley and deepened our understanding of the geological - environmental history of the area.

In terms of the abandonment of the Aguada settlements in the Ambato Valley, the study shows that frequent forest fires might have been a factor that played a role, but due to the regularity of such events in the sediment history it is unlikely to be the only factor in the process of abandonment of the valley.

ACKNOWLEDGEMENTS

This study was financed by FONCyT PICT 2005 - 34558. I am extremely grateful to Alain Viot, Emilio Villafañez, Bernarda Marconetto and Juan Mottino for help with the sampling carried out in the Ambato Valley. I further wish to acknowledge the helpful direction of Bernarda Marconetto in the laboratory during the analysis, and for making various useful comments on this article. I am also grateful for the comments on an earlier draft of this paper presented at the Congreso Nacional de Arqueología Argentina in Mendoza in 2010, and to Débora Kligmann for enabling me to take part in the symposium organized by her and for including me in the publication. I would also like to thank Valeria Ponce and Marcos Gastaldi for help with the graphics, especially Marcos for several helpful comments on the manuscript. Language revision was done by Benjamin Alberti. Finally, I would like to thank the two anonymous reviewers for their helpful comments that improved the paper. Any errors in the article, however, are my own responsibility.

REFERENCES

ABDOUN, F., M. THINON AND M. ALIFRIQUI
2000. La pédoanthracologie, outil paléoécologique des regions sèches. *Libre d´Actes de II Colloque International d´Anthracologie, Paris.*

BLASCO, G., R.L. CAMINOS, O. LAPIDO, A. LIZUAÍN, H. MARTINEZ, F. NULLO, J.L. PANZA, L. SACOMANI, E.L. BARBER, M.A. CHIPULINA AND L. DEL V. MARTINEZ
1994. Hoja Geológica 2966-II, San Fernando del Valle de Catamarca. *Secretaría de Minería de la Nación, Boletín 212*, Buenos Aires.

CABRERA, A.
1976. *Regiones fitogeográficas argentinas. Enciclopedia Argentina de Agricultura y Jardinería, Second Edition*. 2: 1-85.

CARCAILLET, C. AND M. THINON
1996. Pedoanthracological contribution to the study of the evolution of the upper treeline in the Maurienne Valley (North French Alps): Methodology and preliminary data. *Review of Palaeobotany and Palynology* 91(1-4): 399-416.

CLARK, J.S.
1988a. Particle motion and the theory of charcoal analysis: Source area, transport, deposition, and sampling. *Quaternary Research* 30(1): 67-80.

1988b. Stratigraphic charcoal analysis on petrographic thin sections: Application to fire history in Northwestern Minnesota. *Quaternary Research* 30(1): 81-91.

CLARK, R.L.
1982. Point count estimation of charcoal in pollen preparations and thin sections of sediments. *Pollen et Spores* 24: 523-535.

1984. Effects on charcoal of pollen preparation procedures. *Pollen et Spores* 26: 559-576.

DE LA ORDEN, A. AND A. QUIROGA
1997. Fisiografía y vegetación de la Cuenca del Río Los Puestos, Departamento de Ambato, Catamarca. *Revista de Ciencia y Técnica* IV(3).

DI PASQUALE, G., M. MARZIANO, S. IMPAGLIAZZO, C. LUBRITTO, A. DE NATALE AND M.Y. BADER
2008. The Holocene treeline in the Northern Andes (Ecuador): First evidence from soil charcoal. *Palaeogeography, Palaeoclimatology, Palaeoecology* 259(1): 17-34.

GASTALDI, M.R.
2010. *Cultura Material, Construcción de Identidades y Transformaciones Sociales en el Valle de Ambato. Primer Milenio d.C.* Unpublished PhD Dissertation, Facultad de Ciencias Naturales y Museo, Universidad Nacional de la Plata.

GONZÁLEZ, A.R.
1988. *Cultura La Aguada Arqueología y Diseños*. Filmediciones Valero, Buenos Aires.

GONZÁLEZ, A.R. AND J.A. PÉREZ GOLLÁN
1972. *Argentina Indígena. Víspera de la Conquista*. Paidós, Buenos Aires.

GORDILLO, I.
2005. Dimensión temporal del sitio La Rinconada: su interpretación y aportes a la historia del Período Medio. En *La Cultura de La Aguada y sus Expresiones Regionales. V Mesa Redonda La Cultura de la Aguada y su Dispersión*, pp. 159-171. EDULAR, Universidad Nacional de La Rioja, La Rioja.

HABERZETTL, T., M. WILLE, M. FEY, S. JANSSEN, A. LÜKE, C. MAYR, C. OHLENDORF, F. SCHÄBITZ, G.H. SCHLESER AND B. ZOLITSCHKA
2006. Environmental change and fire history of southern Patagonia (Argentina) during the last five centuries. *Quaternary International* 158(1): 72-82.

HEUSSER, C.J.
1987. Fire history of Fuego-Patagonia. *Quaternary South America Antarctic Peninsula* 5, edited by J. Rabassa, pp. 93-109. Taylor & Francis, New York.

1994. Paleoindians and fire during the late Quaternary in southern South America. *Revista Chilena de Historia Natural* 67: 435-443.

1999. Human fourcing of vegetation change since the last ice age in southern Chile and Argentina. *Bamberger Geographische Schriften* 19: 211-231.

2003. Ice Age Southern Andes: A chronicle of paleoecological events. *Developments in Quaternary Science* 3. Elsevier, Amsterdam.

HORN, S.P., R.D. HORN AND R. BYRNE
1992. An automated charcoal scanner for paleoecological studies. *Palynology* 16(1): 7-12.

HUBER, U.M. AND V. MARKGRAF
2003. European impact on fire regimens and vegetation dynamics at the steppeforest ecotone of southern Patagonia. *The Holocene* 13(4): 567-579.

HUBER, U.M., V. MARKGRAF AND F. SCHÄBITZ
2004. Geographical and temporal trends in Late Quaternary fire histories of Fuego-Patagonia, South America. *Quaternary Science Reviews* 23(9-10): 1079-1097.

IVERSEN, J.
1941. Landnam i Danmarks Stenalder: En pollenanalytisk Undersøgelse over det første Landbrugs Indvirkning paa Vegetationsudviklingen. Land occupation in Denmark's Stone Age. *Danmarks Geologiske Undersøgelse* 11(66): 7-65.

KITZBERGER, T. AND T.T. VEBLEN
1999. Fire-induced change in northern Patagonian landscapes. *Landscape Ecology* 14(1): 1-15.

2003. Influences of climate on fire in northern Patagonia, Argentina. In *Fire and Climate Change in Temperate Ecosystems of the Western Americas*, edited by T.T. Veblen, W.L. Baker, G. Montenegro and T.W. Swetnam, pp. 296-321. Ecological Studies 160, Springer, New York.

KITZBERGER, T., T.T. VEBLEN AND R. VILLALBA
1997. Climatic influences on fire regime along a rain forest-to-xeric woodland gradient in northern Patagonia, Argentina. *Journal of Biogeography* 24 (1): 35-47.

LAGUENS, A.G.
2006. Espacio social y recursos en la Arqueología de la desigualdad social. In *Contra la Tiranía Tipológica en Arqueología: una Visión desde Suramérica*, edited by C. Gnecco and C.H. Langebaek, pp. 99-119. Universidad de los Andes, Facultad de Ciencias Sociales, CESO, Ediciones Uniandes, Bogotá.

LINDSKOUG, H.B.
2010. Trás las huellas del fuego. Aportes del análisis de microcarbones a la arqueología del valle de Ambato (Catamarca). Simposio 27: Utilización de indicadores físicos, químicos y biológicos en la resolución de problemas arqueológicos. Tomo *IV Libro de Resúmenes del XVII Congreso Nacional Arqueología Argentina*, Mendoza, pp. 1405-1410.

LINDSKOUG, H.B. AND V. MORS
2010. Donde hubo fuego ¿Cenizas quedan? Residuos de combustión en el sitio Piedras Blancas, Dpto. Ambato, Catamarca. In *La Arqueometría en Argentina y Latinoamérica*, edited by S. Bertolino, G. Cattaneo and A.D. Izeta, pp. 243-250. Editorial de la Facultad de Filosofía y Humanidades, Universidad Nacional de Córdoba.

MARCONETTO, M.B.
2007. Aportes de la Antracología a la cronología del Valle de Ambato. In *Paleoetnobotánica del Cono Sur: Estudios de Casos y Propuestas Metodológicas*, compiled by B. Marconetto, P. Babot and N. Oliszewski, pp. 197-218. Museo de Antropología, FFyH, UNC and Ferreyra editor, Córdoba.

2009. Rasgos anatómicos asociados al estrés hídrico en carbón vegetal arqueológico, Valle de Ambato (Catamarca), fines del primer milenio. *Darwiniana* 47(2): 247-259.

MARCONETTO, M.B. AND A.G. LAGUENS
2013. La trama socio-ambiental de una crisis. In *De las Muchas Historias entre las Plantas y la Gente*, edited by C. Belmar and S. Rojas. British Archaeological Reports, Oxford. In press.

MARKGRAF, V. AND L. ANDERSON
1994. Fire history of Patagonia: Climate *versus* human cause. *Revista do Instituto Geografico de Sao Paulo* 15: 33-47.

MARQUER, L.
2010. From microcharcoal to macrocharcoal: Reconstruction of the "wood charcoal" signature in Paleolithic archaeological contexts. In *The Taphonomy of Burned Organic Residues and Combustion Features in Archaeological Contexts*, edited by I. Théry-Parisot, L. Chabal and S. Costamagno. Proceedings of the Round Table, Valbonne, May 27-29 2008. *P@lethnologie* 2: 105-115.

MILLSPAUGH, S.H. AND C. WHITLOCK
1995. A 750-year fire history based on lake sediment records in central Yellowstone National Park, USA. *The Holocene* 5(3): 283-292.

MISTRY, J., A. BERARDI, V. ANDRADE, T. KRAHÔ, P. KRAHÔ AND O. LEONARDOS
2005. Indigenous fire management in the *cerrado* of Brazil: The Case of the Krahô of Tocantíns. *Human Ecology* 33(3): 365-386.

NICHOLS, G.J., J.A. CRIPPS, M.E. COLLINSON AND A. C. SCOTT
2000. Experiments in waterlogging and sedimentology of charcoal: Results and implications. *Palaeogeography, Palaeoclimatology, Palaeoecology* 164(1-4): 43-56.

PATTERSON, W.A., K.J. EDWARDS AND D.J. MAGUIRE
1987. Microscopic charcoal as a fossil indicator of fire. *Quaternary Science Reviews* 6(1): 3-23.

RHODES, A.N.
1996. *Moorland fire history from microscopic charcoal in soils and lake sediments*. Unpublished PhD Dissertation, University of Newcastle upon Tyne.

1998. A method for the preparation and quantification of microscopic charcoal from terrestrial and lacustrine sediment cores. *The Holocene* 8(1): 113-117.

SCOTT, A.C., J. MOORE AND B. BRAYSHAY
2000. Introduction to fire and the palaeoenviroment. *Palaeogeography, Palaeoclimatology, Palaeoecology* 164(1-4): vii-xi.

SOTTILE, G.
2008. *Dinámica del fuego y de la vegetación en el ecotono bosque-estepa de Santa Cruz durante los últimos 1000 años*. Unpublished Graduate Thesis, Universidad Nacional de Mar del Plata.

TALON, B., C. CARCAILLET AND M. THINON
1998. Études pédoanthracologiques des variations de la limite supérieure des arbres au cours de l'Holocene dans les alpes françaises. *Géographie Physique et Quaternaire* 52(2): 195-208.

THINON, M.
1978. La pédoanthracologie: une nouvelle méthode d'analyse phytochronologique depuis le Néolithique. *Comptes Rendues de l'Académie Des Sciences Paris, Série D* 287: 1203-1206.

TURNER, R.
2007. *Late Quaternary fire histories in the Eastern Mediterranean region from lake sedimentary micro-charcoals*. Unpublished PhD Dissertation, University of Plymouth.

TURNER, R., A. KELLY AND N. ROBERTS
2008a. A critical assessment and experimental comparision of microscopic charcoal extraction

methods. In *Charcoal from the Past: Cultural and Palaeoenvironmental Implications. Proceedings of the Third International Meeting of Anthracology*, Cavallino - Lecce (Italy), June 28th - July 1st 2004, edited by G. Fiorentino and D. Magri, pp. 225-272. BAR International Series 1807, BAR Publishing, Oxford.

TUNER, R., N. ROBERTS, W.J. EASTWOOD, E. JENKINS AND A. ROSEN
2010. Fire, climate and the origins of agriculture: Micro-charcoal records of biomass burning during the last glacial-interglacial transition in Southwest Asia. *Journal of Quaternary Science* 25(3): 371-386.

TUNER, R., N. ROBERTS AND M.D. JONES
2008b. Climatic pacing of Mediterranean fire histories from lake sedimentary microcharcoal. *Global and Planetary Change* 63(4): 317-324.

UMBANHOWAR JR., C.E. AND M.J. MCGRATH
1998. Experimental production and analysis of microscopic charcoal from wood, leaves and grasses. *The Holocene* 8(3): 341-346.

VEBLEN, T.T., T. KITZBERGER, R. VILLALBA AND J. DONNEGANG
1999. Fire history in Northern Patagonia: The roles of humans and climatic variation. *Ecological Monographs* 69(1): 47-67.

WADDINGTON, J.C.B.
1969. A stratigraphic record of the pollen influx to a lake in the Big Woods of Minnesota. *Geological Society of America Special Paper* 123: 263-283.

WHITLOCK, C. AND S.H. MILLSPAUGH
1996. Testing the assumptions of fire-history studies: An examination of modern charcoal accumulation in Yellowstone National Park, USA. *The Holocene* 6(1): 7-15.

WHITLOCK, C., P.I. MORENO AND P. BARTLEIN
2007. Climatic controls of Holocene fire patterns in southern South America. *Quaternary Research* 68(1): 28-36.

ON STEWS AND SEDIMENTS: CONTRIBUTIONS OF EXPERIMENTAL FIELD AND LAB ARCHAEOLOGY TO THE STUDY OF SEDIMENTOLOGICAL MODIFICATIONS

Débora M. Kligmann[1] and Irene J. Lantos[2]

[1] CONICET - Instituto de Arqueología, Facultad de Filosofía y Letras, Universidad de Buenos Aires, [2] CONICET - Museo Etnográfico, Facultad de Filosofía y Letras, Universidad de Buenos Aires

Abstract

This paper presents the geoarchaeological results of an experiment carried out to examine human modifications of sediments resulting from food preparation and consumption activities. We conducted experiments involving the deposition of organic matter coming from food remains in sediments, both in the field and in the lab, and measured their postdepositional decomposition. Our study suggests that while color is not a good indicator of organic matter deposition if the amount deposited is too small, pH, available phosphorus, organic matter concentration, lipids and fatty acids are good indicators of anthropic deposition and postdepositional processes and transformations. We observed that the postdepositional alteration of buried foodstuff is progressive in time, that humid environments accelerate these processes even in the same sedimentary contexts, and that clayey substrates help to better preserve organic matter for longer periods. These results support previous research in human modification of sediments through organic enrichment.

Key words: *Geoarchaeology, Experimental Archaeology, Sediment analyses, Organic matter decomposition*

INTRODUCTION

In this paper we present the results of a geoarchaeological experiment carried out to examine sedimentological modifications as a result of organic matter decomposition. In particular, we examine those modifications originated by human deposition of organic matter in sediments, resulting from food preparation and consumption activities. The main goals are to reproduce decay in sedimentary contexts, to monitor the decomposition processes and their effects, and to create reference data that will enable us to reconstruct past human activities through sedimentological analyses. We are aware that organic matter suffers post depositional changes in time and that, as archaeologists, we don't work with originally deposited and preserved compounds but rather with their degradation products (Kligmann and Díaz País 2012). The experiment here described and discussed contributes to the understanding of the transformations undergone by archaeological sediments and their depositional history.

The questions that guided our experiment were: (i) how much organic matter is left of what is originally deposited anthropogenically in sediments?, (ii) how does it change through time?, (iii) what remains of the organic matter once deposited and what are its degradation products?, (iv) which effects do these transformation processes have in sediment composition?, and (v) how do environmental conditions affect organic matter decomposition?

In order to fulfill our goals we first cooked different stews in field sites located in La Troya Archaeological Study Area (West Tinogasta Region, Catamarca Province, northwest Argentina) using replica ceramic vessels (Figures 1 and 2). We buried the content of each vessel *in situ*, in field sites set up on different texture sediments (sand and clay). We also buried samples of these same stews in open lab sites installed in one of our research institutes (Museo Etnográfico, Facultad de Filosofía y Letras -FFyL-, Universidad de Buenos Aires -UBA-) using identical sediments, collected from the study area field sites (sand and clay). In this paper we analyze two of the stews cooked, named A and B. Open lab site samples were collected after nine and eighteen months. We returned to the field sites after eighteen months and collected samples of the buried stews.

Our experiment was designed using both field sites and open lab sites because we wanted to compare the decomposition of organic matter in two different environments: Catamarca (arid) and Buenos Aires (humid), given that this process is affected by environmental factors such as temperature, moisture and oxygen (Stein 1992). Also, field sites enabled us to observe the transformation products in the natural environmental conditions of our study area, but we were left in the dark as to the agents and processes operating through time (Borrero 1991; Kligmann 2009). Thus, having open lab sites in one of our research institutes allowed us to observe and monitor the intermediate steps of the decomposition process. This would have been impossible in our study area, located 1,500 km away from Buenos Aires, given the high costs of travel and fieldwork. Therefore, installing open lab sites at our research center had both scientific and cost benefits, and allowed us to explore the "black box" of decomposition agents and processes (Borrero 1991; Kligmann 2009). By combining information from both field and open lab sites we attempted to shed light on the products of both biological decomposition and chemical alteration in anthropogenically modified sediments, also called archaeosediments (*sensu* Waters 1992).

Although we acknowledge the complexity of the system of postdepositional alterations in sediments (March *et al.* 2012; Stein 1992), in this paper we only attempt to measure their effects. In order to address each degradation factor, agent and process individually, a lab-controlled experiment should be specially designed and carried out, which falls outside the scope of this work.

Decomposition of Organic Matter in Archaeological Sediments

Organic matter of both natural and anthropic origin in archaeological sediments undergoes different decomposition processes, also defined as diagenetic trajectories or pathways, which act on organic matter that incorporate nutrients into the sediments. They can be classified into three main groups: (i) biological decomposition processes, (ii) chemical degradation processes, and (iii) mineralization processes (Wilson and Pollard 2002). The main interacting factors are humidity, pH, temperature, exposure to light, sediment texture, type of organic matter and time, while the major agents of decomposition are microorganisms such as bacteria and fungi, burrowing or scavenging animals and human activity (Hartley and Ineson 2008; Hayes and Edward 2001; Krull *et al.* 2001; Kumada 1987; Mtambanengwe *et al.* 2004; Sánchez Vizcaíno and Cañabate Guerrero 1998; Sposito 1989; Stein 1992; Van Veen and Kuikman 1990).

It has been demonstrated that organic matter decomposition is dependent on the positive relationship between temperature and humidity that promotes microorganism activity and fosters mineral solubility (Álvarez and Lavado 1998). Carbon content in soils increases according to humidity, and decreases with respect to high temperatures. Thus, carbon content is positively correlated to the precipitation/temperature index, which is the main factor of organic matter mineralization. In arid or semiarid landscapes humidity is normally low and temperatures are commonly extreme, and these are not the best conditions for the formation of humic substances. Nevertheless, monsoonal type rains can produce high availability of rainwater and surface runoff at certain moments of the yearly cycle that, in combination with clayey substrates that absorb water, can be propitious for organic decomposition.

On the other hand, it has been proposed that clayey substrates work as "protectors" of organic matter by creating sealed and anoxic environments that are isolated from microorganisms. It is argued that regardless of climate, clay content promotes carbon concentration because of the protective effect of the matrix on organic compounds (Balesdent *et al.* 2000; Eugène *et al.* 2010; Hassink and Whitmore 1997; Krull *et al.* 2001; Mtambanengwe *et al.* 2004; Oades 1988; Oonk *et al.* 2012; Scott *et al.* 1996; Skjemstad *et al.* 1993; Soliman and Radwan 1981; Van Veen and Kuikman 1990). Thus, with other factors equal, sandy soils are naturally low in organic matter given that it decomposes rapidly while clayed soils are medium to high in organic matter as its decomposition is slow (Brady and Weil 1996).

Regarding the mineralization of organic phosphorus compounds into available mineral phosphorus compounds, both labile and soluble, it has been proposed that this process is favored by neutral or slightly acidic sediment pH values (Eidt 1977; Holliday and Gartner 2007). The decrease in sediment pH when natural conditions tend to alkalinity is directly related to organic acid content as a result of the decomposition of organic compounds. Also, compounds containing phosphorus such as phospholipids and nucleic acids are broken down by microbial activity, resulting in an increase in available phosphorus (Salomonová *et al.* 2003; Soliman and Radwan 1981).

Lipid degradation can occur by a combination of biological and chemical processes. On one hand, the process of acylglycerol hydrolysis breaks the ester bonds between the glycerol molecule and the fatty acid molecules (Hita *et al.* 1996). This reaction can also take place in the presence of some bacteria (Nityananda *et al.* 2006). The result of a total hydrolysis is the complete separation in glycerol molecules and free fatty acid molecules. On the other hand, the oxidative reactions act on the most reactive unsaturated fatty acids, resulting in other more volatile products such as aldehydes, ketones, acids, diacids, hydrocarbons and alcohols (Regert *et al.* 1998; Zamora and Hidalgo 2005). Also, microbial degradation can occur both in aerobic or anaerobic conditions in the burial site. The presence of some fatty acids with odd number of carbons in their alkylic chains is frequent in bacteria and thus can be used as an indication of microbial degradation in archaeological lipid samples (Dudd *et al.* 1998). Degradation of lipids was studied experimentally in archaeological samples to reproduce transformations undergone over long periods of time (Briggs 1999; Evershed *et al.* 2002; Malainey *et al.* 1999; March *et al.* 2012; Pesci 2003; Reber and Evershed 2004). Results show a decrease in lipid concentration and depletion of reactive fatty acids, although some relations between saturated fatty acids are preserved and can indicate lipid origin (Colombini *et al.* 2000; Eerkens 2007).

In brief, current literature agrees that organic matter decomposition is a complex combination of processes and that in order to understand them each variable must be studied separately, despite their interconnection as part of complex feedback systems (Hedges *et al.* 2000). Consequently, in this paper we propose to study a selection of variables that are reliable indicators of organic matter decomposition. In order to better understand this process, we generated hypotheses and expectations for our experimental study.

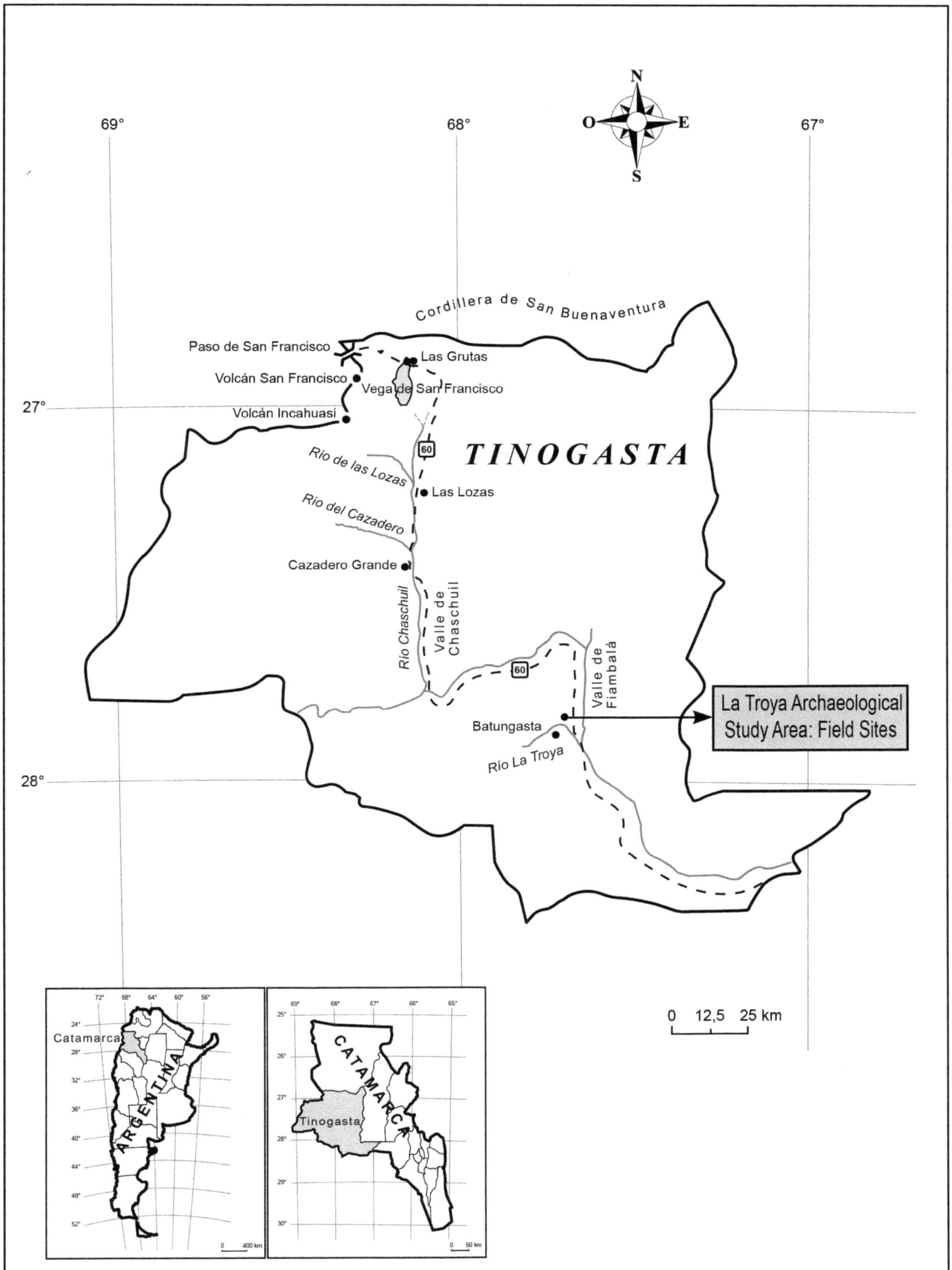

Figure 1. Map of Tinogasta Department, Catamarca Province, northwest Argentina, indicating the area where field experiments were carried out (Adapted from Kligmann 2009).

Figure 2.a. La Troya Valley study area.

Figure 2.b.Lay out of field sites.

Figure 2.c. Stews A and B cooked in the field.

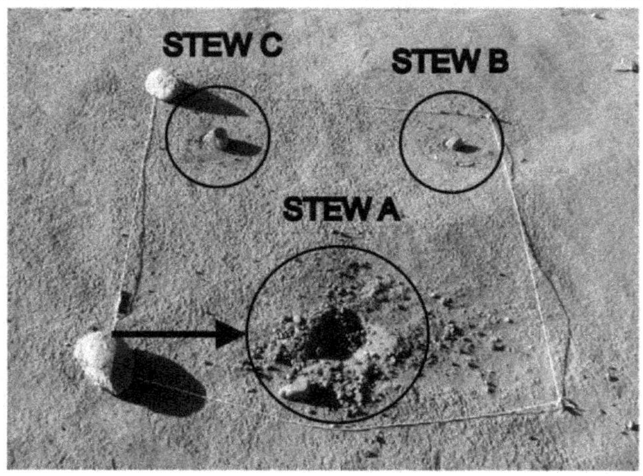

Figure 2.d. Burial in field sites disturbed by burrowing animals.

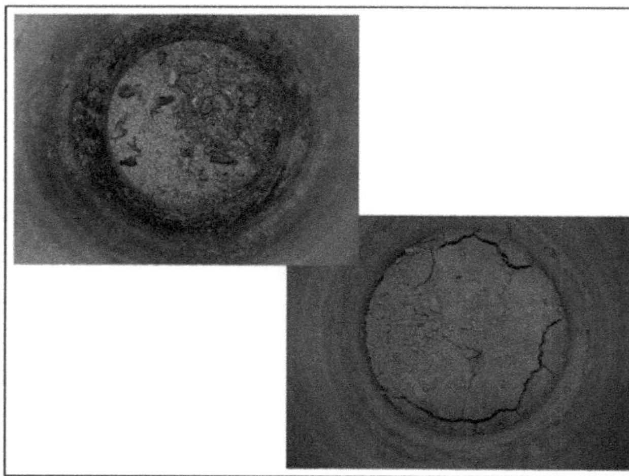

Figure 2.e. Stew B in sand and in clay, open lab site samples.

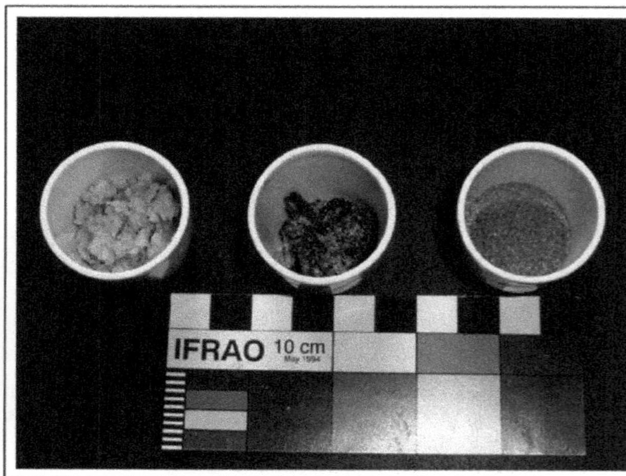

Figure 2.f. Stew A before and after loss-on-ignition, Stew A buried in sand in the lab.

Figure 2. Pictures of the experiment.

GOALS, HYPOTHESES AND EXPECTATIONS

Goals

Our main goal was to reproduce archaeological deposition of organic matter and assess its decomposition in different textured deposits in order to better understand and identify archaeological sediments affected by cooking practices and later postdepositional transformations.

Specific goals were to measure the effects of the decomposition of food remains on sediment attributes such as color, pH, available phosphourus, organic matter, lipids and fatty acids. Also, we wanted to identify differences in decomposition processes in two sediment textures -sand and clay- (or sand/muddy sand and mud/ sandy mud, *sensu* Folk 1980) to observe how grain-size influences organic matter decomposition and preservation. In addition, we also aimed to evaluate differences in decomposition processes depending on the type of organic compounds added to the sediments. Finally, we compared the degradation of organic matter under different environmental conditions.

Hypotheses and Expectations

In accordance with conventional wisdom we propose that once organic matter is deposited, cumulative biological and chemical processes produce the incorporation of nutrients into the sediments through time and therefore older samples are more enriched with decomposed organic content than younger samples. We also suggest that due to the established effect of clay on organic matter preservation, we will find a higher concentration of organic matter in clay than in sand, both for field and open lab sites. Finally, we propose that organic matter decomposition and incorporation into the sediments will be faster in the humid climate of Buenos Aires than in the arid climate of Catamarca.

Considering that experimental samples differ from natural sediments (also called control samples) because the former were affected by both anthropic disturbance and natural decomposition processes, we came up with the following expectations for the experimental samples as compared to the control samples:

1. Darker colors in sediment appearance due to the incorporation of organic matter through humification (browning and Maillard reactions between organic macromolecules, Evershed *et al.* 1997; Hedges *et al.* 2000).
2. Decrease in pH values, given the incorporation of organic acids as a result of organic decomposition (natural sediments in the study area are alkaline).
3. Available phosphorus enrichment due to the breakdown of organic compounds containing phosphorus.
4. Increase in organic matter content due to the incorporation of decomposed food remains into the sediments.
5. Increase in lipid concentration due to the incorporation of lipid-containing food remains into the sediments and changes in fatty acid profiles due to postdepositional degradation processes.

MATERIALS AND METHODS

Materials

Stews: Ingredients and Cooking Methods

Experimental cooking was conducted on replica cooking ceramic pots using recipes and ingredients from local Andean cookery. Recipes were compiled from interviews conducted in Fiambalá and Hualfin regions, Catamarca Province, Argentina. Two stews were boiled in two different pots (Figure 2c). In Pot A, a corn-based recipe called *locro* (Stew A) was made, which is a thick stew consumed traditionally for lunch. Its ingredients were dry kernels of starchy landrace *dentado blanco* corn, dry beans, fresh peppers and fresh beef fat[1]. In Pot B, another corn based recipe called *mazamorra* (Stew B) was made, which is a sweet porridge consumed at breakfast or for dessert. Its only ingredients were dry kernels of sugary landrace *chullpi* boiled extensively in water. Corn landraces were native from Northwest Argentina and were provided by the Maize Germplasm Bank (Instituto Nacional de Tecnología Agropecuaria - INTA-). The other ingredients were bought at a local market near the field sites. Lipids and fatty acids of all ingredients were studied individually both raw and cooked by thin-layer and gas chromatography (Lantos *et al.* 2012). Replica cooking pots were made by an artisan, who applied morphological, volumetric and paste parameters established for local archaeological cooking pots (Feely 2010). Both stews (A and B) were boiled for four hours in an open bonfire built on an experimental field site on the ancient alluvial plain of La Troya River. The bonfire was surrounded with stones to protect from local winds. *Algarrobo* (*Prosopis* sp.) timber was used as the main fuel, and vine (*Vitis* sp.) and *retama* (*Retama* sp.) branches were used as kindling. Before burial, stew samples were collected as control samples (CT 1 and 2, Table 1).

Field Sites at La Troya Archaeological Study Area

La Troya river plain is set in the Fiambalá valley at an altitude of 1,500 m a.s.l. The valley hydrography is formed by seasonal watercourses that flow to the south, and surface deposits are composed primarily of Quaternary alluvium. The La Troya river basin contains the Guanchín-Chaschuil River and multiple temporary streams that flow during seasonal rains from the western border of the Fiambalá range and the eastern border of the Narváez range. Alluvial deposits consist predominantly of clay and sand intermixed with large boulders carried by the high-energy overflowing river during the rainy season (Ratto 2009). Recent geological analysis of the region identified evidence of climatic fluctuations, explosive volcanism and recurrent seismic activity during the Mid-Holocene that has shaped the topography and influenced the habitability of this area (Ratto *et al.* 2012).

The modern climate of La Troya study area is arid with seasonal summer rains, and the basins are predominantly endorheic. The considerable height of the Andes range (4,500 m a.s.l.), running north south and located west of the study area, acts as a geographical barrier forming a rainshadow at the valley bottom. Precipitation varies 150 to 200 mm annually, and the summer rains are torrential and sometimes carry catastrophic consequences such as extensive flooding (Ratto 2009).

The region's phytogeography corresponds to the Central Botanical Province and the Andean *Páramo* (Vervoorst 1951). The climaxic communities of the Botanical District are: (i) the open thorny bush land with predominance of deciduous thorn bushes, and (ii) the open shrub land formed by mycrophyllus evergreens. Also, edaphic communities can be found such as *retamal*, *algarrobal*, *jumeal* and *cachiyuyal* (Morlans and Guichón 1994). It is worth noting that in the past the *algarrobo* (*Prosopis* sp.) woods extended along the La Troya River as a wide gallery forest but due to extensive deforestation is now restricted to sparsely populated areas. This resulted in a loss in total vegetation cover that triggered progressive desertification, more extensive flooding and gully formation (Noetinger 1996).

Four field sites (FS) were established in the alluvial plain of La Troya River, two of them in a sandy deposit and the other two in a clayey deposit of this landscape feature, away from the active channel (Figure 2b). In each site a hole of 5 liters capacity was dug into the ground. We decided to use holes rather than flat surfaces for the deposition of the foodstuff[2]. Sites were marked with camping stakes and geographic positions were recorded with a GPS device. Photographs were taken before and after stew burial. Equal amounts of stew (4 tablespoons, approximately 40 grams) were poured into the holes from each site in the following order: (i) Field Site 1: stew A and sand; (ii) Field Site 2: stew A and clay; (iii) Field Site 3: stew B and sand and (iv) Field Site 4: stew B and clay (Table 1). Each field site was then covered with the same sediment excavated from the appropriate hole and left for the night. The following day we returned to the field sites and recorded photographically the alterations produced during the night. Both Field Sites 1 and 2 containing stew A were burrowed by scavenging animals (Figure 2d), while Field Sites 3 and 4 with stew B remained untouched. The explanation for this could be that stew A had animal fat as one of the ingredients, which is coveted by scavenging animals. On the other hand, stew B, which only contained corn, was not attractive to them. Therefore, we are aware that these events could have affected some of the experimental field sites, given that scavenging animals may have eaten the solid fraction of the stew while leaving the liquid fraction untouched. Finally, field sites were abandoned for eighteen months, after which we returned to collect samples for the analysis discussed in this paper (FS 1 to 4, Table 1).

SITES	SAMPLE NAME	STEW	SEDIMENT TEXTURE	COLLECTION TIME AFTER BURIAL
CONTROL (La Troya archaeological study area, Catamarca)	CT 1	A	N/A	N/A
	CT 2	B	N/A	N/A
	CT 3	N/A	Sand	N/A
	CT 4	N/A	Clay	N/A
FIELD SITES (La Troya archaeological study area, Catamarca)	FS 1	A	Sand	18 months
	FS 2	A	Clay	18 months
	FS 3	B	Sand	18 months
	FS 4	B	Clay	18 months
OPEN LAB SITES (Buenos Aires)	OLS 1.1	A	Sand	9 months
	OLS 1.2	A	Clay	9 months
	OLS 1.3	B	Sand	9 months
	OLS 1.4	B	Clay	9 months
	OLS 2.1	A	Sand	18 months
	OLS 2.2	A	Clay	18 months
	OLS 2.3	B	Sand	18 months
	OLS 2.4	B	Clay	18 months

Table 1. List of samples analyzed.

Open Lab Sites at the University Research Institute

We set up four open lab sites (OLS) in an open terrace at the Museo Etnográfico (FFyL, UBA), an urban setting in Buenos Aires, Argentina. Typical climate is temperate, characterized by hot, humid summers and cool winters, with four distinct seasons and an annual mean temperature of 18° C. Relative humidity is moderately high (64-70%), especially in summer. The city receives 1,242 mm of rainfall per year. Rain can be expected at any time of the year and hailstorms are not unusual (Camilloni and Barros 2010).

The open lab sites consisted of four 5-liter ceramic containers, which were filled to 80% of their capacity.

We decided that using ceramic containers was the best way to reproduce the holes dug in the field sites, while using the same sediments. Two of them were filled with sand and another two were filled the same way with clay. Both types of sediment had been collected and transported from the field sites to ensure identical texture and composition. Samples of both sand and clay sediment types were stored separately to use as control samples (CT 3 and 4, Table 1), while the rest was used to fill the appropriate containers at the open lab sites. Same amounts of stew (4 table spoons, approximately 40 grams) were poured into the containers from each open lab site and covered with sediment in the following order: (i) Open Lab Sites 1.1 and 2.1: stew A and sand; (ii) Open Lab Sites 1.2 and 2.2: stew A and clay; (iii) Open Lab Sites 1.3 and 2.3: stew B and sand and (iv) Open Lab Sites 1.4 and 2.4: stew B and clay (Figure 2e). There were holes in the base of the ceramic containers to help drain the excess moisture in case of heavy rainfall. Stews and sediments were mixed at the beginning of the experiment and samples were taken from the surface down, exposing a fresh surface each time. Samples from each open lab site were taken at 9 and 18 months to monitor gradual decomposition of organic matter and sediment alteration through time (OLS 1.1 to 1.4 and OLS 2.1 to 2.4, Table 1).

Methods

The following variables were selected for analysis in all samples (both control and experimental): color, pH, available phosphorus, organic matter, lipids and fatty acids. Methods are detailed below for each variable.

Color was determined with a Munsell Soil Color Chart, both on dry and wet samples. **pH** was measured with a digital pH-meter (*Denver Instrument*, model UB-10 - UltraBasic Benchtop-), using a 1:1 soil to water ratio. **Available phosphorus** content was determined with a kit designed for farmers and soil specialists (Hanna Instruments, model HI-38073), employing Mehlich 2 extraction method. Color, pH, and available phosphorus analyses were carried out at the Instituto de Arqueología (FFyL, UBA). **Organic matter** was measured by loss-on-ignition at the Instituto de Geocronología y Geología Isotópica (INGEIS), UBA, using a muffle furnace (Verfben).

Lipids and fatty acids were analyzed at Laboratory 8, Facultad de Ciencias Exactas y Naturales (FCEyN), UBA. Samples were prepared for extraction by grinding in a clean porcelain mortar and pestle. All samples were weighed before extraction and the final extract was also weighed to calculate lipid concentration in each sample. Extraction was carried out twice with Folch method (Folch *et al.* 1957). The lipidic composition of all samples was determined before and after derivatisation by thin layer chromatographic analysis against standards of acylglcygerols mixture, cholesterol, n-docosanol, free oleic acid and oleic acid methyl ester (Sigma) with a mixture of cyclohexane:acetone (7:3), and developed under UV (365 nm) and sulfuric acid (10%) plus heat. An aliquot of 5 mg of the extract was saponified with 1 ml of potassium hydroxide solution (4%) in an ethanolic aqueous solution (2:1, vol/vol), at 60° C for 2 h. After cooling at room temperature, the neutral fraction was extracted with 1.5 ml *n*-hexane and the aqueous fraction acidified with 2N HCl solution and extracted with 1.5 ml diethyl ether. The ether phase was evaporated under delicate nitrogen current and immediately 1 ml of boron trifluoride in methanol (20%) was added and heated in a bath at 100° C for 3 minutes. Samples were cooled at room temperature and extracted with 1.5 ml of chloroform and a drop of water in order to extract the fatty acid methyl esters. The extract was transferred to a glass vial and stored in chloroform at -18° C for analysis. Blanks were made for each procedure. The determination of fatty acid composition in all samples was performed on a gas chromatograph (Finnigan Focus chromatographer, Thermo Fisher Scientific, Bremen) equipped with an Omegawax 250 column (Supleco). Standards were used to establish retention times for each fatty acid, which included myristic ($C_{14:0}$), pentadecanoic ($C_{15:0}$), palmitic ($C_{16:0}$), margaric ($C_{17:0}$), stearic ($C_{18:0}$), oleic ($C_{18:1}$), linoleic ($C_{18:2}$) and linolenic acids ($C_{18:3}$) (Sigma).

RESULTS AND INTERPRETATION

Color

Samples can be grouped into 1) the control stews (CT 1 and 2, yellowish), 2) the stews decomposed in sandy sediments (FS 1 and 3, OLS 1.1, 1.3, 2.1 and 2.3, brownish) and 3) the same stews decomposed in clayed sediments (FS 2 and 4, OLS 1.2, 1.4, 2.2 and 2.4, reddish brown) (Table 2). The original colors of the stews were "diluted" in the matrix, with the color of sediment being dominant, i.e., the degraded stews did not significantly change the color of the natural sediments (brown and reddish brown for sand and clay respectively). This is probably related to the small amount of organic matter deposited that is insufficient for humification to significantly alter the color of the sedimentary matrix.

pH

As expected, control stews (CT 1 and 2) were acidic and control sediments (CT 3 and 4) were alkaline. The combination of both stews and sediments (FS and OLS) present intermediate results (slightly alkaline) because extreme values tend to neutralize each other. However, the values obtained are more similar to control sediment samples than to control stews (Figure 3). This is surely related to the small amount of organic matter deposited so that the pH values of the control sediments still dominate the assemblage.

In turn, differences can be seen between the field site samples (FS 1 to 4) and the open lab site samples (OLS 1.1 to 2.4). The former are more alkaline while the latter show values closer to neutrality (Figure 3). Given that the materials used were the same (both stews and sediments), the only difference was the environment where the organic matter decomposed, with Buenos Aires being much more humid than Catamarca province and with heavier rainfall. High humidity accelerates organic matter decomposition. Thus, more organic acids were incorporated into the sediments in the open lab sites than in the field sites, lowering the naturally alkaline pH values. Soil pH also influences the rate of organic matter decomposition because microbial activity is enhanced at neutral or slightly acidic pH ranges (Bååth and Anderson 2003; Rousk *et al.* 2009; Stein 1992).

SAMPLE	HUE, VALUE AND CHROMA (Dry)	COLOR NAME	HUE, VALUE AND CHROMA (Wet)	COLOR NAME
CT 1	10 YR 7/6	Yellow	10 YR 7/6	Yellow
CT 2	2.5 Y 8/3	Pale yellow	2.5 Y 8/3	Pale yellow
CT 3	7.5 YR 5/3	Brown	7.5 YR 4/3	Brown / Dark brown
CT 4	5 YR 5/3	Reddish brown	5 YR 4/3	Reddish brown
FS 1	7.5 YR 5/3	Brown	7.5 YR 4/3	Brown / Dark brown
FS 2	5 YR 5/3	Reddish brown	5 YR 4/4	Reddish brown
FS 3	7.5 YR 5/3	Brown	7.5 YR 4/3	Brown / Dark brown
FS 4	5 YR 5/4	Reddish brown	5 YR 4/4	Reddish brown
OLS 1.1	7.5 YR 5/3	Brown	7.5 YR 4/3	Brown / Dark brown
OLS 1.2	5 YR 4/4	Reddish brown	5 YR 3/4	Dark reddish brown
OLS 1.3	7.5 YR 5/3	Brown	7.5 YR 4/3	Brown / Dark brown
OLS 1.4	5 YR 4/3	Reddish brown	5 YR 3/4	Dark reddish brown
OLS 2.1	7.5 YR 5/3	Brown	7.5 YR 4/2	Brown / Dark brown
OLS 2.2	5 YR 5/4	Reddish brown	5 YR 3/4	Dark reddish brown
OLS 2.3	7.5 YR 5/3	Brown	7.5 YR 4/2	Brown / Dark brown
OLS 2.4	5 YR 5/4	Reddish brown	5 YR 3/3	Dark reddish brown

Table 2. Color of control samples, field site samples and open lab site samples.

Available Phosphorus

The control stews (CT 1 and 2) have higher phosphorus contents than the control sediments (CT 3 and 4). The buried stews show intermediate results, with higher values in the open lab site samples (OLS 1.1 to 2.4) than in the field site samples (FS 1 to 4). The latter are similar to control sediment samples (Figure 4). This is probably related to differential degradation, being faster in Buenos Aires due to higher environmental humidity. In turn, those samples taken 18 months after the beginning of the experiment (OLS 2.1 to 2.4) have higher phosphorus contents that those collected after 9 months (OLS 1.1 to 1.4), showing that as times goes by phosphorus becomes more available (Figure 4).

It is interesting to note that all clay samples show higher values of phosphorus than sand samples (Figure 4). This can be explained by the fact that sediment texture influences the rate of organic matter decomposition and plays a key role in phosphorus fixation. Sediments with high clay contents generally have higher percentages of organic matter, due to soil texture, specific mineral surface area, and soil mineralogy, in addition to secondary parameters like water holding capacity, pH and porosity (Balesdent *et al.* 2000; Eugène *et al.* 2010; Hassink and Whitmore 1997; Krull *et al.* 2001; Mtambanengwe *et al.* 2004; Oades 1988; Oonk *et al.* 2012; Scott *et al.* 1996; Skjemstad *et al.* 1993; Soliman and Radwan 1981; Van Veen and Kuikman 1990).

Organic Matter

Given the differences in the contents of organic matter of the stews and the rest of the samples, two figures are presented for this variable. In Figure 5.1, all the samples analyzed are included. In Figure 5.2, control samples 1 and 2 (the stews) were eliminated and the scale was changed to better appreciate the differences between the samples.

Control stews (CT 1 and 2) are almost 100% organic while control sediments (CT 3 and 4) present very low percentages of organic matter. Decomposed stews, either in the field or in the lab, minimally enhance the organic matter content relative to control sediment samples. The values of open lab site samples are slightly higher than those coming from field site samples (Figure 5.2), showing the same tendency as the phosphorus results. In fact, after 9 months, open lab samples still contained odors related to the stew and some corn kernels were visible whereas after 18 months the samples lacked any odor, and kernels had completely disintegrated.

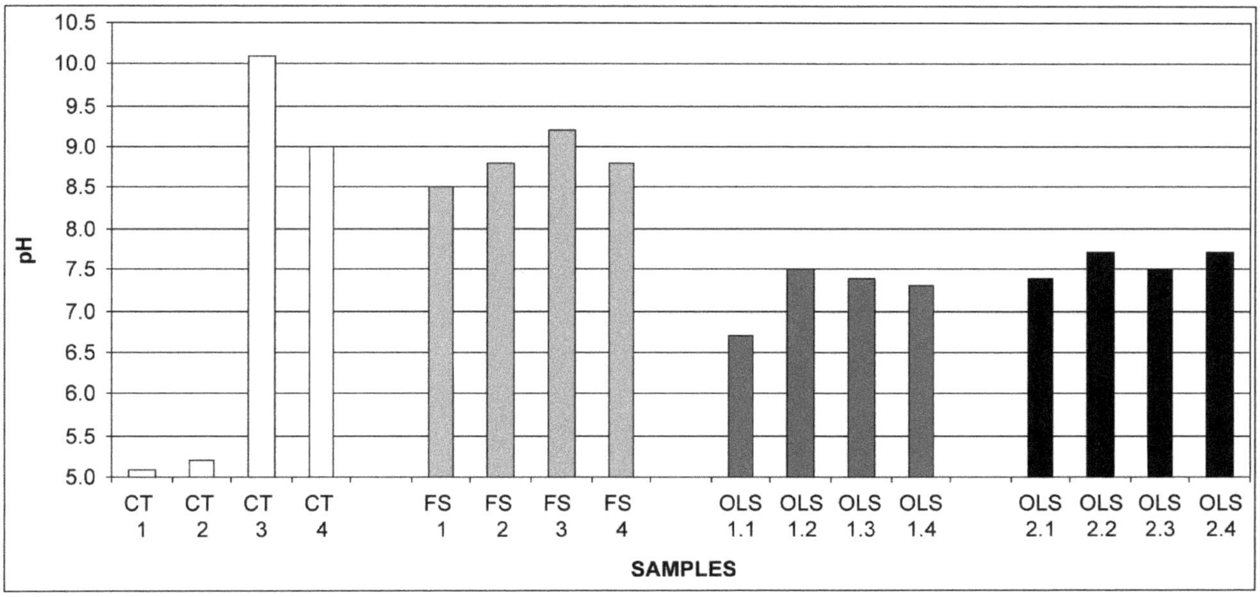

Figure 3. pH of control samples, field site samples and open lab site samples.

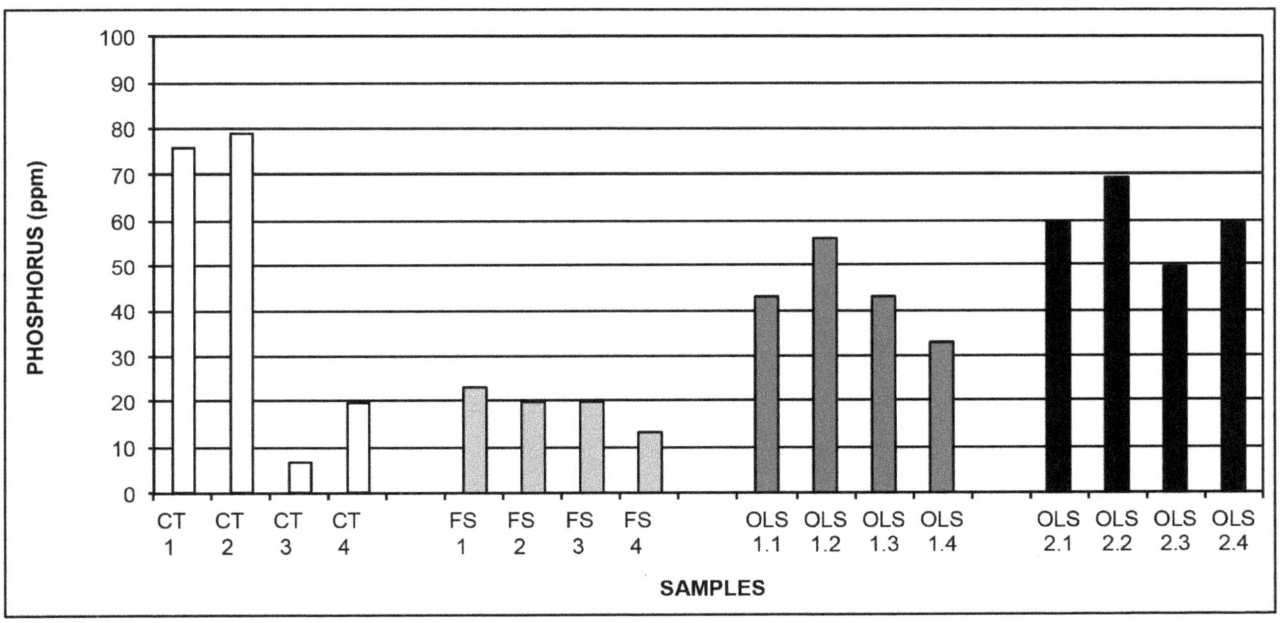

Figure 4. Available phosphorus of control samples, field site samples and open lab site samples.

Once again, all clay samples present higher values than sand samples. This fact can also be appreciated in the natural sediments of the study area (Figure 5.2). Thus, given that soil texture influences other soil properties, it is a key attribute to consider when it comes to organic matter decomposition.

Lipids and Fatty Acids

Total lipid content and relative fatty acid profiles are detailed in Table 3. Given the pronounced differences in the lipid contents of the stews and the other samples, we present two figures: in Figure 6.1, all samples are included while in Figure 6.2 control samples 1 and 2 (the stews) were eliminated and the scale was changed for a better appreciation of the differences between the samples.

Control stew samples had a greater lipid content than the other samples, where CT 1 (Stew A) showed a greater lipid content than CT 2 (Stew B) given that the former had animal fat as one of its main ingredients. On the other hand, natural lipid occurrence in the sediment control samples was extremely low (CT 3: 0,005% and CT 4: 0,008%). Field and open lab site samples present an intermediate situation, showing a substantial decrease in lipid content as compared to control stews, but appeared enriched in comparison to control sediment samples (Figure 6.1). In Figure 6.2 we can observe that in samples from both field and open lab sites, the lipid concentration is always greater in clay than in sand, which could point to a better preservation of organic matter in finer textured sediments.

We observed a decrease in lipid content from 9 to 18 months in the open lab samples, probably due to the prolonged degradation process. Also, when comparing samples at 18 months from field sites and open lab sites, degradation appears to be more intense in the latter samples, possibly due to the increased temperature and humidity that fosters microbial activity in Buenos Aires as compared to Catamarca (Figure 6.2).

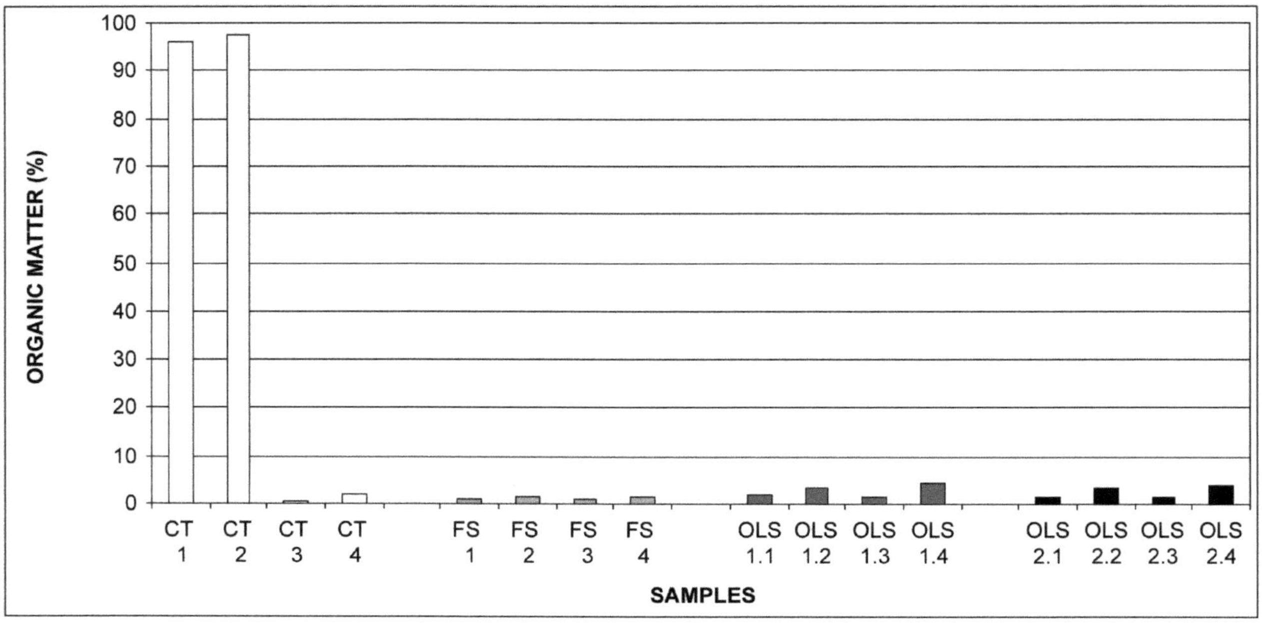

Figure 5.1. Organic matter of control samples, field site samples and open lab site samples (Scale: 100%).

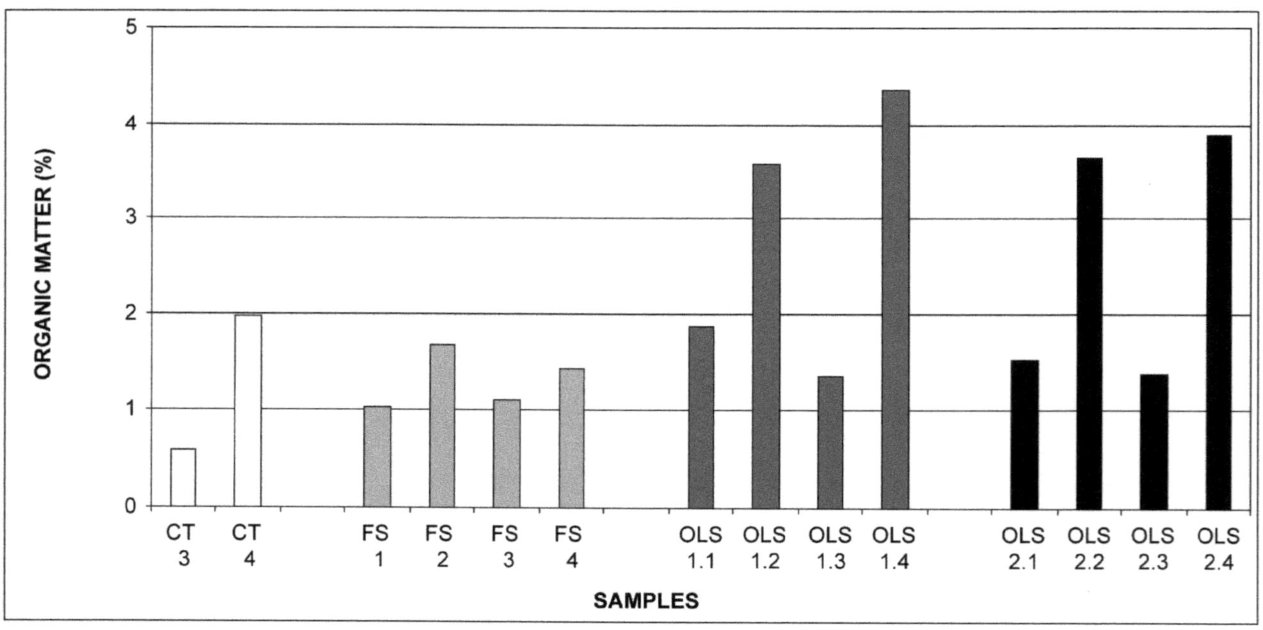

Figure 5.2. Organic matter of control samples (excluding the stews), field site samples and open lab site samples (Scale: 5%).

The lipid compositions determined by thin layer chromatography showed high concentration of triacylglycerols and low concentrations of diacylglycerols, monoacylglycerols and free fatty acids in control stews (CT 1 and 2). The sand and clay control lipid extracts (CT 3 and 4 respectively) were insufficient for thin layer chromatography determination. With this same technique we observed that in open lab and field samples there was a greater concentration of diacylglycerols, monoacylglycerols and free fatty acids due to degradation.

SAMPLE	TOTAL LIPID CONTENT (%)	FATTY ACID RELATIVE PERCENTAGES											
		SATURATED					MONOUNSATURATED		POLYUNSATURATED		SFA	MUFA	PUFA
		C14:0	15:0	C16:0	C17:0	C18:0	C16:1	C18:1	C18:2	C18:3			
CT 1	1.57	3.92	0.00	32.41	0.10	13.95	2.83	25.69	13.98	3.40	50.38	28.52	17.38
CT 2	1.24	0.00	0.00	20.40	0.10	2.44	0.00	27.44	47.32	2.42	22.94	27.44	49.74
CT 3	0.01	0.00	0.00	0.00	0.00	0.00	0.00	0.00	0.00	0.00	0.00	0.00	0.00
CT 4	0.01	0.00	0.00	61.93	0.00	26.77	0.00	11.30	0.00	0.00	88.70	11.30	0.00
FS 1	0.14	6.59	3.24	39.52	2.36	43.00	1.40	3.67	0.20	0.00	94.71	5.07	0.20
FS 2	0.15	7.63	2.40	39.94	2.43	42.33	1.55	3.13	0.59	0.00	94.73	4.68	0.59
FS 3	0.04	13.58	1.50	44.92	1.90	21.75	4.12	9.86	2.36	0.00	83.65	13.98	2.36
FS 4	0.06	16.70	3.81	41.29	2.35	17.75	5.07	10.12	1.77	1.11	81.90	15.19	2.88
OLS 1.1	0.06	4.69	0.00	40.98	1.41	12.59	7.34	26.98	6.01	0.00	59.67	34.32	6.01
OLS 1.2	0.16	1.71	0.00	42.04	0.14	10.97	2.55	30.10	12.04	0.44	54.86	32.65	12.48
OLS 1.3	0.04	2.84	0.00	45.47	0.63	13.03	3.57	25.33	8.57	0.58	61.97	28.90	9.15
OLS 1.4	0.14	0.00	0.00	54.93	0.68	9.44	1.51	25.62	7.36	0.45	65.05	27.13	7.81
OLS 2.1	0.02	4.39	1.31	44.15	1.72	20.56	6.45	18.15	3.28	0.00	72.13	24.60	3.28
OLS 2.2	0.04	7.52	0.08	45.48	4.50	19.82	1.15	18.67	2.96	0.00	77.40	19.82	2.96
OLS 2.3	0.02	0.92	0.00	54.34	1.43	18.64	0.80	20.72	2.81	0.34	75.33	21.52	3.15
OLS 2.4	0.03	3.77	0.00	50.47	1.32	15.60	1.39	21.53	5.06	0.91	71.16	22.92	5.97

Table 3. Lipid and fatty acid composition of control samples, field site samples and open lab site samples. $C_{14:0}$ myristic; $C_{15:0}$ pentadecanoic; $C_{16:0}$ palmitic; $C_{17:0}$ margaric; $C_{18:0}$ stearic; $C_{18:1}$ oleic; $C_{18:2}$ linoleic; $C_{18:3}$ linolenic; SFA: total saturated fatty acids; MUFA: total monounsaturated fatty acids; PUFA: total polyunsaturated fatty acids.

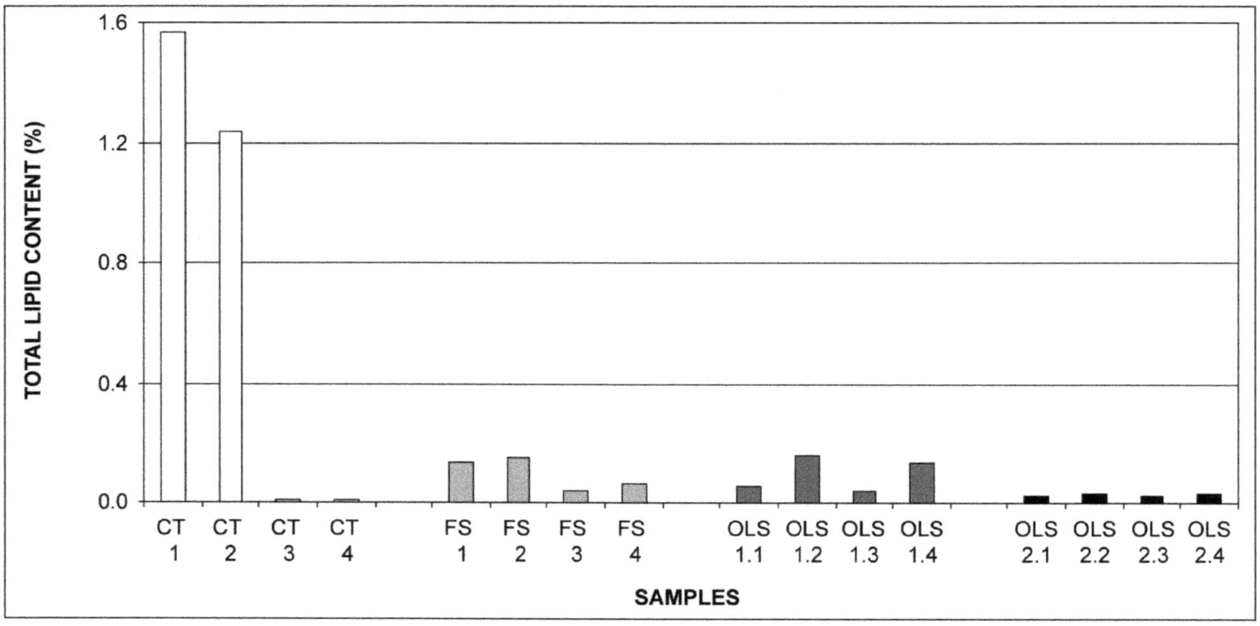

Figure 6.1. Total lipid content of control samples, field site samples and open lab site samples (Scale: 1.6%).

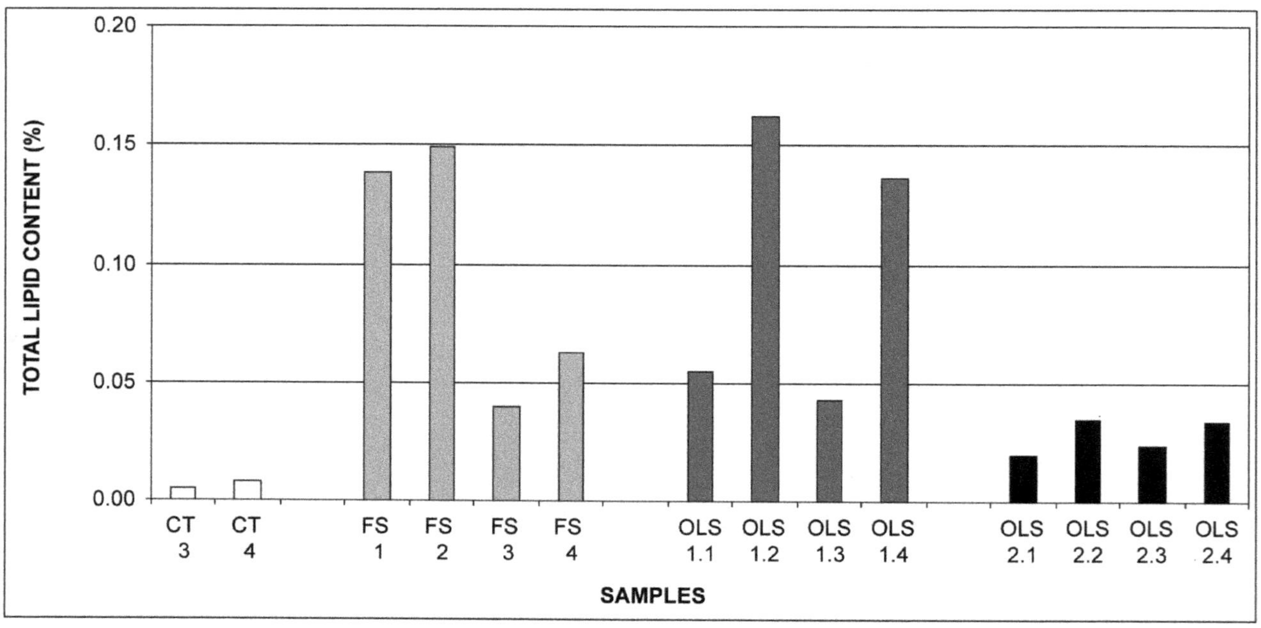

Figure 6.2. Total lipid content of control samples (excluding the stews), field site samples and open lab site samples (Scale: 0.2%).

Fatty acid profiles obtained by gas chromatography were plotted in Figure 7 as the relative proportions of saturated (SFA), monounsaturated (MUFA) and polyunsaturated (PUFA) fatty acids in order to observe the depletion of the more reactive (unsaturated) fatty acids and the consequent increase in the relative levels of the more stable (saturated) fatty acids.

In control samples, CT 1 (Stew A) shows a distribution of SFA, MUFA and PUFA typical of mixtures of animal fats and vegetable oils. On the other hand, CT 2 (Stew B) shows a typical distribution for vegetable oils exclusively. The sand control sample (CT 3) showed no trace of fatty acids, while the clay control sample (CT 4) showed presence of SFA and very low amounts of MUFA, while PUFA were completely absent (Table 3). This profile is usual in degraded organic matter naturally occurring in sediments (Van Bergen *et al.* 1997).

Fatty acid profiles of field site samples show extensive degradation by the increased SFA and depleted MUFA and PUFA. Nevertheless, it is interesting to note that stews buried in clay preserved more MUFA and PUFA than stews buried in sand, possibly due to the protective effect of clay on organic compounds. In the case of open lab sites a similar tendency is detected, but the depletion of MUFA and PUFA is less extensive than in field sites. Progressive decrease of MUFA and PUFA can be noted from 9 to 18 months, pointing towards a process of cumulative loss of unsaturated fatty acids and therefore a relative increase in saturated fatty acids (Figure 7). Even though there was degradation, we could still differentiate stews A and B in field site samples. FS 1 and 2 had a palimitc acid to stearic acid relation of degraded animal fats or degraded mixtures of animal fats and vegetable oils. On the contrary, FS 3 and 4 had a palmitic to stearic acid relation of degraded vegetable oils (Lantos *et al.* 2012). These relationships were not as clear in open lab sites due to factors we have not yet determined.

In addition, the increase of odd numbered fatty acids (pentadecanoic and margaric) in all open lab and field samples are good indicators of microbial activity as agents of biological decomposition (Dudd *et al.* 1998).

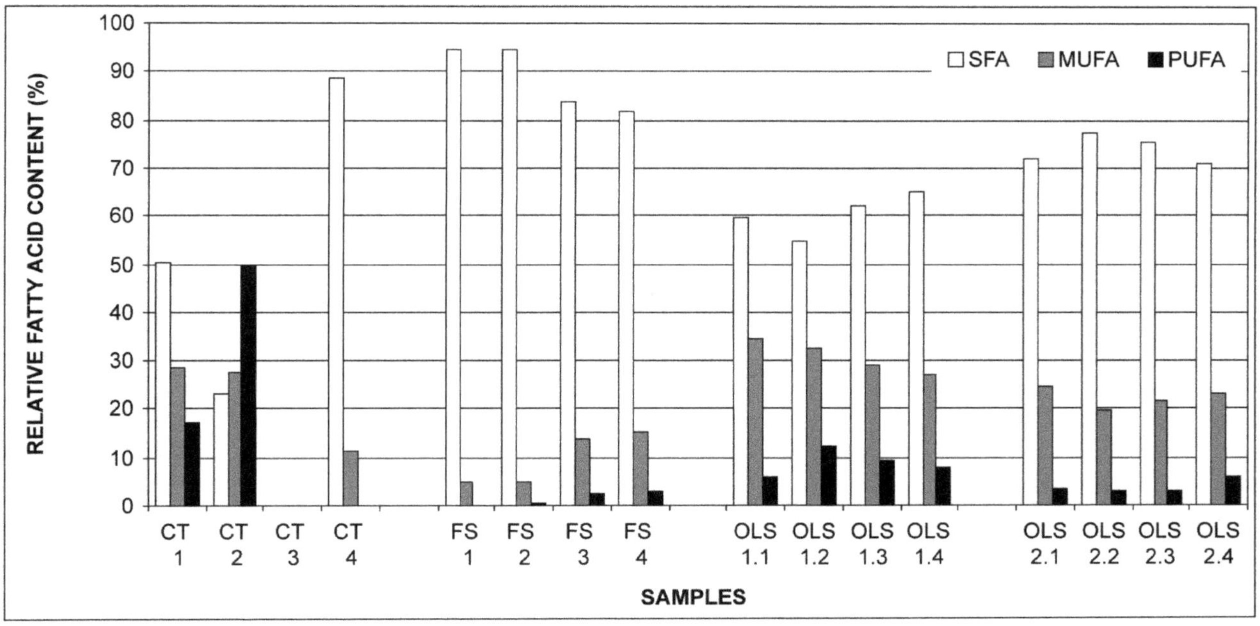

Figure 7. Relative content of saturated (SFA), monounsaturated (MUFA) and polyunsaturated (PUFA) fatty acids in control samples, field site samples and open lab site samples.

CONCLUSIONS

In this paper we presented the preliminary results of an experiment carried out to measure sediment alterations by anthropic disturbance due to organic matter deposition and decomposition. Food remains were buried in holes (in the field) and placed in ceramic containers (in the lab), mixed with sediments, re-examined twice (after 9 and 18 months) and tested for various physical (color) and chemical (pH, available phosphorus, organic matter, lipids and fatty acids) properties. The experiment conducted in holes was took place in an arid environment (Catamarca region) while the experiment conducted in ceramic containers took place in a humid environment (Buenos Aires). The objective was to document how food waste degrades in sediment in two contrasting environments. We compared the results for the two environments, for 9 *vs.* 18 months and for sandy *vs.* clayey sediments.

The preliminary results obtained so far show that color is not a reliable indicator of organic matter deposition if the amount deposited is too small. Although we do not know how much stew is needed to alter the color of the sedimentary matrix, we believe that the spillage produced by only one event of food preparation and consumption will probably not be enough to change the color of the sediments.

However, the same amount of stew was enough to be detected by the other attributes chosen in this study. Thus, pH, available phosphorus, organic matter concentration, lipids and fatty acids may be good indicators of certain types of anthropic disturbance involving the deposition of food products, and of postdepositional processes and transformations of those products in a sedimentary matrix.

By comparing field sites in an arid environment and open lab sites in a humid environment, we could recognize the importance of climate in organic matter decomposition, given that increasing temperature and moisture accelerate degradation processes. We could also single out some general observations that were independent of the climate. For example, degradation is progressive in time, and clayey substrates help preserve organic compounds for longer periods of time.

The results obtained in this experiment are useful for thinking about the archaeological implications of factors such as type of sediment, climate and time, in relation to organic matter preservation in archaeological sedimentary contexts. Thus, sites set in clayey sedimentary contexts could potentially preserve more organic matter than those set in sandy contexts. Humid climate accelerates organic matter decomposition, although in contexts with soil development it may be difficult to differentiate anthropogenic from natural organic matter. In contrast, in arid contexts the presence of organic matter in an archaeological feature can be more indicative of organic matter of human origin. Finally, time is a relevant factor when dealing with ancient deposits given that alteration processes in sediments are progressive and difficult to measure and interpret. An individual event (deposition of stew remains) was sedimentologically visible a few months later. We could also differentiate between the two stews deposited (representing vegetable and/or animal foodstuff), based on the relationships between different fatty acids. However, we do not yet know if this event will still be visible in the sedimentary matrix after several years. Thus, what we measure in archaeological sediments is surely the accumulation of more than one event of foodstuff deposition or spillage.

In summary, the study caried out shows that archaeological sites located in arid environments and set up in clayed deposits will reflect more truthfully what was originally deposited in the past because both of these characteristics tend to protect organic matter, slowering decomposition processes.

Also, the interest to archaeologists is that this type of experiment might lead to the definition of characteristic signatures in sediments for reconstructing certain human activities (e.g. food preparation and/or disposal). Other observations that have archaeological implications include the role of scavenging animals in differential organic matter survival and consequent decomposition.

We are aware that one single experiment cannot provide enough data to allow statistically significant inferences. Our experiment, however, was useful to observe preliminary trends and to generate hypotheses to guide our future research. These preliminary results allowed us to come up with some ideas that would have been impossible otherwise. Further experiments will determine if under the same conditions described in this paper, equal results are obtained.

ACKNOWLEDGEMENTS

We would like to thank Elena Díaz País, Noelia Russo, María de las Victorias Villa, Eduardo Llambias, Marta Maier, Norma Ratto, Raquel Defacio from the Maize Germplasm Bank (INTA Pergamino) and ceramist Daniel Verniers. We would also like to thank the reviewers for their insightful comments that helped improve the paper.

The following research projects provided funding for both fieldwork and labwork: "Haciendo visible lo invisible: Geoarqueología del fuego" (UBACyT 20020100100649, 2011-2014), directed by Dr. Débora Kligmann, "Prácticas sociales en el tiempo: continuidad y cambio en el oeste tinogasteño (Catamarca, Argentina)" (UBACyT F-139, 2008-2011) and "Interacción y cambio: sociedad y ambiente de valle y Puna cordillerana del oeste tinogasteño, Catamarca (*ca.* 5000-500 AP)" (PICT 01539, Agencia Nacional de Promoción Científica y Tecnológica, 2009-2012), both of them directed by Dr. Norma Ratto and "Aislamiento, producción y estudios por espectrometría de masa de metabolitos secundarios bioactivos de origen marino terrestre. Síntesis de análogos de esteroides con actividad antiviral" (PIP 11220090100071, CONICET, 2010-2012), directed by Dr. Marta Maier.

REFERENCES

ÁLVAREZ, R. AND R.S. LAVADO
1998. Climate, organic matter and clay content relationships in the Pampa and Chaco soils, Argentina. *Geoderma* 83(1-2): 127-141.

BÅÅTH, E. AND T.H. ANDERSON
2003. Comparison of soil fungal/bacterial ratios in a pH gradient using physiological and PLFA-based techniques. *Soil Biology & Biochemistry* 35(7): 955-963.

BALESDENT, J., C. CHENU AND M. BALABANE
2000. Relationship of soil organic matter dynamics to physical protection and tillage. *Soil & Tillage Research* 53: 215-230.

BORRERO, L.A.
1991. Experimentos y escalas arqueológicas. *Shincal* 3(1): 142-145.

BRADY, N.C. AND R.R. WEIL
1996. *The Nature and Properties of Soils*. Eleventh edition. Prentice-Hall, New Jersey.

BRIGGS, D.E.G.
1999. Molecular taphonomy of animal and plant cuticles: Selective preservation and diagenesis. *Philosophical Transactions of the Royal Society of London* B (Biological Sciences) 354(1379): 7-17.

CAMILLONI, I. AND V. BARROS
2010. Clima. *Atlas Ambiental de Buenos Aires*. Last Update: 31/12/2010. http://www.atlasdebuenosaires.gov.ar/aaba/index.php?option=com_content&task=view&id=26&Itemid=202&lang=es

COLOMBINI, M.P., F. MODUGNO, E. MENICAGLI, R. FUOCO AND A. GIACOMELLI
2000. GC-MS characterization of proteinaceous and lipid binders in UV aged polychrome artifacts. *Microchemical Journal* 67(1-3): 291-300.

DUDD, S.N., M. REGERT AND R.P. EVERSHED
1998. Assessing microbial lipid contributions during laboratory degradations of fats and oils and pure triacylglycerols absorbed in ceramic potsherds. *Organic Geochemistry* 29(5-7): 1345-1354.

EERKENS, J.W.
2007. Organic residue analysis and the decomposition of fatty acids in ancient potsherds. In *Theory and Practice in Archaeological Residue Analysis*, edited by H. Barnard and J.W. Eerkens, pp. 90-98. BAR International Series 1650, BAR Publishing, Oxford.

EIDT, R.C.
1977. Detection and examination of anthrosols by phosphate analysis. *Science* 197(4311): 1327-1333.

EUGÈNE, N.N., E. JACQUES, T.V. DÉSIRÉ AND B. PAUL
2010. Effects of some physical and chemical characteristics of soil on productivity and yield of cowpea (*Vigna unguiculata* L Walp.) in coastal region (Cameroon). *African Journal of Environmental Science and Technology* 4(3): 108-114.

EVERSHED, R.P., H.A. BLAND, P.F. VAN BERGEN, J.C. CARTER, M.C. HORTON AND P.A. ROWLEY-CONWY
1997. Volatile compounds in archaeological plant remains and the Maillard reaction during decay of organic matter. *Science* 278(5337): 432-433.

EVERSHED, R.P., S.N. DUDD, M.S. COPLEY, R. BERSTAN, A.W. STOTT, H. MOTTRAM, S.A. BUCKLEY AND Z. CROSSMAN
2002. Chemistry of archaeological animal fats. *Accounts of Chemical Research* 35(8): 660-668.

FEELY, A.
2010. *Tradición cerámica y límites sociales en el valle de Fiambalá (Dto. Tinogasta, Catamarca)*. Unpublished PhD Dissertation, Facultad de Filosofía y Letras, Universidad de Buenos Aires.

FOLCH, J., M. LEES AND G.H. SLOANE STANLEY
1957. A simple method for the isolation and purification of total lipides from animal tissues. *Journal of Biological Chemistry* 226(1): 497-509.

FOLK, R.L.
1980. *Petrology of Sedimentary Rocks*. Hemphill Publishing Company, Austin.

HARTLEY, I.P. AND P. INESON
2008. Substrate quality and the temperature sensitivity of soil organic matter decomposition. *Soil Biology & Biochemistry* 40(7): 1567-1574.

HASSINK, J. AND A.P. WHITMORE
1997. A model of the physical protection of organic matter in soils. *Soil Science Society of America Journal* 61(1): 131-139.

HAYES, M.H.B. AND C. EDWARD
2001. Humic substances: Considerations of compositions, aspects of structure, and environmental influences. *Soil Science* 166(11): 723-737.

HEDGES, J.I., G. EGLINTON, P.G. HATCHER, D.L. KIRCHMAN, C. ARNOSTI, S. DERENNE, R.P. EVERSHED, I. KOEGEL-KNABNER, J.W. DE LEEUW, R. LITTKE, W. MICHAELIS AND J. RULLKOTTER
2000. The molecularly-uncharacterized component of nonliving organic matter in natural environments. *Organic Geochemistry* 31(10): 945-958.

HITA, C., E. PARLANTI, P. JAMBU, J. JOFFRE AND A. AMBLIS
1996. Triglyceride degradation in soil. *Organic Geochemistry* 25(1/2): 19-28.

HOLLIDAY, V.T. AND W.G. GARTNER
2007. Methods of soil P analysis in archaeology. *Journal of Archaeological Science* 34(2) 301-333.

KLIGMANN, D.M.
2009. *Procesos de Formación de Sitios Arqueológicos: Tres Casos de Estudio en la Puna Meridional Catamarqueña Argentina*. BAR Publishing - British Archaeological Reports (BAR) International Series 1949, Oxford. ISBN 978 1 4073 0432 8.

KLIGMANN, D.M. AND E. DÍAZ PAÍS
2012. Mirando lo invisible. Primeros análisis físico-químicos de los sedimentos de Tolombón y del Pukara de Angastaco (provincia de Salta). In *Al Borde del Imperio. Paisajes Sociales en Áreas Periféricas del Qollasuyu*, edited by V. Williams and B. Cremonte. In press.

KRULL, E., J. BALDOCK AND J. SKJEMSTA
2001. Soil texture effects on decomposition and soil carbon storage. *NEE Workshop Proceedings*: 103-110.

KUMADA, K. (EDITOR)
1987. Chemistry of soil organic matter. *Developments in Soil Science* 17: 1-241.

LANTOS, I., M. MAIER AND N. RATTO
2012. Recreando recetas: una experimentación con variedades nativas de maíz del Noroeste Argentino. In *Las Manos en la Masa. Arqueologías y Antropologías de la Alimentación en Suramérica*, edited by P. Babot, F. Pazzarelli and M. Marschoff, pp. 533-558. Corintios 13, Córdoba.

MAIER, M.S., D.L.A. DE FARIA, M.T. BOSCHÍN, S.D. PARERA AND M.F. DEL CASTILLO BERNAL
2007. Combined use of vibrational spectroscopy and GC-MS methods in the characterization of archaeological pastes from Patagonia. *Vibrational Spectroscopy* 44(1): 182-186.

MALAINEY, M.E., R. PRZYBYLSKI AND B.L. SHERRIFF
1999. The effects of thermal and oxidative degradation on the fatty acid composition of food plants and animals of western Canada: Implications for the identification of archaeological vessel residues. *Journal of Archaeological Science* 26(1): 95-103.

MARCH, R., A. LUCQUIN, J. DELPHINE, J.C. FERRERI AND M. MUHIEDDINE
2012. Processes of formation and alteration of archaeological fire structures: Complexity viewed in the light of experimental approaches. *Journal of Archaeological Method and Theory*. In press. DOI 10.1007/s10816-012-9134-7.

MORLANS, M.C. AND B. GUICHÓN
1994. Reconocimiento ecológico de la Provincia de Catamarca. *Revista de Ciencia y Técnica* 1: 15-51.

MTAMBANENGWE, F., P. MAPFUMO AND H. KIRCHMANN
2004. Decomposition of organic matter in soil as influenced by texture and pore size distribution. En *Managing Nutrient Cycles to Sustain Soil Fertility in Sub-Saharan Africa*, edited by A. Bationo, pp. 261-275. Academy Science Publishers, Nairobi.

NOETINGER, M.
1996. *Estructura de la vegetación y modelo de depositación dispersión polínica en el Valle de Chaschuil, Catamarca*. Unpublished Final Report, Postdoctoral Fellowship, Consejo Nacional de Investigaciones Científicas y Técnicas, Buenos Aires.

NITYANANDA, R., R. LALITAGAURI AND P. CHATTOPADHYAY
2006. Studys on lipid hydrolysis in a continuous packed bed bioreactor using immobilized lipase. *Indian Chemical Engeneering, Section A* 48(1): 32-34.

OADES, J.M.
1988. The retention of organic matter in soils. *Biogeochemistry* 5(1): 35-70.

OONK, S., E. CAPPELLINI AND M.C. COLLINS
2012. Soil proteomics: An assessment of its potential for archaeological site interpretation. Organic *Geochemistry*. In press.

PECCI, A.
2003. Los alimentos y sus residuos químicos. Arqueología experimental para entender actividades prehispánicas. In *Quaderni di Thule III, Atti del XXV Convegno Internazionale di Americanistica*, pp. 75-83. Perugia.

RATTO, N. (EDITOR)
2009. *Entrelazando Ciencias. Sociedad y Ambiente antes de la Conquista Española*. Eudeba, Buenos Aires.

RATTO, N., C. MONTERO AND F. HONGN
2012. Environmental instability in western Tinogasta (Catamarca) during the Mid-Holocene and its relation to the regional cultural development. *Quaternary International*. http://dx.doi.org/10.1016/j.quaint.2012.09.014.

REBER, E.A. AND R.P. EVERSHED
2004. Identification of maize in absorbed organic residues: A cautionary tale. *Journal of Archaeological Science* 31(4): 399-410.

REGERT, M., H.A. BLAND, S.N. DUDD, P.F. VAN BERGEN AND R.P. EVERSHED
1998. Free and bound fatty acid oxidation products in archaeological ceramic vessels. *Proceedings of the Royal Society of London B* (Biological Sciences) 265(1409): 2027-2032.

ROUSK, J., P.C. BROOKES AND E. BÅÅTH
2009. Contrasting soil pH effects on fungal and bacterial growth suggest functional redundancy in carbon mineralization. *Applied and Environmental Microbiology* 75(6): 1589-1596.

SALOMONOVÁ, S., J. LAMACOVÁ, M. RULÍK, J. ROLCÍK, L. CÁP, P. BEDNÁR AND P. BARTÁK
2003. Determination of phospholipid fatty acids in sediments. *Acta Universitaris Palackianae Olomucensis, Facultas Rerum Naturalium. Chemica* 42: 41-49.

SÁNCHEZ VIZCAÍNO, A. AND M.L. CAÑABATE GUERRERO
1998. *Indicadores Químicos para la Arqueología*. Universidad de Jaén, Jaén.

SCOTT, N.A., C.V. COLE, E.T. ELLIOTT AND S.A. HUFFMAN
1996. Soil textural control on decomposition and soil organic matter dynamics. *Soil Science Society of America Journal* 60(4): 1102-1109.

SKJEMSTAD, J.O., L.J. JANIK, M.J. HEAD AND G. MC CLURE
1993. High energy ultraviolet photooxidation: A novel technique for studying physically protected organic matter in clay and silt-sized aggregates. *European Journal of Soil Science* 44(3): 485-499.

SOLIMAN, A.H. AND S.S. RADWAN
1981. Degradation of sterols, triacylglycerols, and phospholipids by soil microorganisms. *Zentralblatt für Bakteriologie, Parasitenkunde, Infektionskrankheiten und Hygiene* 136(5): 420-426.

SPOSITO, G.
1989. *The Chemistry of Soils*. Oxford University Press, New York.

STEIN, J.K.
1992. Organic matter in archaeological contexts. In *Soils in Archaeology, Landscape Evolution and Human Occupation*, edited by V.T. Holliday, pp. 193-216. Smithsonian Institution Press, Washington.

VAN BERGEN, P.F., I.D. BULL, P.R. POULTON AND R.P. EVERSHED
1997. Organic geochemical studies of soils from the Rothamsted Classical Experiments - I. Total lipid extracts, solvent insoluble residues and humic acids from Broadbalk Wilderness. *Organic Geochemistry* 26(1/2): 17-135.

VAN VEEN, J.A. AND P.J. KUIKMAN
1990. Soil structural aspects of decomposition of organic matter by micro-organisms. *Biogeochemistry* 11(3): 213-233.

VERVOORST, F.
1951. Resultados de un viaje a la Cuenca de la Laguna Verde (Tinogasta - Catamarca). III. Observaciones sobre la vegetación entre Tinogasta y la cuenca. *Actas de la XV Semana de Geografía*, pp. 61-67. Sociedad Argentina de Estudios Geográficos, Facultad de Filosofía y Letras, Universidad Nacional de Cuyo.

WATERS, M.R.
1992. *Principles of Geoarchaeology. A North American Perspective*. The University of Arizona Press, Tucson.

WILSON, L. AND A.M. POLLARD
2002. Here today, gone tomorrow? Integrated experimentation and geochemical modeling in studies of archaeological diagenetic change. *Accounts of Chemical Research* 35(8): 644-651.

ZAMORA, R. AND F.J. HIDALGO
2005. Coordinate contribution of lipid oxidation and Maillard reaction to the nonenzymatic food browning. *Critical Reviews in Food Science and Nutrition* 45(1): 49-59.

---

[1] Since cows were introduced by Europeans, we are aware that beef was not consumed by pre-hispanic populations and that camelids were the most probable source of animal fat. Due to the difficulty in acquiring camelid fat in the region nowadays, we decided to use beef fat instead because it was easily available and degradation processes are similar to those of camelid fat (Maier *et al.* 2007).

[2] Food deposition in archaeological sediments can occur in a number of situations, such as spillage in hearths while stirring and pouring, spillage in eating areas and dumping in discard areas. While some spillage occurs in archaeological living floors, the bulk of food discard occurs in dumpsters or holes in the ground.

www.ingramcontent.com/pod-product-compliance
Lightning Source LLC
LaVergne TN
LVHW071216210926
848482LV00015B/822